Nursing 11/12

Second Edition

EDITOR

Dionne Gibbs, RN, BS, MSN
Fortis College, Norfolk

Dionne Gibbs is a Director of Nursing for Fortis College in the Hampton Roads region of Virginia. In this capacity, she provides direct oversight of the college's practical nursing (PN) and associate degree nursing (ADN) programs. More specifically, she is responsible for regulatory compliance, curriculum development, recruitment, retention and management of a highly qualified nursing education faculty, and most importantly, the development of nurses capable of entering the nursing profession.

Prior to assuming her current role, Dionne was the Corporate Director of Nursing at a four-year private college, where she was responsible for the PN, medical assistant, and dental assistant programs. She has developed and revised ADN and practical/vocational nursing curriculum and started new nursing programs while maintaining the compliance of existing nursing and other medical programs in Virginia, North Carolina, and Florida.

Dionne obtained her MSN in Nursing Education from Walden University and an undergraduate degree from Norfolk State University and Virginia Commonwealth University. Prior to embarking on a career in nursing education, she worked in a variety of nursing specialty areas.

D1315645

ANNUAL EDITIONS: NURSING, SECOND EDITION

Published by McGraw-Hill, a business unit of The McGraw-Hill Companies, Inc., 1221 Avenue of the Americas, New York, NY 10020. Copyright © 2012 by The McGraw-Hill Companies, Inc. All rights reserved. Previous editions © 2006. No part of this publication may be reproduced or distributed in any form or by any means, or stored in a database or retrieval system, without the prior written consent of The McGraw-Hill Companies, Inc., including, but not limited to, in any network or other electronic storage or transmission, or broadcast for distance learning.

Some ancillaries, including electronic and print components, may not be available to customers outside the United States.

Annual Editions® is a registered trademark of The McGraw-Hill Companies, Inc.

Annual Editions is published by the **Contemporary Learning Series** group within the McGraw-Hill Higher Education division.

1 2 3 4 5 6 7 8 9 0 QDB/QDB 1 0 9 8 7 6 5 4 3 2 1

ISBN 978-0-07-351559-5
MHID 0-07-351559-0
ISSN 1558-7886

Managing Editor: *Larry Loeppke*
Developmental Editor II: *Debra A. Henricks*
Senior Permissions Coordinator: *Shirley Lanners*
Marketing Specialist: *Alice Link*
Senior Project Manager: *Joyce Watters*
Design Coordinator: *Margarite Reynolds*
Buyer: *Susan K. Culbertson*
Cover Graphics: *Kristine Jubeck*
Media Project Manager: *Sridevi Palani*

Compositor: Laserwords Private Limited
Cover Images: Jose Luis Pelaez/Getty Images (inset); Ryan McVay/Getty Images (background)

Editors/Academic Advisory Board

Members of the Academic Advisory Board are instrumental in the final selection of articles for each edition of ANNUAL EDITIONS. Their review of articles for content, level, and appropriateness provides critical direction to the editors and staff. We think that you will find their careful consideration well reflected in this volume.

ANNUAL EDITIONS: Nursing 11/12
2nd Edition

EDITOR

Dionne Gibbs, RN, BS, MSN
Fortis College, Norfolk

ACADEMIC ADVISORY BOARD MEMBERS

Preface

In publishing ANNUAL EDITIONS we recognize the enormous role played by the magazines, newspapers, and journals of the public press in providing current, first-rate educational information in a broad spectrum of interest areas. Many of these articles are appropriate for students, researchers, and professionals seeking accurate, current material to help bridge the gap between principles and theories and the real world. These articles, however, become more useful for study when those of lasting value are carefully collected, organized, indexed, and reproduced in a low-cost format, which provides easy and permanent access when the material is needed. That is the role played by ANNUAL EDITIONS.

Nursing has been recognized as a profession for hundreds of years. Although nursing has a long history, it is not a profession that is stagnant, monotonous, unchanging, or lacking adventure. Today, nursing offers more opportunities than at any other time in its history. With all its opportunities however, the nursing profession is also in crisis: the shortage of nurses is staggering and the present workforce is aging. The average age of the working nurse is 47 and this is rising every year. During the last 15 years the number of students choosing to pursue nursing as a career right out of high school has also diminished. There is also a shortage of qualified nursing instructors. As hospitals and other agencies offer higher salaries, benefits, and sign-on bonuses, it has become harder for education to entice the nurse to leave the hospital. If you are reading this book, you have likely decided to pursue the nursing profession. Each of you brings to the profession a rich heritage and hopefully a desire to help, to serve, and to teach those individuals entrusted to your professional care. You may be young and this is your first career, you may be older and finally following your dream to become a nurse, or you may be making a career change. Today's nursing student may be male or female, single or married, childless or a parent, or of any culture or ethnic background. The opportunities for nurses are as varied as the people entering the profession. *Annual Editions: Nursing* provides up-to-date material to supplement any nursing text. Cutting-edge information is presented about nursing practice around the world. *Annual Editions: Nursing* also looks at the diversity of the profession through units on culture, cultural care, and men in nursing.

Annual Editions: Nursing highlights today's new opportunities for using nursing skills. Unit 1 addresses the past, present, and future of nursing—to know where

the profession is headed, nurses must know where it has been. Legal and ethical issues are examined in Unit 2. Unit 3 provides updated information on drugs and medications. Unit 4 highlights new treatments and interventions for diseases. Unit 5 looks at the profession of nursing and the vast opportunities for diversified practice. Unit 6 provides information on weight management and nutrition and on the importance of sound nursing interventions when a patient seeks surgical treatment for weight control. Unit 7 addresses the role of men in the profession. Nursing education is presented in Unit 8, including education's role in alleviating the nursing shortage. Unit 9 examines issues that are affecting the profession and professionalism today. Finally, Unit 10 looks at the diversity of nursing through culture and culturally-proficient care.

Annual Editions: Nursing has been designed to be one of the most useful and up-to-date publications currently available in nursing education. Articles presented here encourage you to learn more about today's nursing profession. A Topic Guide is included to assist the student and the instructor in finding additional articles on a given subject. Critical Thinking questions, located at the end of each article, allow students to test their understanding of key concepts. Internet References are also provided to encourage research or further exploration of a particular topic. This is an evolving work: as a user of *Annual Editions: Nursing,* your input is invaluable. We strive to provide you with material that will enhance your education and assist you in becoming a nurse. Please let us know what you think by filling out and returning the postage-paid article-rating form on the last page of this book.

Dionne Gibbs, RN, BS, MSN
Editor

Contents

UNIT 1
Nursing Past, Present, and Future

The concepts in bold italics are developed in the article. For further expansion, please refer to the Topic Guide.

UNIT 2
Legal and Ethical Issues

The concepts in bold italics are developed in the article. For further expansion, please refer to the Topic Guide.

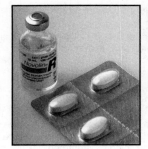

UNIT 3
Drugs, Medications, and Alternative Therapies

UNIT 4
Disease and Disease Treatments

The concepts in bold italics are developed in the article. For further expansion, please refer to the Topic Guide.

The concepts in bold italics are developed in the article. For further expansion, please refer to the Topic Guide.

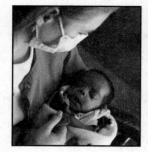

UNIT 5
Nursing Practice Areas/Specialties

UNIT 6
Nutrition and Weight Management

The concepts in bold italics are developed in the article. For further expansion, please refer to the Topic Guide.

UNIT 7
Men in Nursing

The concepts in bold italics are developed in the article. For further expansion, please refer to the Topic Guide.

UNIT 8
Nursing Education

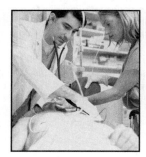

UNIT 9
The Profession and Professionalism

The concepts in bold italics are developed in the article. For further expansion, please refer to the Topic Guide.

UNIT 10
Culture and Cultural Care

The concepts in bold italics are developed in the article. For further expansion, please refer to the Topic Guide.

Topic Guide

This topic guide suggests how the selections in this book relate to the subjects covered in your course. You may want to use the topics listed on these pages to search the Web more easily.

On the following pages a number of websites have been gathered specifically for this book. They are arranged to reflect the units of this Annual Editions reader. You can link to these sites by going to www.mhhe.com/cls

All the articles that relate to each topic are listed below the bold-faced term.

Internet References

The following Internet sites have been selected to support the articles found in this reader. These sites were available at the time of publication. However, because websites often change their structure and content, the information listed may no longer be available. We invite you to visit www.mhhe.com/cls for easy access to these sites.

Annual Editions: Nursing 11/12

General Sources

The American Diabetic Association
www.diabetes.org

American Nurses Association
www.nursingworld.org/ethics/ecode.htm

The American Nurses' Association Center for Ethics and Human Rights provides a look at the revised Code of Ethics for Nurses with Interpretive Statements which was approved in July, 2001, when the Congress of Nursing Practice and Economics voted to accept the new language of the interpretive statements resulting in a fully approved code.

The Center for Disease Control
www.cdc.gov

The CDC in Atlanta, Georgia is the most reliable source for health information. Their website provides up to date information on all areas of health care and treatment.

The Commission on Collegiate Nursing Education
www.aacn.nche.edu/accreditation

CCNE is an autonomous accrediting agency that ensures the quality and integrity of baccalaureate and graduate education programs preparing effective nurses. CCNE assesses and identifies programs that engage in effective educational practices.

National School Nurse Association
www.nasn.org

The Official site for the National School Nurse Association is a source for news articles to advance the delivery of professional school health services to promote optimal health and learning in students.

UNIT 1: Nursing Past, Present, and Future

American Association for the History of Nursing
www.aahn.org

American Association for the History of Nursing fosters the importance of history in guiding the future of nursing.

Barbara Bates Center for the Study of the History of Nursing
www.nursing.upenn.edu

The Barbara Bates Center was founded in 1985 at the University of Pennsylvania School of Nursing to collect, preserve, and make accessible primary sources which document Nursing's history.

Brownson's Nursing Notes
http://members.tripod.com/~DianneBrownson/history.html

This is a site which provides an extensive list of links to other nursing history sites.

Center for Disease Control and Prevention
www.**cdc.gov/vaccines**/vpd-vac/polio/default.htm

The National Immunization Program (NIP)
www.cdc.gov/nip/events/polio-vacc-50th/default.htm

The National Immunization Program (NIP), a part of the Centers for Disease Control and Prevention, provides leadership for the planning, coordination, and conduct of immunization activities nationwide. This portion of the site provides complete coverage of the 50th anniversary of the polio vaccine.

UNIT 2: Legal and Ethical Issues

American Hospital Association
www.aha.org

The American Hospital Association provides a voice to individual members such as Healthcare facilities, networks, patients, and society.

American Nurses Association
www.nursingworld.org/ethics/ecode.htm

The American Nurses' Association Center for Ethics and Human Rights provides a look at the revised Code of Ethics for Nurses with Interpretive Statements which was approved in July, 2001, when the Congress of Nursing Practice and Economics voted to accept the new language of the interpretive statements resulting in a fully approved code.

Legal Eagle Eye Newsletter for the Nursing Profession
www.nursinglaw.com

This online newsletter covers the latest US court decisions and new Federal regulations affecting hospitals and nurses.

Nursing Ethics in Canada
www.registered-nurse-canada.com/nursing_ethics.html

Nursing Ethics in Canada provides information on the role the Registered Nurse plays in ethical decision-making for the patient. Information can also be obtained on nursing standards and ethical practice.

UNIT 3: Drugs, Medications, and Alternative Therapies Internet References

The Center for Disease Control
www.cdc.gov

The CDC in Atlanta, Georgia is the most reliable source for health information. Their website provides up to date information on all areas of health care and treatment.

Institute for Healthcare Improvement: High-Alert Medications
www.ihi.org/IHI/Topics/PatientSafety/MedicationSystems/ImprovementStories/FSHighAlertMeds

The Institute for Healthcare Improvement provides information on improvement of patient care with regard to medication administration.

The Institute for Safe Medication Practices
www.ismp.org/Tools/**highalertmedications**.pdf.

The institute of Safe Medication Practices provides information on medication safety, consulting services, educational programs, professional development, and the ISMP-self assessment.

Internet References

The Joint Commission High Alert Medication and Patient Safety

www.jointcommission.org/sentinelevents/sentineleventalert/sea_11.htm

United States Pharmacopeia

www.usp.org

The United States Pharmacopeia (USP) provides information on dietary supplement, pharmaceuticals, food quality, and developing countries and counterfeit medicines.

UNIT 4: Disease and Disease Treatments

American Diabetic Association

www.diabetes.org/living-with-diabetes/?utm_source=WWW&utm_medium=GlobalNavLWD&utm_campaign=CON

The American Diabetic Association provides information on the treatment & care, complication, planning for a healthy life, for individuals living with diabetes.

Diabetes Monitor

www.diabetesmonitor.com

This is a site created by an M.D. and has professional experts on all area of diabetes serving on an advisory board. The website provides information on economic and political issues for people with diabetes. Commentaries on diabetes, diabetes Q&As, important press release about diabetes, a collection of brief discussions on diabetes topics, recent research findings, and dubious products and how to spot them are included.

Health News Flash

www.healthnewsflash.com/conditions/respiratory_failure.php#8
www.healthnewsflash.com/conditions/pain.php

Health and New Flash provides information on respiratory failure and treatment intervention. Information can also be obtained on other discoveries and medical treatments.

National Heart Lung and Blood Institute

www.nhlbi.nih.gov/health/dci/Diseases/shock/shock_what.html

Provides information on shock, who is at risk, the clinical manifestation, diagnosis, and treatment.

UNIT 5: Nursing Practice Areas/Specialties

Air & Surface Transport Nurses Association

www.astna.org

The webpage for the Air & Surface Transport Nurses Association, also known as the National Flight Nurses Association, strives to advance the practice of transport nursing and enhance the quality of patient care.

APHA Public Health Nursing

www.csuchico.edu

APHA Public Health Nursing Section Web Site. The mission of the Public Health Nursing Section is to enhance the health of population groups through the application of nursing knowledge to the community.

The Hospice Care Network

www.hospice-care-network.org/HospiceWeb/pediatricprogram.cfm

Seriously ill children and their families need special care. The Hospice Care Network website strives to educate the pediatric health care community about these needs.

The International Association for Forensic Nurses

www.iafn.org

The webpage for the International Association for Forensic nurses provides links to news, related resources, and information on certification.

Mercy Ships

www.mercyships.org

Mercy Ships, a global charity, has operated a growing fleet of hospital ships. Following the example of Jesus, Mercy Ships brings hope and healing to the poor in port areas around the world. This is its official webpage and presents information on current and future voyages and opportunities for service.

National School Nurse Association

www.nasn.org

The official site for the National School Nurse Association is a source for news articles to advance the delivery of professional school health services and to promote optimal health and learning in students.

Pfizer

www.pfizercareerguides.com

The nursing profession has evolved significantly in the past 100 years. This site supplies links to numerous professional organizations.

UNIT 6: Nutrition and Weight Management

National Institute of Children and Human Development

www.nichd.nih.gov/health/topics/Diet_and_Nutrition.cfm

The National Institute of Children and Human Development provides health Information of human development topics, clinical research, clinical trial, health education, and publication material.

Nutrition.gov

www.nutrition.gov/nal_display/index.php?info_center=11&tax_level=1

Nutrition.gov provides information on a variety of food and nutrition. Information is provided on smart Nutrition, life stages, weight management, and dietary supplement.

Obesity Help

www.obesityhelp.com

Obesity Help provides information about obesity treatment, professional referrals, and resources.

United States Department of Agriculture

www.mypyramid.gov

Mypyramid.gov provides information on the food groups, dietary guidelines, interactive videos, and dietary guidelines.

WebMD Medical Reference

http://my.webmd.com

WebMD Medical Reference provides information in collaboration with The Cleveland Clinic Guide to weight loss and bariatric surgery.

UNIT 7: Men in Nursing

The American Assembly for Men in Nursing

http://aamn.org

AAMN supports male nurses in their professional development and provides a framework for them to join together with all nurses in strengthening and humanizing health care.

Internet References

Men in Nursing History
www.geocities.com/Athens/Forum/6011

A site which contains a chronological survey of the contribution of men to the profession of nursing.

Texas Health Resources
www.minoritynurse.com

Texas Health Resources is the career resource for minority nursing professionals.

UNIT 8: Nursing Education

The Commission on Collegiate Nursing Education
www.aacn.nche.edu/accreditation

CCNE is an autonomous accrediting agency that ensures the quality and integrity of baccalaureate and graduate education programs preparing effective nurses. CCNE assesses and identifies programs that engage in effective educational practices.

The National League for Nursing
www.nln.org

The NLN is a leader in the advancement of quality nursing education. This site contains listings of job and career opportunities, practice assessments, and lifelong learning programs.

The National Student Nurses' Association
www.nsna.org

The National Student Nurses' Association is a membership organization representing students in ADN, Diploma, Baccalaureate, generic Masters and generic Doctoral programs preparing students for Registered Nurse licensure, as well as RNs in BSN completion programs.

UNIT 9: The Profession and Professionalism

The American Nurses Association
www.nursingworld.org

The official website of the American Nurses Association.

National Center for Health Workforce Analysis
http://bhpr.hrsa.gov

National Center for Health Workforce Analysis, US Department of Health and Human Services provides up to date information on nurse shortage 2000–2020.

The National Council State Boards of Nursing
www.ncsbn.org

NCSBN supplies the latest information on licensing examinations and on public policy regarding nursing, nurse aides, and nursing regulation.

University of North Carolina Chapel Hill
http://nursing-research-editor.com

The website brings the advantages of the Web to Nursing Research authors and readers. It provides expanded content on selected Nursing Research. It serves as an archive of Nursing Research contents.

UNIT 10: Culture and Cultural Care

Cultural Diversity in Healthcare
www.ggalanti.com/cultural_profiles

This site is designed to introduce the viewer to issues of diversity in health care settings, as well as to provide information on caring for patients from different cultures.

Cultural Diversity in Nursing
www.culturediversity.org

Provides basic concepts related to cultural care and case studies.

Diversity Rx
www.diversityrx.org

This site promotes language and cultural competence to improve the quality of health care for minority, immigrant, and ethnically diverse communities.

Transcultural Nursing Society
www.tcns.org

Official website of the Transcultural Nursing Society. Promotes training of nurses in transcultural nursing in order to meet the needs of all clients.

University of Washington Harborview Medical Center
http://ethnomed.org

The EthnoMed site contains information about cultural beliefs, medical issues, and other related issues pertinent to the health care of recent immigrants to Seattle or the US, many of whom are refugees fleeing war-torn parts of the world.

UNIT 1

Nursing Past, Present, and Future

Unit Selections

1. **Mary Breckinridge,** Suzanne Ridgway
2. **Hospitals Were for the Really Sick,** Clancy Strock
3. **Jane Delano,** Suzanne Ridgway
4. **An End to Angels,** Suzanne Gordon and Sioban Nelson
5. **Delores O'Hara,** Suzanne Ridgway
6. **Shots Heard 'Round the World,** Daniel J. Wilson
7. **Linda Richards,** Suzanne Ridgway
8. **Susie Walking Bear Yellowtail,** Suzanne Ridgway
9. **Lillian Wald,** Suzanne Ridgway

Learning Outcomes

After reading this unit, you should be able to:

- List the personal tragedies that caused Mary Breckenridge to devote her life to nursing.

- Identify some of the changes brought on by a hard-to-kill virus in hospitals.

- Describe Jane Delano's contribution to nursing during the Spanish American War.

- Discuss how promotion of traditional images of nursing has reinforced traditional stereotypes of the profession.

- Identify some of the responsibilities of the NASA's nurses.

- State the differences between the Salk Polio Vaccine and the Oral (Sabin) vaccine.

- Explain the nursing practice developed by Linda Richards that continues to be utilized by health-care professionals for every patient in hospitals today.

- Discuss how Susie Walking Bear Yellowtail improved the health-care system on the Indian reservation.

- Describe the legacy left by Lillian Wald to nursing.

Student Website

www.mhhe.com/cls

Internet References

American Association for the History of Nursing
www.aahn.org
Barbara Bates Center for the Study of the History of Nursing
www.nursing.upenn.edu
Brownson's Nursing Notes
http://members.tripod.com/~DianneBrownson/history.html
Center for Disease Control and Prevention
www.**cdc.gov**/vaccines/vpd-vac/**polio**/default.htm
The National Immunization Program (NIP)
www.cdc.gov/nip/events/polio-vacc-50th/default.htm

Nursing has a rich past that has made it one of today's top professions. In order to understand the future of nursing, individuals considering pursuing a career in nursing as their life's work must first understand where the profession of nursing has evolved from, before there is a through comprehension of where the future is taking the profession. This unit reflects the journey of nursing pioneers such as Susie Walking Bear Yellowtail and Lillian Wald. Each of their personal contributions has made nursing what it is today. Mary Breckinridge, Jane Delano, Delores O'Hara, and Linda Richards contributed to the profession through their individual efforts. Mary Breckinridge and Lillian Wald were responsible for the creation of rural and visiting nurse services respectively. Jane Delano added to the work of Clara Barton on behalf of the American Red Cross. Linda Richards was the first nurse in this country to complete formal nursing education. These accomplishments help to raise the standards of the nursing profession. Susie Walking Bear Yellowtail was the first Native American to become a registered nurse. Finally, Delores O'Hara was the first nurse to work for NASA.

This unit also looks at changes that have occurred in the delivery of care. During the early and mid-twentieth century, the majority of the sick were cared for at home by family members, or occasionally by a practical nurse. Patients were hospitalized for occurrences such as surgery, treatment of serious injuries, or to have a baby. Doctors cared for the family from birth to death. Specialists were unheard of except in large metropolitan hospitals. "Miracle drugs" like today's antibiotics had not been developed and technology was very primitive.

Although the changes have been profound, for example, increasing our life expectancy and improving our overall health, with all the health care delivery and a technology change, the nurse has not been replaced. Rather, these changes require nurses to be more knowledgeable and have driven nursing to embrace life-long-learning. "Take away direct patient care from a registered nurse and vital knowledge affecting the health of patients is lost." (Gordon & Nelson, 2005, p. 65).

Finally, the unit looks at an important contribution to medical care: the discovery of the polio vaccine to treat one of the most devastating diseases to affect children. It has been over fifty years since a safe polio vaccine was made. Today, children can enjoy all forms of water activity without the fear of developing

© Royalty-Free/CORBIS

paralysis or other deadly symptoms due to the poliovirus. The history of nursing and health care has been remarkable. Today, the shortage of nurses is presenting a crisis, but the future provides many opportunities for those who seek a career in this honored profession.

Reference

Gordon, S., & Nelson, S. (2005, May). An end to angels. *American Journal of Nursing* 105 (5), 62–69.

Mary Breckinridge

Suzanne Ridgway

Mary Breckinridge was born in 1881 into a distinguished and influential family; her grandfather was the vice president of the United States and her father, a congressman and diplomat. Educated at a Swiss boarding school, she moved in elite circles and married, only to be widowed at age 26.

Before she married again, she attended nursing school in New York and became an RN. Her second marriage produced two children, but her daughter died as an infant, then she lost her four-year-old son two years later. Grief-stricken, she devoted herself to nursing as a way to honor their memory. She believed she had a calling "to work directly for little children now and always because that is the work I can do best, in which my health and enthusiasm and happiness do not fail."

Separated from her second husband, Breckinridge worked as a public health nurse during World War I and afterward went to France under the auspices of the Red Cross as part of the American Committee for Devastated France. There she set up a medical assistance program for pregnant and nursing women and their children.

Believing that the rural poor in America could benefit from similar programs and midwife services, she got additional nursing training and midwife certification. She knew that the people of the Appalachians in southeast Kentucky were impoverished and underserved. With poor roads and lack of reliable transportation, they had limited access to healthcare.

In 1925 Breckinridge began work setting up a decentralized system to deliver services in this region, an organization originally called the Kentucky Committee for Mothers and Babies (later, the Frontier Nursing Service). There was one central hospital and one physician, with nursing outposts scattered throughout the region. Horses could traverse most types of terrain, so nurses on horseback were dispatched to remote areas to provide preventive and curative healthcare services and to deliver babies. In this way, patients were never more than six miles from help.

The availability of nurse-midwives in this area caused a substantial decline in the rate of childbirth-related and neonatal deaths. A story on the FNS website reports that parents in this territory would tell their children that babies were brought not by the stork, but in the nurse's saddlebag.

The Committee became the Frontier Nursing Service in 1928, which Breckinridge funded entirely from her personal funds for several years. Starting with only three nurses in 1925, the staff grew to 30 by 1928. Within five years, FNS nurses were visiting a 700-square-mile area with 1,000 rural families.

The FNS started its own graduate school of nurse-midwifery in 1939 when the British-trained midwives that it had previously relied upon became scarce at the start of World War II. Today the educational program offers a Masters of Science in Nursing in the specialties of midwifery and family nurse practitioner.

Mary Breckinridge died in May 1965, but the FNS has continued to thrive. In 1998, it had registered a quarter of a million patients, delivered 25,000 babies, and its nurses still visit 35,000 homes a year. Its non-profit hospital in Hyden, Kentucky, is now called the Mary Breckinridge Hospital. She was inducted into the American Nursing Association Hall of Fame in 1982.

Critical Thinking

1. What personal tragedies caused Mary Breckenridge to devote her life to nursing?

Suzanne Ridgway is a freelance writer with extensive experience in the health care field, including administrative roles at Cedars-Sinai Medical Center and Orthopedic Hospital.

From *Working Nurse*, March 14–April 4, 2005, p. 30. Copyright © 2005 by Suzanne Ridgway. Reprinted by permission of Suzanne Ridgway.

Hospitals Were for the Really Sick

Before today's miracle drugs, specialists and high-tech procedures, you stayed at home unless the diagnosis was dire.

CLANCY STROCK

It's a safe bet that you or someone close to you has been hospitalized already this year. But there was a time when lots of people lived long healthy lives without ever setting foot in a hospital. My own dad didn't become a patient until he was 78.

Not that folks were made of sturdier stuff years ago. It's just that there was not much hospitals could do for you that couldn't be handled just as well at home.

Remember, now, that "miracle drugs" had yet to be invented. "They say she has double pneumonia" was pretty much an automatic death sentence.

My Aunt Grace, who became a nurse in the '20s and stayed at it for a half century, says people who were hospitalized in her era were there because they needed surgery, had been in a bad accident or were having a baby. (Dr. McCandless, our family doctor, brought over 4,000 babies into the world . . . quite a few of them at home.)

Most sick people just stayed home and were cared for by the family. Or, if lucky enough to be well-off, they hired a "practical nurse," who put in 12-hour days providing skilled care.

Back when Aunt Grace started out, private rooms were only for the very rich. Most patients were in 10-bed wards tended by a single nurse. Some big-city hospitals even had 20-bed wards with two nurses sharing the chores.

It's hard to imagine one woman trying to look after 10 patients. But she took temperatures and blood pressure, brought meals and changed bed linens, scrubbed floors, kept the charts and administered the best medicine available . . . doses of TLC (tender loving care).

I jokingly asked Aunt Grace what she did in her spare time. She didn't think it was much of a joke.

"If and when we had spare time, one of our jobs was to sharpen the hypodermic needles," she said.

She explained that needles were used over and over again and eventually became dull. Nurses used a sharpening stone to restore them to a fine point.

Some Sharp Improvements

Several decades ago, an especially hard-to-kill virus swept through many hospitals. That led to the widespread use of disposable needles, sheets, hospital gowns and dozens of other items that might be contaminated with germs. It added a bit to the daily hospital bill but was a major step ahead in patient care.

In recent years, ingenious gadgets have come along to spare nurses much of the legwork that once was so common. To cite one example, monitors at central nursing stations now chirp away, sounding an instant alarm when something goes wrong.

Back in Dr. McCandless' day, doctors were physician/surgeons, at least in small-town America, where I grew up. Specialists were unheard-of. Your family doctor was a jack-of-all-trades, providing total care for the entire family from birth through adulthood. He scheduled surgery in the morning and made the hospital rounds. In the afternoon, he would come home late in the evening and pile eggs, chickens, canned items on the kitchen table. It was the only way many rural people had to pay his dad.

A lady I know was the daughter of an Omaha doctor in those lean times. His office was on the first floor of their home. He worked long days, often getting home late at night. When he passed away, the family discovered that he had nearly $60,000 in uncollected fees on the books.

"Exploratory surgery" was a common—and dreaded—phrase until recent times. It mean't "We can't figure out what's going on inside Uncle Fred, so we're going to cut him open for a look-see."

Nowadays, there are huge—and hugely expensive—machines that snoop around and put pictures right up there on a TV screen for your viewing pleasure. Yup, there's your heart pumping away. No guesswork necessary. And what a wonderful feeling of relief when the doctor smiles and says that everything looks okay.

However, some things stay the same, as I found out a few months ago. Nurses still wake you from a sound sleep to administer a sleeping pill. Then, while you are snoring away at 2 A.M., they flip on the overhead lights to wake you up so they can check your temperature and blood pressure. Soon they rouse you while it's still dark outside to bathe you . . . even though breakfast won't show up for another 3 or 4 hours, by which time the make-believe scrambled eggs and lumpy oatmeal are as cold as the gelatin that comes with every meal.

Nurses still wake you from a sound sleep . . .

Yes, I know . . . no one checks into a hospital to savor the food. That's why restaurants were invented.

One thing has changed, although I'm not sure whether it's for the better. No matter what the hour, there's always the television—now often with cable—to keep you amused. In earlier times, there wasn't much to do but stare at the wallpaper.

The Biggest Pain of All

Finally, can someone explain to me why, despite all these wonderful advances in technology, those hospital gowns have not changed in more than a century?

It's impossible to prowl the halls without worrying about what's on display for the amusement of those following you. I can't believe that something easier to tie and more comfortable to wear can't be invented. Haven't they heard of Velcro and zippers, or just plain snaps?

Is this some sort of secret medical amusement? In short what's, uh, *behind* this aggravation?

But I'm being picky. In few areas of life have profound changes been more evident than in medical care. As a result, we live longer and healthier lives. And as a nice bonus, even the needles are sharper and sting less.

I know . . . I was there.

Critical Thinking

1. What were some of the changes brought on by a hard-to-kill virus in hospitals?

Jane Delano

Suzanne Ridgway

Although Clara Barton founded the American Red Cross in 1882, there was no formal structure within the organization to recruit or maintain a permanent nursing force. Jane Delano changed that. In 1909, Delano founded the Nursing Service Corps of the American Red Cross. Her mission was to develop and maintain an efficient reserve of Red Cross nurses for military service.

Born in New York in 1862, Delano chose to be a nurse not, she insisted, for any "romantic" or "sentimental" reasons about alleviating suffering, but because "the nurse's profession is a fine one, and I like it."

She displayed a heartfelt devotion to the American Red Cross when she went to work for them full time without pay, saying, "I would rather live on a crust and serve the Red Cross than do anything else in the world."

Delano graduated from the Training School for Nurses at Bellevue Hospital in 1886. When the Red Cross arrived in Jacksonville, Florida, to help with a yellow fever epidemic, a physician from a local hospital brought in Jane Delano to take charge of the volunteer nursing effort there.

Later, during the 1898 Spanish-American War, Delano worked with the Red Cross as a recruiter of volunteer nurses. The aid the Red Cross was able to provide to the military was put together under emergency conditions, and, while it achieved notable results, the war demonstrated that the Red Cross needed to have a peacetime corps of nurses trained and ready for wartime emergencies.

She was already Superintendent of the Army Nursing Corps in 1909 when Delano was appointed Chair of the new National Committee on Red Cross Nursing Service. By coordinating with hospitals, training schools and various nurses' associations, she developed a comprehensive program for enrolling nurses nationwide who would be prepared to act in times of war or disaster.

Only well-trained nurses with high professional standing were accepted. In order to occupy nurses between emergencies, Delano developed valuable peacetime programs. These included classes for women in home nursing, and elementary hygiene and training courses for nurses' aides.

Delano's strategies for preparing nurses for national emergencies were successful: at the outbreak of World War I in 1914, there were 8,000 Red Cross nurses ready. By the end of the war, 20,000 had volunteered for service in the U.S. and abroad.

Delano was on official Red Cross business in France inspecting base hospitals in 1919 when she became ill and passed away at the age of 57. After temporary interment at Saveney, France, she was eventually laid to rest in the Nurse's Corner at Arlington Cemetery.

In 1990, The National Nursing Advisory Committee of the Red Cross formed the Jane Delano Society. The goals of this society named in Delano's honor are to promote the professionalism of nursing, to assist with recruitment of nurses, and to contribute policy deliberations regarding critical nursing issues.

Critical Thinking

1. What was Jane Delano's contribution to nursing during the Spanish American War?

Suzanne Ridgway is a freelance writer with extensive experience in the health care field, including administrative roles at Cedars-Sinai Medical Center and Orthopedic Hospital.

An End to Angels

Moving away from the 'virtue script' toward a knowledge-based identity for nurses.

SUZANNE GORDON, BA AND SIOBAN NELSON, PhD, RN

Nurses often disagree on the causes of and possible solutions to the current nursing shortage. Mandatory staffing ratios versus Magnet hospitals? Sign-on bonuses for nurses versus more unionization of RNs? The aging of the nursing workforce versus working conditions? Still, most nurses agree that the profession needs a contemporary image to attract new recruits and reinforce the idea that nursing is a profession grounded in science, technology, and knowledge. To present a modern image and solve the crisis, dozens of different groups have produced advertising campaigns and promotional messages to attract new recruits to the profession.

A close analysis of the words and images used in these campaigns reveals that, instead of creating a modern, accurate version of today's nurse, many simply repackage nursing's traditional stereotype of women born to be good, kind, and self-sacrificing—not educated to provide care based on science and practical skill. Although many studies—conducted by nursing, medical, and public health researchers—have documented the links between nursing care and lower rates of nosocomial infections, falls, pressure ulcers, deep-vein thrombosis, pulmonary embolism, and death, most promotional campaigns are conspicuous for their failure to promote these data. Even when ads feature a mix of men, women, and minorities, what is often communicated is a sexist, archaic message: nursing is "virtuous" work.

Learning from the Past

Nursing's image has been studied in past decades, but what's rarely addressed is why and how nurses themselves reinforce traditional images of their work.[1] Similarly, the historical origins of nurses' choices of verbal and visual images have been poorly explored. It's crucial that, in a period of rampant cost cutting to health care services, nurses convey their central importance to hospital and health service managers, insurers, policymakers, politicians, and new recruits to the profession. To do so, nurses must reexamine the history of their image as "virtuous workers" and understand the power that what we call the "virtue script" has over the nursing profession.

The virtue script bases the presentation of nursing on characteristics such as kindness, caring, compassion, honesty, and trustworthiness, attributes associated with "good women." This script sentimentalizes and trivializes the complex skills, including caring skills, nurses must acquire through education and experience—not simply individual inclination. Only when freed of the virtue script can nursing assert its identity as a knowledge-based profession that is critically important to patient care.

The Angel Image: Nursing's Beginnings

In the mid-19th century, urbanization and industrialization helped to eliminate many small family farms, and factories began to provide many of the goods that women had previously produced in the home. There was now a generation of women who needed to support themselves by working outside their homes, but those who wanted to work as nurses confronted a pressing problem: moving unchaperoned in public places and working for money were frowned upon; how could women remain respectable members of society and work for a living? Before the efforts of nursing reformers, such as Florence Nightingale, nursing was considered the domain of religious women and servants.

While religious women were portrayed as angels and thus protected from the stigma of working with sick strangers (most of whom were male), nonreligious nurses were not considered to be respectable. Charles Dickens's depictions of drunken, heartless nurses Sarah Gamp and Betsy Prig in *Martin Chuzzlewit* (originally published in the 1840s) both encapsulated and solidified the stigma of secular nursing. Nurse reformers thus faced a challenge—to make nursing safe for respectable lower- and middle-class women who would not be attracted to its ranks otherwise. In a society in which gender roles were very rigid, they also needed to make it safe for female nurses to work with male medical students, surgeons and physicians, hospital managers, and boards of governors. Whether nurses practiced on battlefields or in hospitals or homes of the rich and poor, they were confronted with unpleasant, unladylike realities such as poverty and disease. In an era that prized blushing innocence, nurses' knowledge of anatomy and their experience of the world outside the domestic sphere threatened their respectability.

For centuries, before the arrival of nurse reformers like Nightingale, nursing care was delivered mainly by religious women whose vows and veils—and self-sacrificing, altruistic mission—protected them from the stigma of hard work caring for the sick. By borrowing this religious template as well as traditional Victorian notions of charity, nurse reformers, who were, like Nightingale, often very pious women, made it possible for thousands of women to find purposeful work immersing themselves in the gritty realities of nursing. It also allowed nursing to become the first social activity of women outside the home that gained acceptance among respectable classes and ultimately facilitated women's work in other professions.[2]

Equally important, the foundation of nursing as virtuous work—as opposed to knowledge-based work—furnished nurses who were trained in secular institutions with an important arm in their battles with medicine and male physicians over what was to become, by the late 19th century, the highly contested terrain of the contemporary hospital. Until well into the 20th century, these hospitals were charitable foundations, which frequently had religious affiliations.

The Battle for Reform

As scientific medicine consolidated its authority, many physicians felt threatened by the movements for women's suffrage and education. These movements included not only women who wanted to be physicians, but also nurses who wanted more training, education, and authority. Nightingale fought for a woman's right to do purposeful work outside the home. Elizabeth Blackwell, the first female physician, was also a staunch advocate of nursing education. And outspoken English nurse reformer Ethel Bedford Fenwick, who championed nursing registration and founded the *British Journal of Nursing* in 1893, and her American colleague Lavinia Lloyd Dock, author of *Materia Medica for Nurses,* one of nursing's first textbooks, were prominent advocates of women's suffrage.

When the movement for nursing reform began in the 1860s, these women initiated the long journey that transformed nursing from lowly work performed by servants or women of religious affiliation into a secular profession. Key North American reformers included Isabel Adams Hampton Robb, the ANA's first president; Dock; and M. Adelaide Nutting, who became the first nurse ever appointed to a university professorship. Key women in the United Kingdom included the Nightingale Nurses, trained at St. Thomas's training school for nurses, as well as Fenwick. This new breed of women formed nursing associations and nursing schools, and they were determined to turn nursing into a profession. They published journals, such as *AJN* and the *British Nursing Journal,* and exerted their leadership over rapidly developing groups of trained nurses. Between 1900 and 1920, nursing registration was introduced in many parts of the world.

Improvements to nursing. From their earliest days, reformers struggled to place nursing authority in control of the nursing workforce. They wanted a matron—not a physician—to be in charge of nurses and to decide where nurses worked. They wanted nurses to be accountable to the matron's authority and insisted that students have a formal program of training and experience. To gain this sphere of influence, Nightingale and her counterparts across the Atlantic fought a series of battles with medical and hospital administrators. As a result, hospitals gained better-trained, better-educated nurses, but physicians lost direct control of the nursing workforce (although they still controlled much of the practice of nursing).

The movement to improve nursing was infused with the politics of women's emancipation. Medicine was a male bastion that quickly began defending its territory against the incursions of female physicians and contemporary nurses.[3] Newspapers and medical journals of the time were full of debates, letters, and squabbles over the proper relationship between medicine and nursing—and one of the fundamental issues was power over nurses' knowledge and practice. Physicians and their political supporters insisted that nurses were no more than physicians' servants, dependent on physicians' knowledge. In an unsigned and untitled 1880 editorial, the *Times of London* presented the point clearly: "nursing is merely one of the means of cure, . . . and . . . it can only be rightly carried out under absolute and unconditional subjection, in every principle and detail, to the doctor who is responsible for the case."

Fighting the opposition. Faced with medical opposition and patriarchal traditions, nursing reformers, who as women had no political, legal, or economic power, had to transform nursing into a profession respectable enough to attract middleclass women and yet not be a threat to male medical authority. This class issue was critical for nursing reform. For better-educated, middle-class women—who were used to employing and directing servants—to be attracted to nursing, many changes were needed. Traditional nurses had been accommodated on the wards, directly off stairwells, and no family of standing would allow a daughter to live in such precarious lodging. This prompted the creation of cloister like group homes for nurses that had chaperones and did not permit male visitors. It was a first step in attracting a higher class of women to train as nurses.

To assuage physicians, nurse reformers also downplayed nurses' knowledge and skills and emphasized their virtue and ethics. Like Nightingale, reformers in Britain, the United States, Australia, Canada, and France exploited the Victorian view that women possessed a superior moral power and essential female virtues that could be used for the common good. The very success of nurse reformers in creating the first mass profession for women put nurses in the paradoxical position of playing an important role in health care, while sentimentalizing and trivializing the very critical role they played. Taught in religious orders to "say little, but do much," the only way nurses could say more was to couch their description of their work in charitable, devotional, or altruistic terms. (For more on this topic, see *Say Little, Do Much: Nursing, Nuns, and Hospitals in the Nineteenth Century,* University of Pennsylvania Press, 2001, written by Sioban Nelson, one of the authors of this article.)

Pursuing Professionalism

Even as women have gained greater social, economic, legal, and political power in the late 20th and early 21st centuries, nurses and their political supporters still rely on the virtue script. This is apparent in many campaigns that support the nursing profession, including advertisements, videos, brochures, articles, and newsletters, which are targeted to those in the profession and the public.

"Nursing: The ultimate adventure." This video, produced by the National Student Nurses' Association, promotes the career of registered nursing to junior and senior high school students. The video makes references to learning and knowledge, but what nurses learn and know is never specified. What is instead emphasized is the public warmth and love nurses receive. Beverly Malone, then-president of the ANA, declares: "The public loves me as a nurse and they don't even know my name, but if I say I'm an RN, there's affection and warmth and an experience that means so much to me." A young woman in the video further encourages students to choose nursing by saying that nursing is "a job where people will love you."

Nursing profile. In October 2001, this Michigan nursing magazine ran an article on the accomplishments of an African-American nurse, Birthale Archie, RN, who also has MSN and BS degrees.[4] Rather than alerting the reader to the fact that this nurse was "educated to help others," the headline on the cover of the magazine proclaimed she was "Born to Help Others."

Nurses week. In 2002 the ANA chose the tagline "Nurses Care for America" for its biennial convention and Nurses' Week theme; in 2003 its Nurses Week slogan was "Lifting Spirits, Touching Lives." And in 2002, to celebrate National Nurses Week, Ohio Health Systems produced a brochure with a gauzy picture of nurses wheeling a sick patient. The copy read as follows:

People believe there are beings
That come to you in your darkest hour
Guide you when your life hangs in the balance

Cradle you.
Calm you.
Protect you.
Some people call them *guardian angels.*
We call them nurses.

Perpetuating the Angel Image

These messages are so pervasive that they create a social feedback loop that reinforces and then reproduces the 19th-century view that nurses are sentimental workers who may even act as agents of a higher power (God or the physician). Through a complex historical process nurses inherit these virtuous images. Nurses then stress these images when they discuss their work. Nursing departments and executives may approve these traditional images for use in promotional copy. Not realizing they are reinforcing traditional stereotypes about the profession, they may also suggest the use of these images to public relations departments. Once the virtue script is relayed by nurses to other health care professionals, the public, patients, and the media, these groups broadcast the messages to an even wider audience. This audience then closes the social feedback loop when the idea is projected back on its source—nurses who then have to "live" the ideal.

These messages are so pervasive that nurses themselves believe them. They then try to pass them on to other health care professionals, patients, and media.

Some of the most critical participants in the social feedback loop are the mass media. In stories or headlines about nurses in newspapers and magazines, nurses are often portrayed as self-sacrificing, self-effacing angels of mercy.

In 2001 the *Toronto Star* ran a story about a nurse who had founded a community health care center dedicated to serving children and adolescents. The nurse, Ruth Ewert, after identifying a glaring lack of adolescent health care services, raised money for a center to provide the needed health care. The headline, instead of reflecting her knowledge, courage, and persistence, introduced her as an "angel in our midst."[5]

In the spring of 2003, the *New York Times* ran an article, "Premature Births Rise Sharply, Confounding Obstetricians," on the rising number of premature births.[6] The article featured a photograph of a nurse reaching tenderly toward a premature infant in a neonatal ICU in what could be considered a nurturing act reserved for mothers. The lengthy article is filled with quotes from numerous physicians and demonstrates their scientific knowledge on the subject—but there is only one quote from a nurse and nurses never comment on the science involved in caring for premature infants. The article further supports physicians' quest for knowledge by saying: "Doctors can save most premature babies, but they haven't found a way to stop premature births." Such depictions of nurses have serious effects. Nurses are excluded from the process of scientific curiosity and discovery and from the acts of rescuing babies from complications and saving their lives—which is precisely what nurses who work with such babies do. If health care administrators and policymakers are allocating scarce resources, to whom will they give the money—the "tenders" or the "savers"?

Given the virtue script, it is not surprising that patients whose lives have been saved by good nursing care also seem unable to recognize the knowledge and skill it requires. In his book *Still Me,* the actor Christopher Reeve described in great detail the extraordinary activities of the physicians who saved his life after his 1995 equestrian accident. Of his ICU nurses, he had this to say[7]:

The nurses were so gentle. I still remember their sweet southern voices, trying to strike the correct balance between being sympathetic and being straightforward. One morning a favorite nurse, Joni, arranged for me to be taken up on the roof of the hospital to watch the sunrise.

While this is certainly part of good nursing care, Reeve had made it the totality of nursing care.

This social feedback loop has influenced one of the most expensive contemporary campaigns designed to recruit nurses, address the nursing shortage, and change the profession's public image. The Johnson & Johnson company has spent more than $20 million on its Campaign for Nursing's Future. With the help of nurse advisers, it has produced television spots, videotapes (one in which nurses talk about their work and another in which patients testify to the importance of nurses), and brochures about nursing work.

The campaign's television advertisements are accompanied by voiceovers and a soundtrack. A female nurse appears and says, "The art and science of medicine combined with awesome nursing care can perform miracles." The soundtrack has the following ditty:

There are some who live for caring with all they have to give
There are some who have comfort to share
They dare to care
They dare to cry
They dare to feel
They dare to try . . .
There are some who dare to care

The jingle accompanying the "Patient Perspectives" video includes the following lines:

You're always there when someone needs you
You work your magic quietly
You're not in it for the glory
The care you give comes naturally

In the campaign's recruitment video, a nurse says: "Being a nurse is about holding someone's hand. Being a nurse is about giving a really good shot to a six-year-old who's terrified. It's about putting an ice pack and making it better on someone . . . or getting the wrinkles out of the back of a sheet that's causing someone to be uncomfortable who has to lay on the bed. . . . And sometimes, you know, just rubbing someone's back is the answer to all their prayers." Many nurses have lauded the campaign as a welcome recognition of the importance of nursing—which it certainly is—but have failed to recognize its problematic aspects. Thus nurse Melissa Fitzpatrick, former editor-in-chief of *Nursing Management,* wrote: "These ads showcase diversity, intelligence, competence, and caring—the essence of nursing. They make me even more proud than usual to be a nurse and thrilled that our profession is getting prime airtime that millions of viewers worldwide see each day."[8]

One is struck by the similarity of these campaign messages to those of Hallmark cards for Nurses Week. "What is a nurse?" one of its 2002 Nurses Week cards asks. "A nurse," according to the card's answer, "is a special person, an angel in disguise, with tenderness in every touch, and caring, watchful eyes." These messages are so pervasive that nurses themselves believe them. They then pass them on to other health care professionals, patients, and the media.

Nurses and Public Opinion

Today's nurses are under increasing pressure to concretely connect nursing practice and patient outcomes. It is thus difficult to understand why nursing and nurses appear to have such a limited vocabulary when discussing and promoting their own work. Even more difficult to understand is why, when there is a great deal of data documenting the critical importance of nursing in patient care, nursing groups and their political supporters make so little use of it.

One reason nurses may rely so heavily on the virtue script is that many believe this is their only legitimate source of status, respect, and self-esteem. For the past 150 years nurses have been told that only physicians really need scientific training. Deprived of status and respect that stems from a standard university education, nurses were taught the way to gain social respect was to establish themselves as the most devoted, altruistic, and trustworthy members of the health care team. The polarized view of women and nurses holds that they are either good or evil, Madonnas or whores. Nurses may also feel that these images help to counteract depictions of the cruel nurse, like Nurse Ratched, or the nurse as porn star.

Opinion polls reinforce the belief that nurses are generally prized for their virtues, not their knowledge. Highly publicized polls conducted in North America by Harris Interactive and the Gallup Organization give nurses very high marks for being more ethical, honest, and trustworthy than physicians and many other professionals (nurses topped the Gallup trustworthiness poll last year). But when it comes to "knowledge," nurses survey ratings may plummet. When the Harris poll asked respondents whether they would consult a nurse on a variety of health care issues, on which nurses clearly have great expertise, very few said they would.[9] Those who felt nurses were ethical and honest apparently felt they had only the most limited knowledge and would not ask them questions about women's health, osteoporosis, or sexually transmitted diseases.

Moving Forward

Given the persistence of institutional restrictions on nurses, it is also understandable that so many focus on their virtues. Nurses are too frequently denied a voice on issues of relevance to patient care or their concerns are restricted to the boundaries of caring and then dismissed by hospital administrators.[10] They may face a closed door when it comes to claiming a legitimate voice in the scientific and medical management of patients. They may often be prevented or discouraged from speaking to the media on medical treatment or research and may be excluded from consultations with politicians and policymakers about "medical" policies and issues in which nurses play a critical role.

Although much changed for women in the 20th century, nurses continue to rely on images of hearts and angels and appeals based on references to closeness, intimacy, and holism. These images are a fundamental part of nurses' claim that they have a superior connection to patients—and are a humanizing presence in an increasingly impersonal health care system.

While Johnson & Johnson and others who sponsor recruitment advertisements insist their efforts are attracting new candidates to nursing, it is far from clear that new nurses will remain at the bedside once they discover the realities of nursing work and contemporary working conditions. A focus on nurses' knowledge in the context of the realities of contemporary health care might actually help retention as well as recruitment. For example, in the Johnson & Johnson recruitment video, the nurse could have spoken about the ways in which nurses make sure patients don't develop fatal post surgical complications, and that they pay attention to small details, such as smoothing out wrinkles on a sheet to reduce the risk of a patient developing an excruciating and costly pressure ulcer. Or he could have mentioned the fact that when nurses talk with patients they discover important facts, such as whether patients understand how to take their medications, whether they have support at home, and whether they are frightened and anxious.

Nursing groups need to promote professional data in recruitment and Nurses Week campaigns—and in hospital promotional literature about nursing staff. An ad used by the British Columbia Nurses Union is an excellent example of how these data can be used in a clever way. In Australia, the Australian Nursing Federation has a great slogan: "Nurses, you can't live without them." And United American Nurses has a wonderful comment in one of its brochures: "Saving Lives Is Serious Business."

The crisis in recruitment and retention of nurses demands that the value of nursing knowledge and work, as opposed to the sentimental valuing of nurses, be made clear to the public and employers.

References

1. Buresh B, Gordon S. *From silence to voice: what nurses know and must communicate to the public.* Ithaca, NY: Cornell University Press; 2000.
2. Nelson S. *Say little, do much: nursing, nuns, and hospitals in the 19th century.* Philadelphia: University of Pennsylvania Press; 2001.
3. Peterson MJ. *The medical profession in mid-Victorian London.* Berkeley: University of California Press; 1978.
4. Boyd T. Born to be a nurse. *Nursing Profile.* 2001. www.nursingprofile.com/issue/200108/coverstory.shtml.
5. Taylor B. "Angel in our midst." *Toronto Star,* 9 February 2002, T1.
6. Brody JE. "Premature births rise sharply, confounding obstetricians." *New York Times,* 8 April 2003, D5.
7. Reeve C. *Still me* New York City: Random House; 1998.
8. Fitzpatrick MA. "I" is for image. *Nurs Manage* 2002;33(6):6.
9. Dickenson-Hazard N. New Harris poll is sobering wake-up call for profession. *Excellence in Clinical Practice* 2000;2.
10. Weinberg DB. *Code green: money-driven hospitals and the dismantling of nursing.* Ithaca, NY: Cornell University Press; 2003.

Critical Thinking

1. How has the promotion of traditional images of nursing reinforced stereotypes of the profession?

Suzanne Gordon is the author of Nursing Against the Odds: How Health Care Cost Cutting, Media Stereotypes, and Medical Hubris Undermine Nurses and Patient Care (Ithaca, NY: Cornell University Press, 2005). She lives in Arlington, MA. Sioban Nelson is head of the School of Nursing at the University of Melbourne, Melbourne, Victoria, Australia. Contact authors: Suzanne Gordon, lsupport@comcast.net; Sioban Nelson, siobanmn@unimelb.edu.au.

Delores O'Hara

SUZANNE RIDGWAY

Today there are more than 100 astronauts at the Kennedy and Johnson space centers in Florida and Houston who are attended by more than a dozen aerospace nurses. But when Delores O'Hara was recruited to be NASA's first nurse in 1959, she did not even know what an astronaut was.

It had only been one year since the National Aeronautics and Space Administration had been created, a civilian space agency charged with putting a man into space. Dozens of military test pilots endured rigorous physical and psychological testing before the first seven astronauts were selected in April 1959. Air Force 2nd Lt. Dee O'Hara, RN, became their nurse that November.

In an interview for NASA Quest, an educational website, she says it was "the most ideal job in the world." Dee had the opportunity to travel, meet celebrities, and associate with many famous and interesting people.

Dee had graduated from nursing school and was 22 or 23 when her roommate suggested they join the Armed Forces. They did, and Dee was sent to Patrick Air Force Base in Florida. As a child from a low-income family, she had not had many special opportunities. Her selection as the Mercury Program nurse was, she said, because she "happened to be in the right place at the right time." When offered the job of astronaut nurse, she accepted despite the fact that she wasn't sure what it was.

Dee found that she would be responsible for setting up the Aeromed Lab, the astronauts' exam area, working with Dr. Bill Douglas, head of medicine for the Mercury Program. This was at a time when the effects of g-forces and weightlessness on the human body were still unknown.

America and the Soviet Union were in a space race at the time, with America falling behind. Great importance was attached to the scientific and public relations value of the race and the possible military surveillance opportunities in space. It was a completely new field with much at stake and much to be learned.

When the Mercury astronauts would leave their base in Langley, Virginia, and come to Cape Canaveral for the launches, Dee would help with preflight physicals. When the Mercury Program relocated to the newly built Johnson Space Center, she had to resign her Air Force commission in order to follow the Program, and in 1964 Dee went to Houston to set up the Flight Medicine Clinic there. After the Mercury program ended, she stayed on to work with the Gemini, Apollo, and Skylab programs.

Dee saw all the launches, from the earliest Mercury flights to the first shuttles, and she was close to the astronauts and their families for many years. Dee considers herself lucky to have had such an exciting career and is grateful to have been a part of space history.

Critical Thinking

1. What are some of the responsibilities of NASA's nurses?

SUZANNE RIDGWAY is a freelance writer with extensive experience in the health care field, including administrative roles at Cedars-Sinai Medical Center and Orthopedic Hospital.

Shots Heard 'Round the World

MICHAEL DOSS

"Paralytic polio is not a disease that one recovered from easily. Its physical and emotional consequences were long-lasting and profound, even when polio survivors appeared to have overcome their disability."

—Daniel J. Wilson

Tuesday marks the 50th anniversary of the announcement that a safe polio vaccine had been developed. Before that, epidemics of polio were common and greatly feared because the disease left many patients paralyzed for life. Since the first vaccine, given by injection, was introduced by American researcher Jonas Salk and the second, given orally and developed by another American researcher, Albert B. Sabin, polio has been nearly eliminated.

What is it?: Poliomyelitis is a virus that invades lymphoid tissue, enters the bloodstream and then may infect cells of the central nervous system. As it spreads, it destroys cells.

How is it spread?: Poliovirus is mainly passed through person-to-person contact or contact with waste matter, especially in places with poor sanitation. Only one in 200 people infected develops polio paralysis or other symptoms of polio infection, but it can be carried even by those without symptoms. For that reason, the World Health Organization considers a single confirmed case of polio paralysis to be evidence of an epidemic.

Symptoms (in adults): Usually occur seven to 14 days after infection and may include fever, bad headache, stiff neck or back, deep muscle pain, pins and needles feeling, weakness in limbs and/or paralysis, difficulty swallowing.

Treatment: Vaccines are believed to have eliminated polio in the United States in the 1980s. The World Health

Pioneer in Medicine

Jonas Salk worked in the field of preventive medicine. In addition to his work developing the polio vaccine, he made significant contributions to the understanding of influenza, which can be a severe infectious disease. The vaccine was first tested in 1954 with a mass trial involving 1,830,000 schoolchildren. It was approved for widespread use the next year. Salk got many honors, including a citation from President Eisenhower and a congressional gold medal. He refused all cash awards and returned to his lab to improve the vaccine.

Poliovirus Cases

Global cases of poliovirus, 2004: **1,263**
Polio-endemic countries, 2004
Nigeria **789**
India **136**
Pakistan **53**
Niger **25**
Afghanistan **4**
Egypt **1**
Global cases of poliovirus to date, 2005: **68**
Cases to date, 2005
Nigeria **32**
Sudan **18**
India **12**
Pakistan **4**
Cameroon **1**
Ethiopia **1**

Organization resolved in 1988 to eliminate polio worldwide. Since then, cases have declined but have seen a small resurgence in recent years, especially in parts of Africa and Asia.

How the vaccines work: Two types are available, an inactivated (killed) vaccine and a live attenuated (weakened) oral vaccine. The inactive vaccine works by producing protective antibodies in the blood, preventing the spread of poliovirus to the central nervous system. The oral vaccine also produces antibodies in the blood, but has the additional benefit of limiting the multiplication of the poliovirus. In addition, the oral vaccine is more easily administered and can be produced more inexpensively.

Critical Thinking

1. What are the differences between the Salk Polio Vaccine and the Oral (Sabin) vaccine?

Linda Richards

SUZANNE RIDGWAY

Melinda Richards, known as Linda, learned to be a nurse while she was a child caring for her tubercular mother. Later, the village doctor would take her on house calls and show her how to do simple tasks to care for the sick. She became known as someone who could help if the doctor were unavailable.

As a young woman, Linda wanted a career in nursing and applied for work at Boston City Hospital. At that time "nursing" posts were usually given to charwomen who could also do heavy scrubbing and lifting. What Linda really wanted was formal instruction such as that offered in England at the Nightingale Training School for Nurses, but she knew that no such program was available in the U.S. However, she was determined to get some experience as a nurse, so she took the job. But when she discovered that Boston City Hospital considered the cleaning more important than the patients, she did not stay long.

In 1872 she learned that the New England Hospital for Women and Children would be starting a nurse-training program and she enrolled immediately. In addition to attending lectures and witnessing surgeries, Linda and the four other students cared for patients from 5:30 A.M. to 9:00 P.M., then were on call all night. Linda's suggestion that the nurses take turns working and sleeping was quickly implemented.

Linda was the only one of the five students to graduate and so became "America's first trained nurse" in 1873. Several job offers followed.

She accepted a position as night superintendent at Bellevue Hospital. There a new administrator, Sister Helen Bowdoin, was trying to turn the institution around. As was common at the time, this hospital for the indigent was staffed by drunken criminals who were actually working off their sentences!

Six wards had been set aside for the experiment of using trained nurses instead, in order to determine whether this would facilitate the doctors' work and hasten the patients' healing. Here Linda created the idea of patient charts, tying "penny notebooks" to each patient's bed so that nurses on different shifts could record the patient's condition and graph their vital signs over time.

Subsequent jobs at Massachusetts General and back at Boston City Hospital, supervising their new nurses' training programs, gave Linda more opportunities to implement change while she developed regular classroom instruction. She hired scrubwomen to do the heavy cleaning, established nursing shifts, and set up patient charts. Although many of the physicians were hostile to the training school and its pupils, when they saw results in their patients from the proper utilization of trained nurses, some were won over.

Over the next 20 years, she served as superintendent of training at five more hospitals in the northeastern U.S. and established training courses at three mental hospitals. She even went to Japan to train nurses working in a missionary hospital.

When Linda Richard retired at the age of 70, hundreds of nursing schools had been established, modeled after those she developed. She had succeeded in establishing the importance of training for nurses and bringing nursing out of the realm of menial servant to that of professional caregiver.

Critical Thinking

1. What nursing practice developed by Linda Richards continues to be utilized by health-care professionals for every patient in hospitals today?

SUZANNE RIDGWAY is a freelance writer with extensive experience in the health care field, including administrative roles at Cedars-Sinai Medical Center and Orthopedic Hospital.

Susie Walking Bear Yellowtail

SUZANNE RIDGWAY

The first native American to become a registered nurse, Susie Walking Bear Yellowtail devoted her life to improving healthcare conditions on Indian reservations. Susie Walking Bear was born on a Montana Crow Indian reservation in 1903. She was the child of a Crow father and Sioux mother, but was orphaned at age 12. She later went east to attend the Boston City Hospital School of Nursing. After graduation, she returned home to work on the Crow Reservation in Montana.

Susie married Tom Yellowtail in 1929, a Crow spiritual and tribal leader, and they raised three children of their own and adopted two. She became an activist on behalf of her people when she became aware of sub-standard healthcare practices affecting Native Americans.

While working with white doctors at the Bureau of Indian Affairs Hospital at the Crow Agency, she saw sterilization surgeries performed on Crow women without their consent. Later, she worked for the U.S. Public Health Service, traveling to many reservations, assessing the health, social and education problems plaguing the Native American populations. In one region, she discovered that acutely ill children would sometimes die en route to far-scattered medical facilities while their mothers carried them 20 to 30 miles to seek medical care.

She sought improvement in the healthcare systems on Indian reservations by participating in many councils and committees. She served on the Crow Tribal Education and Health Committees, the U.S. Department of Health, Education and Welfare's Council on Indian Health, and the President's Special Council on Aging. She was a director of the Montana Advisory Council on Vocational-Technical Education and was appointed to the President's Council of Indian Education and Nutrition. In 1962 she received the President's Award for Outstanding Nursing Health Care.

She practiced her native religion and wore native dress throughout her life. She was an artist who worked in Native American beadwork as well. She shared her heritage with non-Indians and became an ambassador for her culture, not only within the United States, but throughout the world. In the 1950s she and her husband were part of a group sent by the U.S. State Department to Europe and the Middle East to promote an understanding of American Indian cultures overseas.

She founded the Native American Nurses Association (now The National Alaska Native American Indian Nurses Association or NANAINA), which named her the "Grandmother of American Indian Nurses," and she helped to win government funding to assist Native Americans enter the nursing profession.

Susie Walking Bear Yellowtail died on Christmas Day 1981 after a lifetime of service to her people as an ambassador, political activist and healthcare giver. Posthumously, she had her picture placed in the "Outstanding Montanans Gallery" at the state capitol in Helena as a tribute to her years of public service.

Critical Thinking

1. How did Susie Walking Bear Yellowtail improve the healthcare system on the Indian reservation?

SUZANNE RIDGWAY is a freelance writer with extensive experience in the health care field, including administrative roles at Cedars-Sinai Medical Center and Orthopedic Hospital.

Lillian Wald

Suzanne Ridgway

Lillian Wald was born into a comfortable Jewish family in 1867, but chose to work in the tenements of New York City. She coined the phrase "public health nursing" and is considered to be the founder of that profession.

Lillian was educated at a private boarding school. She had graduated from a two-year nursing program and was taking classes at the Women's Medical College when she became involved in organizing a class in home nursing for poor immigrants on New York's Lower East Side.

Lillian, distressed by the conditions in the multi-story walkup, cold-water flats, moved to the neighborhood and, along with her classmate and colleague, Mary Brewster, volunteered her services as a visiting nurse. With the aid of a couple of wealthy patrons, the operation quickly grew in size. The Henry Street Settlement (otherwise known as the VSN, or Visiting Nurses Society) grew from 2 nurses in 1893 to 27 in 1906, and to 92 in 1913.

The nurses educated the tenement residents about infection control, disease transmission, and personal hygiene. They stressed the importance of preventative care, but also provided acute and long-term care for the ill. They received fees based on the patient's ability to pay.

The organization also eventually incorporated housing, employment, and educational assistance and recreational programs as well.

In 1912, Wald helped found the National Organization for Public Health Nursing, which would set professional standards and share information. She served as its first president.

Her other accomplishments included:

- Persuading President Theodore Roosevelt to create a Federal Children's Bureau to protect children from abuse, especially exploitation such as improper child labor.
- Lobbying for health inspections of the workplace to protect workers from unsafe conditions and encouraging employers to have nursing or medical professionals on-site.
- Convincing the New York Board of Education to hire its first nurse, which lead to the standard practice within in the U.S. of having a nurse on duty at schools.
- Persuading Columbia University to appoint the first professor of nursing in the country, and initiating a series of lectures for prospective nurses at Columbia's Teachers College. This became the basis a few years later for the University's Department of Nursing and Health and caused nursing education to shift away from solely hospital-taught training to university courses augmented by hospital fieldwork.

Wald wrote two books about her experiences, *The House on Henry Street,* and *Windows on Henry Street.* She died in Westport, Connecticut, on September 1, 1940.

Wald's legacy is seen in the lasting good of her many accomplishments in the areas of public health and social services, not the least of which is her founding of the VSN. The New York Visiting Nurses Association continued to grow and thrive, increasing to 3,000 employees, with the number of people served annually now totaling 700,000. The original VSN is still a model for the 13,000 visiting nurse groups which exist today.

Wald said, "Nursing is love in action, and there is no finer manifestation of it than the care of the poor and disabled in their own homes."

Critical Thinking

1. What was Lillian Wald's legacy to nursing?

Suzanne Ridgway is a freelance writer with extensive experience in the health care field, including administrative roles at Cedars-Sinai Medical Center and Orthopedic Hospital.

UNIT 2

Legal and Ethical Issues

Unit Selections

Learning Outcomes

After reading this unit, you should be able to:

• Discuss the importance of integrating harm reduction when promoting health and well-being for patients with illicit drug use.

• List the ethical and legal considerations for patients who chose to discontinue dialysis.

• Describe the nurse's ethical responsibility to the patient during research participation.

• Explain the ethical considerations of treating the mentally handicapped patient who is on death row.

• Identify the nurse's role when treating a patient who lacks decision-making capability.

Student Website

www.mhhe.com/cls

Internet References

American Hospital Association
www.aha.org
American Nurses Association
http://nursingworld.org
Legal Eagle Eye Newsletter for the Nursing Profession
www.nursinglaw.com
Nursing Ethics in Canada
www.registered-**nurse**-canada.com/**ethical**_issues_in_**nursing**.html
www.nursingworld.org/mods/mod820/ptprivfull.htm

The area of ethics is being discussed by the medical, government and political communities. Legal and ethical issues take on a variety of topics when they are addressed in the context of nursing. The ethical, legal, and social context of harm reduction is essential when promoting ethical practice and education about illicit drug use, harm reduction, and health promotion. It is essential to promote health and well-being so that individuals can obtain their optimal health.

Furthermore, the debate of ethics continues as healthcare professionals interact daily with patients and engage in the decision-making process. The concept of prolonging life or prolonging dying in patients who make the decision to discontinue dialysis treatment should be viewed from the eyes of the patient. It is important that healthcare providers be able to differentiate their own belief from the patient's end-of-life decision and work collaboratively with their patient and other members of the healthcare team to ensure quality of life is achieved.

On the other hand, healthcare providers and patients have an ethical and legal right to refuse to participate in research. Additionally, there are basic principles in research that must be disclosed to participants such as personal autonomy, maintenance of one's dignity, non-coercion, privacy, confidentiality, non-maleficence consent, and informed consent. A primary role of the nurse during research is to act as a patient advocate and protector of the patient's rights. Controversy can occur over ethical, legal, and professional issues when working with the mentally ill on death row. A number of the mentally handicapped prisoners receive psychiatric treatment to assess if they are competent enough to

© PhotoLink/Getty Images

have their sentence carried out. However, what roles, if any, do healthcare professionals play as advocates for these individuals?

Finally, every competent adult has the legal right to refuse medical treatment; however, do older adults whose mental capacities are impacted by diseases such as dementia poses the same legal right to receive and refuse medical treatment? The nurse plays a vital role in ensuring all patients receive respect as they exercise their right to refuse medical treatment. All of the mentioned scenarios require the nurse to act beneficently as she/he advocates for the patient's rights.

The Ethical, Legal and Social Context of Harm Reduction

BERNADETTE PAULEY ET AL.

I llicit drug use is associated with multiple drug-related harms, including high rates of HIV, viral hepatitis, bacterial infections, overdoses, decreased immunity, addiction and violence (Hunt, 2003). From the perspective of those who are street involved, illicit drug use is a primary source of stigma and discrimination that limits access to health care, negatively affects the quality of care and contributes to further marginalization (Butters & Erickson, 2003; Pauly, 2005). Understanding the legal, political and organizational conditions that perpetuate stigmatization and discrimination in health-care practice is essential (Rodney, Pauly & Burgess, 2004).

The purpose of this article is to examine the ethical, legal and social context of harm reduction and to provide guidance to nurses to enhance safe, competent and ethical care for people who use illicit drugs. The importance of integrating harm reduction as part of a broader ethical commitment to promoting health and well-being for individuals, groups and communities is discussed.

The origin of the stigma and discrimination associated with drug use is deeply rooted in the history of societies experience and response to people who use alcohol, tobacco and drugs (Escohotado, 1996). Boyd (1991) argues that the current criminalization of specific psychoactive drugs is part of our "cultural script" and reflects a historical legacy in which certain psychoactive drugs, especially those used by socially and economically marginalized groups, have been branded as eroding morality in society. According to Wood et al. (2003), a primary focus on law enforcement as the means of limiting the supply of illicit drugs has failed and has created a public health crisis. Although ineffective, the continuing focus on law enforcement in drug policy contributes to a "war on drugs" mentality that seeks to punish those who use illicit drugs and the increasing numbers of individuals with problematic substance use in the justice system (Elliott, Csete, Palepu & Kerr, 2005). Nurses may uncritically absorb negative societal attitudes if they are unaware of the history of drug use and the influence of responses to drug use locally, nationally and globally.

When nurses work in ethical climates in which negative attitudes and judgments prevail toward people who have substance use problems, the delivery of health care may be adversely affected. For example, working in a milieu in which it is the norm that those suspected of misusing substances are labelled as drug seeking, lacking in personal responsibility or undeserving of care can have a negative impact on the development of health-care relationships and may manifest in punitive treatment of individuals. The invisible effects of such judgments are lost opportunities for providing timely health care, lack of attention to the underlying social conditions that contribute to poor health and further marginalization (Pauly, 2005).

Harm Reduction

As a philosophic approach, the key principles of harm reduction are pragmatism; humanistic values; focus on harms; balancing costs and benefits; and hierarchy of goals (British Columbia Ministry of Health, 2005; Canadian Centre on Substance Abuse, 1996; Hilton, Thompson, Moore-Dempsey, & Janzen, 2001; Hunt, 2003). The impact of HIV/AIDS and knowledge of the role of injecting drug use in the transmission of the virus prompted implementation of harm reduction strategies (Hilton et al.). Harm reduction is part of a continuum of potential responses to problematic drug use that range from safer use to abstinence (International Harm Reduction Association, 2006; MacPherson, Mulla, Richardson, & Beer, 2005). Strategies that aim to reduce the harms of illicit drugs include needle exchange, heroin and methadone maintenance, distribution of safer crack kits, provision of information on safer injecting practices, supervised injection of illicit drugs in designated environments and drug policy reform.

The adoption of a harm reduction philosophy and these practices have primarily been confined to street outreach, inner city health-care centres, needle exchange programs and, more recently, supervised injection sites. Existing societal values and organizational norms can act as barriers to the adoption of harm reduction strategies where organizational policies are not in place. For example, nurses working in an environment that has no needle exchange policy may feel morally conflicted over their duty to prevent the harms associated with injection drug use. Providing clean needles in such situations may result in feelings of vulnerability, alienation from colleagues or even censure. Conversely, nurses may feel uncomfortable and morally distressed when they act in ways that are consistent with discriminatory attitudes and practices in the presence of cultural norms that stigmatize those who use illicit drugs.

Ethical Reflection

Examining the values and accompanying responsibility statements found in CNA's *Code of Ethics for Registered Nurses* (2002) can guide nurses in their ethical reflection and provide insights into ethical practice in the context of illicit drug use and harm reduction.

Examining the values and accompanying responsibility statements found in CNA's Code of Ethics for Registered Nurses (2002) can guide nurses in their ethical reflection and provide insights into ethical practice in the context of illicit drug use and harm reduction.

Safe, Competent and Ethical Care

The code calls on nurses to "base their practice on relevant research findings and acquire new skills and knowledge in their area of practice throughout their career". Needle exchange programs have been shown to be safe, effective and cost efficient in reducing drug-related harms (Hunt, 2003; Wodak, 2006; Wodak & Cooney, 2005). Supervised injections sites, such as Vancouver's Insite, have been shown to reduce overdoses; limit blood-borne diseases and soft tissue infections; facilitate referrals to other health and social services; and reduce public disorder (Wood et al., 2004; Wood, Tyndall, Montaner, & Kerr, 2006). Currently, there is a disturbing trend in Canada toward disregarding the scientific evidence, potentially jeopardizing public health initiatives designed to improve the health of those who use illicit drugs and protect the health of the community at large.

Health and Well-Being

The code directs nurses to value health promotion and assist individuals to achieve optimum levels of health relative to their own situation. In February 2002, in response to growing concern about the health and well-being of their clients, an inquiry from nurses at the Dr. Peter Centre in Vancouver to the Registered Nurses Association of British Columbia resulted in the statement that "providing clients with evidence-based information to more safely give themselves intravenous injections is within the scope of registered nursing practice" (Wood, Zettel, & Stewart, 2003). The code directs nurses to "recognize the need for a full continuum of accessible health services . . .". Incorporation of a harm reduction philosophy and specific harm reduction strategies helps to shift the health-care culture away from negative judgments, enhances the development of trusting relationships and improves access to

health care for those who are street involved (Pauly, 2005). However, promotion of health and well-being cannot be achieved solely through provision of clean needles and safer injecting practices. Enhancing health and well-being requires attention to the underlying root causes of ill health such as homelessness, poverty, racism and violence.

Dignity

The code states, "Nurses recognize and respect the inherent worth of each person and advocate for respectful treatment of all persons". Additionally, nurses have a particular responsibility to "advocate for health and social conditions that allow persons to live and die with dignity". Such duties are consistent with the principles of harm reduction and a non-judgmental approach that is central to working with all people, including those who use illicit substances.

Choice

Although nurses have a duty to build trusting relationships, foster informed choices, respect choices of individuals and continue to provide opportunities for individuals to make choices, they "are not obligated to comply with a person's wishes when this is contrary to law". Technically, if nurses are found in possession of used syringes that contain drug residue, they can be charged with possession (Wood, Zettel, et al., 2003). However, prosecution is unlikely, because the intent is to prevent overdose deaths, reduce injection-related harm and promote health and well-being for individuals and the community at large. To date, there has not been a documented case of arrest or conviction of a health-care provider in Canada who is providing harm reduction services (Gold, 2003).

Justice

Nurses have a responsibility to participate in "the development, implementation and ongoing review of policies and procedures designed to provide the best care for persons with the best use of available resources given current knowledge and research". The code directs nurses to advocate for policies consistent with current knowledge and research. Nurses can play a key role in fostering the development of organizational policies that include harm reduction strategies such as in-patient needle exchange programs and in-patient safer injecting education.

Understanding the nature of addiction and theories of change can assist nurses in recognizing that frequent visits, relapses or a return to drug use are symptomatic of a chronic relapsing illness and a necessary part of the process of change (Prochaska & DiClemente, 1986). Rather than labelling such individuals as "frequent flyers," it may be more appropriate to keep the door open and implement a revolving door policy (Malone, 1996; Pauly, 2005; Payne, 2007). This approach is more consistent with the needs of those experiencing addiction because it recognizes the cyclical nature of addiction and facilitates access to appropriate and timely health care that maximizes opportunities for disease prevention and health promotion.

According to the code of ethics, "an ideal system of law would be compatible with ethics, in that adherence to the law should never require the violation of ethics. There may be situations in which nurses need to take collective action to change a law that is incompatible with ethics". We contend that criminalization of illicit drugs is such a law. Public health officers in British Columbia (Health Officers Council of British Columbia, 2005), the King County Bar Association (2005) in Seattle and the Transform Drug Policy Foundation (2005) in the U.K. are among many groups who have created policy documents recommending regulation of all currently illegal drugs. Nurses should join those who recognize the failure of prohibition and its impact on public health as part of their ethical responsibilities within a broader commitment to social justice. Action on drug policy reform must be part of a broader agenda to enhance social justice that seeks to take action on the underlying conditions that produce poor health such as homelessness, violence, poverty and racism.

> **Action on drug policy reform must be part of a broader agenda to enhance social justice that seeks to take action on the underlying conditions that produce poor health such as homelessness, violence, poverty and racism.**

Conclusion

Societal values, the criminalization of some psychoactive drugs and the lack of organizational policies endorsing harm reduction contribute to the stigmatization of, and discrimination toward, those who use illicit drugs, creating a context that challenges nurses in their ability to provide safe, competent and ethical care. Failure to embrace a harm reduction philosophy and the lack of policies and strategies may further marginalize individuals who are already disadvantaged in society and can result in serious consequences for the health and well-being of those who use illicit drugs and of the community at large. It is incumbent on nurses to insist on a harm reduction approach for such individuals in all settings, to advocate for changes in drug policy and to take action on the social determinants of health. The lives of many depend on it.

References

Boyd, N. (1991). *High society: Legal and illegal drugs in Canada.* Toronto: ON Key Porter.

British Columbia Ministry of Health. (2005). *Harm reduction. A British Columbia community guide.* Retrieved January 12, 2006, from www.health.gov.bc.ca/prevent/pdf/hrcommunityguide.pdf

Butters, J., & Erickson, P.G. (2003). Meeting the health care needs of female crack users: A Canadian example. *Women and Health, 37*(3), 1–17.

Canadian Centre on Substance Abuse. (1996). *Harm reduction: Concepts and practice.* Retrieved November 21, 2006, from www.ccsa.ca/pdf/ccsa-006491-1996.pdf

Canadian Nurses Association. (2002). *Code of ethics for registered nurses* [Electronic version]. Available from www.cna.aiic.ca

Elliott, R., Csete, J., Patepu, A., & Kerr, T. (2005). Reason and rights in global drug control policy. *Canadian Medical Association Journal, 172*(5), 655–656.

Escohotado, A. (1999). *A brief history of drugs: From the stone age to the stoned age.* Rochester, VT: Park Street Press.

Gold, F. (2003). *Supervised consumption: A professional practice* [Video recording] Vancouver, BC: B.C. Centre for Excellence in HIV/AIDS. Retrieved Oct. 24, 2006, from www.cfenet.ubc.ca/video.php?id=23&sid=33&cat=1

Health Officers Council of British Columbia. (2005, October). *A public health approach to drug control: A discussion paper.* Victoria, BC: Author. Retrieved June 7, 2007, from www.healthcare.ubc.ca/mathias/Framework_for_Drug_Control.pdf

Hilton, B.A., Thompson, R., Moore-Dempsey, L., & Janzen, R. (2001). Harm reduction theories and strategies for control of humen immunodeficiency virus: A review of the literature. *Journal of Advanced Nursing, 33*(3), 357–370.

Hunt, N. (2003). *A review of the evidence-base for harm reduction approaches to drug use.* Available from http://pubs.cpha.ca/PDF/P37/23837.pdf

International Harm Reduction Association. (2006). *What is harm reduction?* Retrieved January 17, 2007, from www.ihra.net/Whatisharmreduction

King County Bar Association. (2005). *Effective drug control: Toward a new legal framework.* Seattle. WA: Author. Retrieved from www.kcba.org/ScriptContent/KCBA/druglaw/pdf/EffectiveDrugControl.pdf

MacPherson, D., Mulla, Z., Richardson L., & Beer, T. (2005). *Preventing harm from psychoactive substance use.* Vancouver, BC: City of Vancouver. Retrieved January 12, 2006, from www.city.vancouver.bc.ca/fourpillars/pdf/PrevHarmPsychoSubUse.pdf

Malone, R. (1996). Almost 'like family': Emergency nurses and 'frequent flyers.' *Journal of Emergency Nursing 22*(3), 176–183.

Pauly, B. (2005). Close to the street. *The ethics of access to health care.* Unpublished doctoral dissertation, University of Victoria, Victoria, BC.

Payne, S. (2007). Caring not curing: Caring for pregnant women with problematic substance use in an acute-care setting: A multidisciplinary approach. In S. Boyd & I. Marcellus (Eds.), *With child: Substance use during pregnancy: A woman-centred approach* (pp. 56–69). Halifax, NS: Fernwood.

Prochaska, J.O., & DiClemente, C.C. (1986). Toward a comprehensive model of behavior change. In W.R. Miller & N. Heather (Eds.). *Treating addictive behaviors: Processes of change.* New York, NY: Plenum Press.

Rodney, P., Pauly, B., & Burgess, M. (2004). Our theoretical landscape: Complementary approaches to health care ethics. In I. Storch, P. Rodney, & R. Starzomsk (Eds.), *Toward a horizon. Nursing ethics for leadership and practice.* (pp. 77–97). Toronto, ON: Pearson.

Transform Drug Policy Foundation, (2006). *After the war on drugs— options for control,* Bristol, U.K.: Author.

Wood, E., Kerr, T., Small, W., Li. K. Maish, D. C. Montaner. J.S. et al (2004) Changes in public order after opening of a medically supervised safer injecting facility for injections drug users. *Canadian Medical Association Journal, 171*(17), 731–734.

Wood, E., Kerr, T., Spitial, P.M. Tyndall, M. W., O'Shaughnessy, M.V., & Schecnter, M.T. (2003). The health care and fisal costs of the drug use epidemic. The impact of conventional drug control strategies, and the potential of a comprehensive approach. *BC Medical Journal, 45*(3), 128–134.

Wood, E., Tyndall, M.W., Montaner, J.S., & Kerr, T. (2006). Summary findings from the evaluation of a medically supervised safer injection facility *Canadian Medical Association Journal, 175*(1), 1399–1404.

Wood, R.A., Zettel, P., & Stewatt, W. (2003). Harm reduction nursing. *Canadian Nurse, 99*(5), 20–24.

Wodak, A., & Cooney, A. (2005). Effectiveness of sterile needle and syinge programmes. *International Journal of Drug Policy, 168,* S31–S44.

Wodak, A. (2006). *Global harm reduction efforts to control HIV among infecting drug users. Current status.* Prenary address to the XVI. International AIDS Conference Toronto, ON. Retreved November 14, 2006, from www.kaisernetwork.org/health_cast/hcast_index.cfm?display=deta&hc=1801.

Critical Thinking

1. Why is it important to integrate harm reduction when promoting health and well-being for patients with illicit drug use?

BERNADETTE PAULY, RN, PHD, Professor, School of Nursing, University of Victoria, British Columbia. **IRENE GOLDSTONE,** RN, BN, MSc, is Director of Professional Education and Care Evaluation, B.C. Centre for Excellence in HIV/AIDS, St. Paul's Hospital, Providence Health Care, Vancouver, British Columbia. **JANE McCALL,** RN, BScN, MSN, is Nurse Educator, St. Paul's Hospital, HIV/AIDS Program, Providence Health Care. **FIONA GOLD,** BA, RN, is Project Coordinator, B.C. Centre for Disease Control Street Nurse Program, Vancouver. **SARAH PAYNE,** RN, MA, is Senior Practice Leader Fir Square Combined Care Unit, B.C. Women's Hospital, Vancouver.

Acknowledgment—The authors thank Jan Storch, RN, PhD, professor emeritus at the University of Victoria, for her feedback and comments on the initial draft of the manuscript.

Dialysis: Prolonging Life or Prolonging Dying? Ethical, Legal and Professional Considerations for End of Life Decision Making

Y. WHITE AND G. FITZPATRICK

Introduction

Discontinuation of dialysis is the second most frequent cause of death in dialysis patients in Australia (22%)[1], which is similar to that reported in the USA [2]. Because the decision to discontinue dialysis is a major life choice, collaborative decision making should be encouraged between the patient, their family and the health care team, and the patient needs assurances of the continuation of care and kindness, a palliative care plan, and the alleviation of suffering.

In Australia there are 7,674 people on dialysis and new patients entering dialysis programs are increasing in age, with the mean age being 59.3[1]. With increasing age come the consequences of ageing on the human body, which has implications for the management of end-stage renal disease (ESRD). Of those who withdrew from dialysis most were in the elderly age group, and 26% had diabetes mellitus[1].

It is suggested that on diagnosis of ESRD, prior to initiation of dialysis, discussion commences with the patient and their family in relation to advanced health care planning. Issues, which should be addressed in these discussions, are shown in Table 1. The outcomes of these discussions should be documented and reviewed during a patient's time on dialysis, as their medical condition changes or deteriorates.

Reasons behind the discontinuation of dialysis are varied and can be either patient or medically based. Patients decide to discontinue dialysis because of an unacceptable quality of life, depression and a chronic failure to thrive[3]. A study reported by Cohen and Germaine[4] stated that although depression was associated with mortality it was not a major factor in decisions to withdraw from dialysis. Medical reasons to discontinue dialysis include: acute, hopeless and intercurrent illness with no hope of recovery or survival, those with a hopeless disability, a patient's response to marked disability, a malignancy that is unresponsive to therapy, and cases of a severe neurological deficit[3]. However, there are many situations where patients decide to discontinue dialysis for other personal reasons.

Decision Making in Relation to the Discontinuation of Dialysis Treatment

Ethical Considerations

Ethical considerations are those, which should be considered in all health care decisions: autonomy, respect for the individual, beneficence, non-beneficence, and justice[5]. The decisions should be patient focused and may require that health care personnel should step back from their own beliefs and should not judge a patient's decision, in relation to their own beliefs. Ethical decisions should be a shared responsibility, 'based on a com-

Table 1 Health Care Planning Issues to Be Discussed Prior to the Initiation of Dialysis

Treatment options including dialysis, transplantation, or no dialysis. (8)

What are the patient's wishes in cases of extraordinary medical conditions and need for resuscitation measures? (8)

What goals, values and other factors does the patient want considered by a substitute decision maker and the health care team? (8)

Who does the patient wish to nominate as their substitute decision maker, and does that person freely agree to be this person. (8)

mon understanding of the goals of treatment, risks and benefits, and the values and preferences of the patient'[6].

A review of the literature identified issues in relation to ethical aspects associated with the discontinuation of dialysis[6–9]. These issues are listed in Table 2.

Because of the continuing stress of dialysis treatments and the associated consequences, it could be expected that patients would welcome the opportunity to discuss end-of life decisions [10]. Studies have reported that 17% of patients on haemodialysis have discussed end of life care with their physician, and 21% have completed a written advanced directive[2], and only 7% a living will[9]. These findings are consistent with the authors' clinical experience, in that very few dialysis patients discuss end of life issues, and in many cases when this subject is mentioned the patient's body language indicates that they do not want to hear (e.g. averting eyes, moving body away from the person speaking). A further study reported 'that patients want to continue to receive the life-saving treatment of haemodialysis until the treatments no longer work, and to receive resuscitation efforts, making the need for advance directives unnecessary'[10]. Ethical decision making should be based upon what the patient and their family perceive as best for them. However, the providing of information by the health care team will foster informed decision making by the patients and their families.

In Australia, New South Wales Health has developed guidelines[11,12] for advanced health care planning, which provides advice to health professionals on how to commence end of life care planning with patients and their families. These documents strongly recommend that patients be actively encouraged to discuss their wishes, with their families and health care team, for those situations where medical treatment becomes futile, or where they may become incompetent and unable to make decisions regarding their health care. Decisions will be reliant upon the individual's own beliefs, and will be influenced by previous life experiences, cultural and religious/spiritual belief systems. Clinical experience would suggest that the patient's physician or dialysis nursing staff will initiate discussions in relation to discontinuing dialysis in most cases, as a response to observed overt suffering of the patient.

Legal Considerations

Legal considerations are also dynamic and any decision to discontinue dialysis may be open to lawful challenges. Patients do have a right to self-determination, even if this should result in their death. There have been several cases in Australia, which highlighted the confusion in law in relation to withdrawing treatment. A judgement made in the USA in a withdrawal of treatment case[6] found that 'Declining treatment may not be properly viewed as an attempt to commit suicide as the refusal merely allows the disease to take its natural course and the subsequent death would be the result primarily of the underlying disease and not the result of a self-inflicted injury'. Decisions to discontinue medical treatment in the cases of a non-competent person can be very difficult from an ethical and legal standpoint. In these situations a substitute decision maker should be nominated (this is where advanced directives can make decision making easier), and this person should have Power of Attorney or be given Enduring Guardianship. When the patient or their families demand the initiation of 'inappropriate' treatments, a very grey area ensues. However, a physician would be under no obligation to treat a patient where there is irrefutable evidence to support that any such treatment would be futile[11]. Renal health care professionals 'should recognise when dialysis changes from being a measure that prolongs life to one that merely prolongs dying [this is] the point to help patients and their families accept the futility of further treatment', and make the dying process comfortable and dignified[13]. It is imperative that any advanced health care directives, or discontinuation of futile medical treatment should be documented unambiguously and includes what specific treatment is to be commenced or continued, what specific treatment is to be stopped and what treatment is not to be given.

Table 2   Issues to Be Addressed in Decision Making Regarding the Discontinuation of Dialysis

Does the patient really understand the implications of stopping dialysis? (6)

Does the patient mean what they say or is it an attempt for attention? (6)

Can any changes be made to treatment to increase the patient's perceived quality of life to allow further dialysis? (6)

Are there other professionals to whom the patient would like to discuss their decision? (6)

What is the overall psychological state of the patient? (8)

Has the patient been assessed for depression? (8)

What is the cognitive ability of the patient to understand not only the options but also the consequences of any decisions? (8)

How does/or did the patient define quality of life verbally with the renal team and their family? (8)

Does the burden of continued treatment outweigh the benefit based on the patient's wishes of factors they would want considered? (8)

What is the patient's prognosis with and without dialysis? (3)

Has the patient been assured that they will not be abandoned and that their death will be comfortable and dignified? (3)

Has the patient been informed of what death may be like following the build up of uraemic toxins? (3)

Has time been allowed for the patient and their family to reach an informed decision? (3)

Has the patient been assured that the level of care will remain the same with the goals of care changing from the prolongation of the patient's life to comfort and palliation? (3)

Practical Tips for the Care of the Person after the Discontinuation of Dialysis

It has been found that death occurs within 5 ± 3 days (median 5 days) after the last dialysis, in those who discontinue dialysis and have an acute medical complication[2]. In those without an acute medical condition death occurs within 9 ± 8 days (median 6 days) after stopping dialysis therapy ($p = 0.03$)[2]. Therefore the palliation period in the case of ESRD is very brief. However, if there have been free and open discussions between the patient and their physician most patients with ESRD come to this period in readiness for death[2].

Renal nurses should be assessed for their knowledge and skills in relation to palliative care so they can optimise a patient's care and dying experience[7].

There is a paucity of information with regard to palliation in those with ESRD who have discontinued from dialysis. The fear of dying, pain, suffering, and abandonment that a patient and/or their family may perceive as being associated with death may create barriers to decisions to discontinue with dialysis treatments. Therefore health care personnel (in particular doctors and nurses) should provide information with honesty to allow patients to predict their quality of life and death. Support for the patient and family during the dying period should be multidisciplinary, with clear and timely communication between all members of the team. All health care team members should be aware that the decision to discontinue dialysis is a very courageous one with the patient making an overt decision to die. Studies[14, 15] found that the patient and family defined a 'good death' as pain free, peaceful and brief, and that they should have the choice as to the site of care (hospital, hospice and home). These studies found the 'best' of deaths provided time to communicate, the reliving of meaningful past experiences, laughing and crying, and an orderly closure of personal and business affairs[14, 16].

Initiating the Option to Withdraw from Dialysis

Initiation of discussions in relation to discontinuing dialysis should only be undertaken following a detailed assessment of the patient, which should involve both physical and psychological aspects. Clinical experience would suggest that if a patient initiates this discussion, they will approach a member of the renal care team, in whom they have sincere trust. Any such approach by a patient or a family member should be treated with respect.

Collaborative Management Following the Discontinuation of Dialysis

Because of the complexity of ESRD, patients will have many symptoms associated with this disease, which have to be managed during the palliative stage. During the terminal phase of ESRD there will be a rapid accumulation of uraemic toxins and fluid, which will exacerbate symptoms. The major problems, which should be addressed, are hyperkalaemia (cardiac arrhythmias and cardiac arrest—lethal at approximately 8 mmols./litre), fluid overload (generalised oedema) and acidosis[3]. It is the degree of acidosis, which has been postulated as the primary cause of the uraemic coma, which occurs following discontinuation of dialysis[3]. Uraemic symptoms, which require special consideration for those with ESRD in the terminal phase of their life, include:

Thirst

Due to the high solute load from the build up of uraemic toxins, thirst is always an issue for those with ESRD and this continues during the terminal phase. Therefore, any care should involve a review of any medications or sustenance (high salt foods), which may exacerbate this thirst. Diligent mouth care should be undertaken with the use of frequent mouth toilets and the use of mouth sprays.

Pulmonary Oedema

Although there may be generalised oedema, the development of pulmonary oedema is very distressful for the patient, and families. This may be controlled with the use of salt and fluid restrictions, and if necessary the use of selective ultrafiltration to remove excess fluid to ensure the patient is comfortable.

Uraemic Taste in the Mouth

This is related to the build up of uraemic toxins, and the breakdown of urea by bacteria in the mouth[3]; this can be relieved with diligent mouth care. Associated with this is the development of a fetid oral odour and family members should be informed of the reason behind this, as clinical experience would suggest that the family may find this very distressing.

Hiccoughs

These are most likely the result of irritation of the phrenic nerve, and can be relieved by the administration of chlorpromazine[3, 17].

Restless Legs

This condition is common in people with ESRD, and is most likely related to the degree of peripheral neuropathy, which occurs in ESRD. The use of Clonazepam and opiates may be helpful[3,17].

Pruritus

Pruritus in ESRD is related to a complex interaction of neuropathies and hormones (parathyroid hormone) and resultant calcium and phosphate imbalances. Keeping the skin clean, cool and moist can relieve the itching. Antihistamines may be of some use[3,17].

Lethargy and Irritability

Although patients with ESRD report a severe degree of fatigue and lethargy[18], this increases in the terminal phase. This increase in lethargy and the onset of irritability is associated with the increasing uraemic effects on neurological functioning.

Clear explanations explaining to the family why the patient is becoming sleepier and less responsive will be paramount[3].

Convulsions and Coma

As the uraemia worsens so do the neurological signs and symptoms. There may be myoclonic jerks, isolated muscle contractions and asterixis[19]. As the uraemia increases patients become more confused, disorientated, and very sleepy which will ultimately result in a coma. Treatment with Dilantin at this time is beneficial in controlling seizure like activity and decreasing anxiety in the patient and their family members[3].

Pain

Pain is unusual in a patient who has withdrawn from dialysis however there may be pain associated with comorbidities (e.g. neuropathy, arthritis), and as with any person undergoing palliation the control of pain is vital. An in-depth pain assessment should be undertaken and the most appropriate analgesia commenced. Morphine is probably the most effective agent in this situation although its use is now being debated (see Murtagh et al in this journal).

It should be cautioned that medications, which may be given to the patient with ESRD in the terminal phase, should be reviewed, as dosages may have to be modified. The excretion of metabolites from many medications is impaired due to loss of renal function and so may lead to the development of toxicity[17]. This is especially relevant to the administration of morphine.

Spiritual and Psychological Needs

It should be recognised that in this twilight period there is often a heightened need for comfort and reassurance. The balance between needing company (such as with family members) and quiet time (for peaceful contemplation) needs consideration. Dying patients need to be respected and cared for, yet for a dignified ending they should be neither smothered with attention nor abandoned. Helping manage this balance in palliative care situations can require tact and firmness to ensure the patient's needs are given priority and the family's needs are appropriately supported. Knowing what is right in each situation requires careful assessment of each individual situation before the discontinuation of dialysis. Finally, it should also be recognised that at this time staff may well need additional support. Fostering this support within the team (such as through open and honest team meetings) can help give staff emotional strength in this difficult work. A supportive staff environment will also benefit and support patients and their families with these difficult decisions.

Conclusion

This paper has presented a review of current literature with an aim of providing practical tips for collaborative management of the person with ESRD who has discontinued dialysis and requires palliative care. The discontinuation of dialysis is the second most common cause of death in ESRD patients in Australia, and so presents a challenge to health care services to provide palliative care services for these people. The identified issues of use of advanced health care directives and the ethico/legal considerations in decision making are still controversial and can create conflict for health care professionals. The use of advanced health care planning is useful in decision making as it provides direction for health care personnel in the care of those with ESRD. However, it must be stressed that those who choose to discontinue dialysis are comprehensively assessed prior to dialysis being ceased. End of life care for those with ESRD provides opportunities for further research.

Personal Comment by Author

I have been involved in decisions to withdraw dialysis treatment. The decision making has been a collaborative effort between the renal team, the patient, and their family. However where there have been difficulties, these have in the main related to the inability of some health care personnel to accept the patient's right to decide to withdraw from treatment. Some of these have been based upon the personal religious beliefs of the staff concerned, and short term contact with the patient and their family. There has also been difficulty in communicating patient and family wishes to health care professionals over a large geographical area where the patient has chosen to die in their local environment which may be many kilometres from the parent dialysis centre. However through committed renal care staff, these issues can be overcome. It has been my experience that some patients who discontinue dialysis do so because of fatigue. Some have had many illness events and associated hospitalisations, and are 'just too tired to fight anymore'.

Summary

There are over 7,000 people on dialysis in Australia and this is predicted to increase due to the ageing population and the high incidence of diabetes mellitus. Discontinuation of dialysis is the second most frequent cause of death in dialysis patients in Australia. Risk factors for the discontinuation of dialysis include: co-morbidities (especially diabetes mellitus) and being older. Because the decision to discontinue dialysis is a major life choice, collaborative decision making should be encouraged, and the patient needs assurances of the continuation of care and kindness, a palliative care plan, and the alleviation of suffering. Patients decide to discontinue dialysis because of an unacceptable quality of life, depression and a chronic failure to thrive. Health professionals need to support end of life decision making using an ethical decision framework. A review of current literature was undertaken and revealed a paucity of information in regard to palliation in those with end stage renal disease who had discontinued dialysis. The fear of dying, pain, suffering, and abandonment that a patient and/or their family may perceive as being associated with death may create barriers to decisions to discontinue with dialysis treatments. Therefore health care personnel should provide information with honesty to allow patients to predict their quality of life and death. Support for the patient and family during the dying period should be multi-disciplinary, with clear and timely communication between all members of the team.

References

1. Excell L and MacDonald S. **The 27th Report Australian and New Zealand Dialysis and Transplant Registry (ANZData).** Adelaide: Australian and New Zealand Dialysis and Transplant Registry, 2004

2. Holley J. L. A single centre review of the death notification form: discontinuing dialysis before death is not a substitute for discontinuation from dialysis. **American Journal of Kidney Diseases** 2002; 40: 525–530.

3. DeVelascoR and Dinwiddie LC. 1998. Management of the patient with ESRD after discontinuation from dialysis. **American Nephrology Nurses Journal** 1998; 25: 611–614.

4. Cohen L.M., Germain M.J. The psychiatric landscape of discontinuation. **Seminars in Dialysis** 2005; 18: 147–153.

5. Berglund, C. **Ethics for Healthcare.** Oxford: Oxford University Press, 1998.

6. Moss Ah, Holley J.L, Davison S.N., Dart R.A, Germain M.J, Cohen L. and Swartz R.D. Core Curriculum in Nephrology: Palliative Care. **American Journal of Kidney Diseases** 2004; 43: 172–185.

7. Johnson A. and Bonner A. Palliative Care challenges: implications for nurses' practice in renal settings. **Contemporary Nurse** 2004; 17: 95–101.

8. Perkins Loftin, L. and Beumer C. Collaborative end of life decision making in end stage renal disease. **American Nephrology Nurses Journal** 1998; 25: 615–618.

9. Singer P A, Thiel E.C., Naylor D., Richardson R.M.A., Llewellyn-Thomas H., Goldstein M., Saipoo C., Uldall P.R., Kim D. and Mendelssohn D.C. Life sustaining treatment preferences of haemodialysis patients: implications for advance directives. **Journal of the American Society of Nephrology** 1995: 1410–1417.

10. Olivier-Calvin A. Haemodialysis patients and end-of-life decisions: a theory of personal preservation. **Journal of Advanced Nursing** 2004; 46: 558–566.

11. New South Wales Department of Health. (a). **Guidelines for end-of-life care and decision making.** New South Wales Department of Health. North Sydney 2005. (www.health.nsw. gov.au).

12. New South Wales Department of Health. (b). **Using Advanced Care Directives.** New South Wales Department of Health. North Sydney 2005. (www.health.nsw.gov.au).

13. Oreopoulos D.G. Discontinuation from dialysis: when letting die is better than helping to live (Commentary). **The Lancet** 1995; 346: 3–4.

14. Cohen L.M., Germain M.J., Poppel D.M., Woods A.L., Pekow P.S. and Kjellstrand C.M. Dying well after discontinuing the life support treatment of dialysis. **Archives of Internal Medicine** 2000; 160: 2513–2521.

15. Steinhauser K.E., Christakis NA., Clipp EC., McNeilly M., Mcintyre L., and Tulsky JA. Factors considered important at the end of life by patients, family, physicians, and other care providers. **Journal of the American Medical Association** 2000; 284: 2476–82.

16. Cohen L.M., McCul J.D., Germain M. and Kjellstrand C.M. Dialysis discontinuation: a good death? **Archives of Internal Medicine** 1995; 155: 42–47.

17. Cole R. and Formby F. Symptom **Control in Palliative Care.** Frenchs Forest: Adis International, 2000.

18. White Y. and Grenyer BFS. The biopsychosocial impact of end-stage renal disease: the experience of dialysis patients and their partners. **Journal of Advanced Nursing** 1998; 30: 1312–1320.

19. Nicholls A.J. **Nervous System.** In: Daugirdas, JT and Ing, TS. (eds). Handbook of dialysis (2nd edn). Boston: Little, Brown and Company, 1994.

Critical Thinking

1. What are the ethical and legal considerations for patients who chose to discontinue dialysis?

From *EDTNA/ERCA Journal,* April/June 2006, pp. 99–103. Copyright © 2006 by EDTNA/ERCA (European Dialysis and Transplant Nurses Association/European Renal Care Association). Reprinted by permission via Wiley-Blackwell.

Being a Research Participant: The Nurse's Ethical and Legal Rights

PAULINE GRIFFITHS

This article explores the rights and responsibilities of nurses and patients who are participants in research projects conducted in clinical settings. Although there is a substantial body of literature that discusses the role of the researcher and the ethical principles involved in nursing research (Johnson, 1992; Merrell and Williams, 1994; Holloway and Wheeler, 1995; de Laine, 2000), there is need for the rights of the nurse as a research participant to be detailed. Nurses in clinical practice have the additional responsibility of a professional duty of care when their patients are used as data sources during a research project. As the Nursing and Midwifery Council (NMC, 2004 p4) note:

'You are personally accountable for your practice. This means that you are answerable for your actions and omissions, regardless of advice or directions from another professional.'

While teaching research ethics to nurses during research appreciation courses the author found that nurses without a research background often lacked an understanding of the questions to ask of would-be researchers. Sometimes nurses indicated that they were not aware that they could refuse to participate in research. This article provides an overview of the key legal and ethical principles that the nurse should insist be satisfied before any researcher conducts data collection involving the individual nurse, or within the nurse's clinical area. Firstly an overview of the path to gaining access to conduct research within a clinical area is discussed.

Gaining Access

If a research project is to be conducted on NHS premises and/or seeks to involve human subjects a research proposal must first have been presented to the Local Research Ethics Committee (LREC). The researcher will have completed an in-depth research proposal in a standardized format that is used across the UK (Central Office for Research Ethics, 2005). The role of the LREC is to judge whether proposed research will be conducted in an ethically appropriate manner (Royal College of Nursing (RCN), 2004).

LRECs have been criticized as being too geared to quantitative research and lacking understanding of qualitative research (Dolan, 1999). This is relevant for research in clinical settings as such research will often be qualitative and, as such, the correct way to proceed ethically is not always as clear-cut as in quantitative studies. For example, if a new drug is being tested a quantitative study could be designed. A suitable sample of research subjects will be recruited and after explanation a signed consent will be obtained that explains the purpose of the study, the risks and benefits involved, and that the subject can withdraw from the study at any time. A quantitative trial is set up, the findings analysed and written up (Polit and Hungler, 1999).

Most qualitative research approaches, however, do not follow such an explicit path. The emergent design of most qualitative studies means that many decisions about who will be observed or asked for interview will be made while the research is in progress (Hammersley and Atkinson, 1995). Data collection methods in qualitative research include, most commonly, participant and non-participant observation and semi-structured and unstructured interviews (Morse and Field, 1996), plus other methods such as diary writing or focus groups (Bowling, 2002). These qualitative data collection methods may then require particular and contextual approaches to issues such as gaining consent (Madjar and Higgins, 1996; Johnson and Long, 2003). However, once a research proposal has been given ethical approval by the LREC, the researcher, even in qualitative research, must not change the research approach fundamentally without gaining fresh approval from the LREC.

Example 1

A researcher approaches the ward manager and asks if she can conduct data collection using participant observation of the nurses working on the ward. The ward manager then asks to view the research proposal and the letter giving ethical consent from the LREC. The ward manager ensures that the researcher has gained permission to conduct the research from the Trust manager and the Trust's own Research Committee. These criteria being satisfied, the ward manager would then ensure that full information is given to potential participants and that it is made clear that participation is voluntary.

Nurses can be guided to what is appropriate research conduct, both legally and ethically, by professional advice (RCN, 2004; 2005) and by research governance. In the UK, research governance has been developed to promote high standards in the conduct of research (Department of Health (DH), 2001; Wales Office of Research and Development for Health and Social Care, 2001). This governance was developed from existing guidance in place to prevent any recurrence of the human atrocities perpetuated by healthcare workers in the name of scientific research during World War II. After the war, the Nuremberg Code on research involving human subjects was agreed and developed by the Tokyo Declaration revised in 2004 (World Medical Association, 2005). Within this guidance, which acts as a reference for all human research ethical codes, it is noted that:

> 'The right of research subjects to safeguard their integrity must always be respected. Every precaution should be taken to respect the privacy of the subject, the confidentiality of the patient's information and to minimize the impact of the study on the subject's physical and mental integrity and on the personality of the subject.'

Such underlying principles are translated into the legal and ethical considerations to which researchers adhere. These are listed in *Table 1*.

Using these principles as a guide, the ethical and legal questions that the nurse should ask of the researcher will be discussed next.

Autonomy, Consent and Informed Consent

The ethical principle of autonomy (or self-rule) is a basic right of the individual. Autonomy respects our right to make decisions based on our own deliberations (Gillon, 1994). From this principle, develops the individual's right to informed consent and an absence of deceitfulness from others. This means that any individual who is asked to take part in a research project can say no without fear of negative effect from the refusal.

Example 2

A lecturer from the local college of nursing wishes to interview nurses working on your ward. Your manager has told the

Table 1 Ethical and Legal Principles in Research

- Personal autonomy
- Informed consent
- Dignity
- Non-coercion
- Privacy
- Confidentiality
- Non-maleficence (do no harm)

Source: Department of Health (2001).

Table 2 Consent and Informed Consent

Consent
- Common law recognizes that the individual has the right to have his bodily integrity protected against invasion by others
- The giving of consent is a defence against a charge of battery for which damages can be awarded

Informed consent
- Consent that is informed is based on knowledge of the nature, consequences and alternatives associated with a proposed therapy or research undertaking
- Purpose of the research, the benefits to the participant and society, the risks involved, and alternatives available must be explained

Source: Mason and McCall Smith (1994).

lecturer that her staff will cooperate. Do you have to be involved in these interviews? The answer is no. No one can consent for you and you have the right to agree or refuse without fear or favour.

As in clinical practice the legal requirements of consent and informed consent are not one and the same. Consent gives a defence to a legal charge of battery whereas informed consent takes this further and requires full disclosure of information from the researcher (see *Table 2*). As Holloway and Wheeler (1995 p224) note:

> 'Informed, voluntary consent is an explicit agreement by the research subjects, given without threat or inducement and based on information that any reasonable person would want to receive before consenting to participate.'

When giving your consent it is useful to check that:

- Full information has been given that is easy to understand. If not, seek more information or refuse to give consent
- Sufficient information has been given so that an informed consent can be given
- It is clearly stated that the participant is free to consent or not; also that consent can be withdrawn at any time
- It is clear what the data will be used for and who will be able to access them.

However, since World War II, despite clear ethical guidance, there have been shocking abuses of individuals' rights in the name of science. For example, in 1943, a study was developed in the US to study the effects of syphilis on a group of poor, African-American men. These men had agreed to take part in the study but the participants were not told that a treatment for syphilis (penicillin) was being withheld from them on purpose so that the long-term physical effects of the disease could be studied. This study contradicted ethical guidance that was in place, as well as being morally wrong, but this study, known as the Tuskegee Study, continued up until the 1970s (Davies, 2001). Thus, ethical principles alone are not enough to ensure ethically appropriate behaviour; other safe-guards must be in place, like LRECs and the vigilance of others.

Informed consent requires that participants not only give their consent to be involved in the research but also that they understand what it is that they are agreeing to. The RCN (2005 p7) have suggested that:

'Informed consent is an ongoing agreement by a person to receive treatment, undergo procedures or participate in research, after risks, benefits and alternatives have been adequately explained.'

So the nurse, like the rest of society, has a clear right to informed consent, as has the patient in the nurse's care. As the World Medical Association (2005) notes, all subjects must be volunteers and informed participants in research projects. Moreover, the gaining of consent should not be considered a once and for all event as researchers are expected to ensure that participants continue to agree to data collection (Merrell and Williams, 1994) and participants must be kept informed of any changes in the study (RCN, 2005).

Complex Areas When Gaining Consent

There are circumstances where gaining consent is particularly challenging. However, this discussion is beyond the scope of this article (see *Table 3*). The reader is guided to the references used in this article that will give further information on these situations. In summary, the situation is, that unless the would-be participant is an adult (over 18 years of age), clearly has the mental ability to give informed consent, and that the consent can be given freely and without coercion (RCN, 2005), then the validity of the given consent must always be queried.

Confidentiality and Privacy

The data generated in qualitative research are:

'. . . personal, identifiable and idiosyncratic material [so] that questions of confidentiality and anonymity are raised in particularly sharp form' (Mason, 2002 p201).

Therefore, the researcher should explain what procedures are in place to protect the individual's right to confidentiality and privacy (RCN, 2004) (*Table 4*).

Table 3   Complex Areas When Gaining Consent

- Emergency situations: delayed consent might be appropriate
- Children: proxy consent, such as from parents may be used
- Mental incapacity: best interests arguments and personal wishes and values might be used

Source: Royal College of Nursing (2005).

Table 4   The Right to Confidentiality: A Checklist of Questions

- How will data collected be store?
- What steps have they in place to ensure confidentiality?
- What will be done with the data after the study is completed?
- Who else will access the data?
- Are you able to view data collected from yourself and ask for data to be removed?

Source: Royal College of Nursing (2004)

Data Protection Act 1998

An area of special responsibility for the nurse regarding patients and research relates to the Data Protection Act (1998) which protects the individual's right to privacy of information. Therefore, researchers are not automatically permitted access to the names of potential participants, such as NHS patients. To comply with the Data Protection Act researchers are sometimes required to use nurses as intermediaries to recruit patients to research studies (Redsell and Cheater, 2001) Acting as an intermediary, places a special ethical responsibility on the nurse. The nurse must satisfy herself that the research is ethically sound and that the correct access permissions have been obtained, such as LREC sanction and written consent to proceed gained from Trust managers. Approval is also likely to be needed from the in-patient's medical consultant. Care must be exercised to guard the patient's privacy and rights to confidentiality. The patient might agree more readily to participate in the research when an approach comes from a nurse whom they know than if they were approached directly by the researcher. Patients may wish to please the nurse or feel that they cannot say no, so the use of the nurse as an intermediary may, without care, lead to a degree of coercion (Redsell and Cheater, 2001).

Dual Roles

If the researcher is undertaking data collection using participant observation (Morse and Field, 1996), they may present themselves in the clinical setting as a healthcare worker. This is common form of data collection in ethnographic studies and action research. It is important that patients then understand that the person caring for them has a dual role as a practitioner and as a researcher (Holloway and Wheeler, 1995). Not to clarify this compromises the patient's right to privacy and autonomy. For example, if the patient consents to a nurse researcher providing care, this does not mean that they have consented to be part of the researcher's data collection. The researcher, who touches a patient or client without consent, as discussed earlier, commits a battery upon the person (Mason and McCall Smith, 1994). Rights to confidentiality can also be compromised if the patient tells the researcher private details believing that they are confiding in a nurse. Ethically sound research practice would consider that any form of covert (hidden) research is seldom justified (Hammersley and Atkinson, 1995).

Non-Maleficence (Do No Harm)

Any research project must ensure that any potential harm is minimized (DH, 2001) and this is a key aspect that the LREC will have addressed. However, the harm that can occur does not only relate to physical harm, as would be found in human subject experiments. Harm experienced could also be psychological or related to the participant's professional standing. Indeed, poorly thought out or carelessly conducted research causes harm in that it wastes limited time and resources (Mason, 2002).

Example 3

Staff nurse Mary has been asked to participate in a research project related to caring for terminally ill patients in the community. Mary consents, as this is an area in which she has an interest. During the interview, Mary finds that she becomes very tearful and finds that the questions have probed areas that she had not before examined herself.

The researcher should acknowledge that his questions might cause the participant to become distressed; this is particularly true in sensitive areas such as in the above example. The researcher should have given information to the nurse before the interview explaining the purpose of the interview and the use that the data will be put to. It is useful to note if the researcher in this information has considered that such a reaction is possible and what they will do should it occur. Astedt-Kurki et al (2001) suggest that the researcher, while not a counsellor, does have a moral responsibility to his participants and this may involve suggesting referral to other agencies, such as counselling services. Hammersky and Atkinson (1995) warn researchers to be mindful of such issues and that a careful ethical line must be drawn between sensitive issue research and exploitation.

Problematic Care Situations

The researcher should acknowledge what her response will be if she witnesses care delivery that is less than perfect, or indeed harmful. For instance, in a study that used covert participant observation on a forensic mental health unit, poor care was described in some detail that was not challenged at the time (Clarke, 1996). Such research can be criticized for the lack of intervention on the part of the nurse researcher. If the researcher is a nurse, then she carries a duty of care and is professionally bound to intervene if witness poor or harmful care is witnessed. Moreover, just to witness poor care and not to intervene is increasingly being seen as unjustifiable, even if this will negatively affect continued data collection. Researchers who do not query issues when in the field but merely write them up in their research reports can be judged as acting in an exploitive manner (Kleinman and Copp, 1993). If the researcher discusses observations of care practice that she finds troublesome then the issue can be addressed, or at least the participant can be given an opportunity to explain and justify why such actions were taken. However, while no one can ever condone practice that causes harm to patients or is illegal, there are situations in clinical practice which are ambiguous, such as the 'gallows' humour' that makes fun of life-threatening, disastrous or terrifying situations and is used by nurses to relieve stress.

Example 4

A researcher has been conducting an ethnographic study using participant observation of an intensive care unit. After a horrific road traffic accident several members of the same family are admitted to the unit. Tragically at the site of the accident one of the family members was decapitated. At handover, the unit nurses discuss this and someone, to relieve the tension, makes a joke. The researcher uses this example in an article published in a nursing journal from his research findings but subsequently national newspapers then pick up this story and run banner headlines criticizing 'cold-hearted nurses'.

As research participant you have the right to ask that the researcher does not include sensitive issues in the research report, for example, you can ask for a copy of the transcript of any interview that you are involved in and ask for removal of statements. However, as in example 4, even though the researcher was conducting overt (not hidden) participant observation the nurses could not view the data collected as in an interview transcript. Researchers should agree to discuss potentially sensitive issues with the participants and reach a consensus on how they will present such data (Mason, 2002). As in the example above, how the incident is reported and the interpretation given to it will make the difference between sensitive insights as opposed to sensational reporting.

Conclusion

Potential research participants have the right to seek assurances that their autonomy will be respected and that their informed consent will be sought. Questions must be asked about how the researcher will ensure the privacy and the right to confidentiality of the research participant. Is there any aspect of the research study that has the potential to cause harm and if so what safeguards are in place? If there is a risk that participating in the research may lead to harm to the participants the researcher must acknowledge any risk factors and strategies to minimize or deal with any such risk. The nurse within her duty of care must also ensure that researchers who wish to collect data involving patients in the nurse's care have fulfilled the required ethical and legal requirements.

Being a participant in a research project can be very interesting, and for the nurse new to the research, an exciting way of gaining insights into the research process by personal experience. Properly planned and scrutinized research proposals would never condone illegal or ethically incorrect practice. However, those nurses working in clinical settings where research is being conducted must monitor the conduct of data collection. By understanding basic principles the nurses can protect their own rights and those of their patients.

Key Points

- Nurses as potential research participants must ensure that their rights are not compromised.
- All nurses require an understanding of the legal and ethical requirements placed on researchers.

- Key principles in ethically conducted research are autonomy, informed consent, confidentiality, and non-maleficence (do no harm).
- The correct approaches, ethically and legally, to research undertaken in the clinical setting require special considerations.
- The duty of care that nurses carry for their patients includes ensuring that patients' rights are not compromised during research projects.

References

Astedt-Kurki P. Paavilainen E, Lehti K (2001) Methodological issues in interviewing families. *J Adv Nurs* **35**: 288–93.

Bowling A (2002) *Research Methods in Health.* 2nd edn. Open University Press, Buckingham.

Central Office for Research Ethics Committees (COREC) (2005) *About Us.* www.corec.org.uk/applicants/about/about.htm. Accessed 21 March 2006.

Clarke L (1996) Participant observation in a secure unit: care, conflict and control. *NT Research* **1**: 431–40.

Davies CA (1999) *Reflexive Ethnography: A Guide to Researching Selves and Others.* Routledge, London.

de Laine M (2000) *Fieldwork, Participation and Practice.* SAGE Publications, London.

Department of Health (2001) *Research Governance Framework of Health and Social Care.* DH, London.

Dolan B (1999) The impact of local research ethics committees on the development of nursing knowledge. *J. Adv Nurs* **30**: 1009–10.

Gillon R (1994) Medical ethics: four principles plus attention to scope. *Br Med J* **309**: 184–8.

Hammersley M, Atkinson P (1995) *Ethnography: Principles in Practice.* 2nd edn. Routledge, London.

Holloway I, Wheeler S (1995) Ethical issues in qualitative research. *Nurs Ethics* **2**(3): 223–31.

Johnson M (1992) A silent conspiracy? Some ethical issues of participant observation in nursing research. *Int J Nurs Stud* **29**(2): 213–23.

Johnson M, Long T (2003) Rigorous governance, yes—but let's get the controls right. *Nurs Res* **11**: 4–6.

Kleinman S, Copp MA (1993) *Emotions and Fieldwork.* SAGE Publications, Newbury Park.

Madjar I, Higgins I (1996) Of ethics committees, protocols, and behaving ethically in the field: a case study of research with elderly residents in a nursing home. *Nurs Inq* **3**: 130–7.

Mason J (2002) *Qualitative Researching* 2nd edn. Sage Publications, London.

Mason JK, McCall Smith RA (1994) *Law and Medical Ethics.* 4th edn. Butterworths, London.

Merrell J, Williams A (1994) Participant observation and informed consent: relationships and tactical decision-making in nursing research. *Nurs Ethics* **1**(3): 163–72.

Morse JM, Field P A (1996) *Nursing Research: The Application of Qualitative Approaches* 2nd edn. Chapman and Hall, London.

Nursing and Midwifery Council (NMC) (2004) *The NMC Code of Professional Conduct: Standards for Conduct, Performance and Ethics.* NMC, London.

Polit DF, Hungler BP (1999) *Nursing Research.* 6th edn. Lippincott, Philadelphia.

Redsell SA, Cheater FM (2001) The Data Protection Act (1998): implications for health researchers. *J Adv Nurs* **35**(4): 508–13.

Royal College of Nursing (2004) *Research Ethics: RCN Guidance for Nurses.* RCN, London.

Royal College of Nursing (2005) *Informed Consent in Health and Social Care.* RCN Research Society, London.

Wales Office of Research and Development for Health and Social Care (2001) *Research Governance Framework for Health and Social Care in Wales.* National Assembly for Wales, Cardiff.

World Medical Association (2005) Declaration of Tokyo 2004. www.wma.net/e/ethicsunit/helsinki.htm.

Critical Thinking

1. What is the nurse's ethical responsibility to the patient during research participation?

PAULINE GRIFFITHS is Senior Lecturer, School of Health Science, University of Wales, Swansea.

From *British Journal of Nursing-BJN,* 15:7, April 13, 2006, pp. 386–390. Copyright © 2006 by MA Healthcare Limited. All rights reserved. Reprinted by permission.

Exploring Ethical, Legal, and Professional Issues with the Mentally Ill on Death Row

Jessica E. Plichta, MS, APRN-PMH, P-CS

World Opinion on Capital Punishment versus U.S. Practice

In order to establish an appreciation of the ethical dilemmas involving current execution practices in the United States, it is essential to understand our views in contrast to the international position on capital punishment practices. In 1999, a worldwide moratorium on executions was passed by the United Nations Human Rights Commission. The resolution suggested that those countries that have not abolished the death penalty restrict its use to a limited number of offenses and discontinue the execution of juveniles (Amnesty International, 2005). Further, only 76 of 196 countries retain use of the death penalty. Ironically, the United States shares the practice of capital punishment not with nations who are considered allies or similar in belief systems, but with countries such as China, Cuba, Iran, Iraq, Saudi Arabia, Kuwait, and Nigeria, to name only a few. Moreover, 94% of the 3800+ executions in 2004 were in the United States, Vietnam, and Iran (Amnesty International, 2005). In mid 2003, the United States (the only country other than Somalia) refused compliance with the *United Nations Convention on the Rights of a Child* because the treaty required that juvenile offenders could not be killed (Prejean, 2005).

Clearly, America's practices conflict with most world opinion regarding the death penalty. Although the United States has not yet discontinued use of this form of punishment, international influences may promote changes in the near future. According to Dieter (2003), there are at least three prominent reasons for the anticipated recognition and implementation of international directives in the United States: (1) there is a growing need for international cooperation and respect for laws in other democracies, (2) international human rights issues such as with extradition, foreign terrorists, and disputes in relation to the execution of foreign nationals from other nations such as Paraguay, Germany, and Mexico have become more frequent, and (3) at present, views are almost unanimous in condemning use of the death penalty in certain groups, such as juveniles.

New developments in trends toward abolishment of the death penalty are being considered with the emergence of new information and concerns, including but not limited to racial disparities, wrongful convictions, prediction of future dangerousness, jury decision-making, public opinion, costs, and methods of execution (Acker et al., 2003). Currently, mentally ill or retarded inmates are the only groups legally protected from capital punishment in the United States.

Law Versus Practice

In *Ford v Wainwright* (1986), the Supreme Court ruled the Eighth Amendment proscription against cruelty prohibits killing the insane but provided no definition of insanity. As a consequence, mental illness continues to be ignored and often times these individuals fall through the cracks of the law and onto death row. According to the National Mental Health Association (NMHA, 2005), common mental illnesses seen on death row include Bipolar Disorder, Schizophrenia, Post-Traumatic Stress Disorder, and Schizoaffective Disorder. It is estimated that at least 10% of U.S. death row inmates are mentally ill (NMHA, 2005).

One example is Kelsey Patterson who was executed in Texas on May 18, 2004, despite his suffering with severe paranoid schizophrenia. He was diagnosed 10 years prior to his offense and for many years was declared mentally incompetent to stand trial. The day before his execution, the Pardon and Paroles Board met and recommended clemency to the governor, which was denied (Death Penalty Information Center, 2005).

Another case is Sammy Perkins who was executed by lethal injection in North Carolina on October 8, 2004. He was the 933rd murderer to be executed in the United States and the 32nd to be executed in North Carolina since 1976. He was diagnosed with Bipolar Disorder after his offense and was previously unaware that the symptoms that he had been living with were treatable. A jury never heard about his mental illness (Clark Prosecutor, 2005).

In 2002, *Atkins v Virginia* declared it unconstitutional to execute mentally retarded persons. The ruling was one that the United Nations Commission on Human Rights had called for on numerous occasions (Dieter, 2003). Prior to *Atkins v Virginia*, 35 mentally retarded prisoners were executed in the United States. However, despite the law and an improvement in the situation, mentally retarded offenders continue to be sentenced with capital punishment (Prejean, 2005).

A Discussion of Ethical Issues
How Do the Mentally Handicapped End Up on Death Row?

While there are many aspects of executing mentally ill or mentally retarded prisoners that deserve debate, there is one obvious question that must be addressed first. Why are the criminally insane or mentally retarded in prison, particularly on death row, rather than in psychiatric treatment facilities? The answer is not simple.

It should be noted that there are safeguards within the system that serve to "ensure that the serious mental health needs, including those related to developmental disability and addictions are identified." (NCCHC, 2005). Every inmate should be screened and evaluated for mental conditions upon booking to the detention facility. The standard is clearly stated by the National Commission on Correctional Health Care (NCCHC, 2005). This ensures that the inmate receives proper psychiatric treatment, as well as documentation of the diagnosis prior to judgment and/or sentencing. The process should eliminate errors that result in improper sentencing of a detainee.

However, many state laws include a condition that if the mentally ill prisoner on death row is restored to sanity through therapy or medications, the state can then execute him. Furthermore, on February 10, 2003, a federal appeals court ruled that administering psychotropic medications to an unwilling death row inmate was not a violation of the US Constitution (Hausman, 2003). Clearly, the 1986 ruling against executing mentally ill inmates in *Ford v Wainwright* has been buried under exceptions and loopholes in the interest of death penalty supporters.

American culture plays a significant role in reasons for ambivalence regarding the management of mentally handicapped who may have committed serious crimes. According to Phoebe Ellsworth, Distinguished University Professor of Law and Psychology and expert on death penalty issues, most Americans are now in favor of capital punishment (Wadley, 2004).

A closer look at reasons for this may be of profound concern. Studies conducted by Barkan and Cohn (Broom, 2004) indicate that racial prejudices are a major reason for such overwhelming support of the death penalty. The studies revealed that more than 80% of those on death row are African–American or Hispanic. After four studies that explored reasons for supporting the death penalty, results consistently showed a "strong link between punitive attitudes and racial prejudice." The studies conclude that law-makers should not base decisions on public opinion if they are tinged by racial prejudice. It is possible that empathy for people who are different and commit crimes may be much less than for those who are better understood.

Another view (Zimring, 2003), suggests that a highly diverse country will inevitably have highly diverse cultural values and division concerning decisions about what is right or wrong. Zimring suggests that the death penalty violates our legal system's highest principles of fairness and due process. Conversely, some citizens see the executions as a means of local control and a safeguard to the public. His observation that most executions occur in the southern states where lynching was most common is worthy of some consideration. Is it not the same horrible form of punishment if by electrocution or lethal injection? Is the outcome and principle of the action any different from a lynching that most would consider barbaric in today's culture?

A legal analysis by Zimring (1999) suggests that there is an evident and remarkable conflict between the operational needs in a system that performs executions and the fundamental principles and procedures of the American legal system. If capital punishment is to be a normal state of conduct in our country, then our legal system and social system must redefine and "reimage" the basic principles of fairness in criminal justice. Thus, one might conclude from the stated arguments that the conflicting nature of the law when dealing with the mentally handicapped is a product of diversity of opinion, ignorance concerning mental diagnoses, insecurity with a need to feel safe, and legislation torn between the fundamental values of our forefathers and pressure to change (even if without a conscious decision to do so) our principles based on current public opinion and levels of violence in America today.

To Treat or Not to Treat?

The issues are particularly complex for psychiatric providers who work on death row. Correctional health care organizations responsible for setting standards and guidelines for inmate treatment, such as the NCCHC and the American Correctional Health Services Association (ACHSA), state that only if an inmate is a danger to himself or others, can he or she be forcibly, but humanely, treated with medication or other necessary interventions (NCCHC, 2005). Moreover, both agree that inmates have a right to refuse treatment of any kind, unless, of course, dangerous behavior warrants involuntary treatment. However, neither of these correctional facility organizations addresses forced treatment to restore sanity in death row inmates. This is not surprising since, in theory, there should be no mentally ill prisoners on death row. Although the situation occurs regularly, and new twists in the law provide avenues to execute the mentally ill, the problem is one that is not discussed formally.

There are many documented cases of inmates who have been forced to take psychotropic medications and were subsequently executed. An example is Charles Singleton of Arkansas who was forcibly medicated to restore his sanity and consequently executed in 2004. An appeal to the U.S. Supreme Court resulted in a vote of 6 to 5 in favor of Arkansas' contention that it should be allowed to medicate Singleton against his will. This case has been cited and discussed by courts and the medical community since its occurrence. American Psychiatric Association President, Paul Appelbaum, M.D., stated that he disagreed with the ruling and that, "Physicians violate their ethical obligations as healers when they treat condemned prisoners for the purpose of restoring competence to be executed" (Hausman, 2003).

It is essential to recognize the ethical predicament for psychiatric practitioners caring for death row inmates. If the provider treats the prisoner, and the prisoner is "cured," he will be sentenced to death. Even if the client is voluntarily treated, due to cognitive disability or altered thought processes, the outcome is not in the best interest of the inmate. Participating in such practice is a clear violation of the Code of Ethics, which is a key element of a provider's decision-making process from the first day of training.

Professional and correctional organizations alike are in agreement as to health care providers' involvement in the execution process, unquestionably opposing their participation in capital punishment. The American Medical Association (AMA), American Nurses Association (ANA), and the American Public Health Association released a joint statement indicating that involvement in executions is a serious violation of professional ethical standards. The release (ANA, 1996, p. 2) states that since "these ethical codes are integral parts of the state medical, nursing, and other health professional practice and licensing acts, participation in execution violates state law." Furthermore, correctional health care organizations observe the same outlook. The ACHSA published a Code of Ethics specific to detention facilities. The code states that health care providers "are not to be involved in any aspect of the death penalty" (ACHSA, 1990). The idea calls for good judgment and professional responsibility. Is restoring the sanity of a mentally ill inmate for the purposes of imposing the death penalty considered involvement in the execution process?

The Maturana case plainly depicts a psychiatric provider's ethical difficulty when treating death row inmates. Claude Maturana of Arizona was declared mentally incompetent related to severe psychosis. He was then transferred to a state psychiatric hospital, where his psychiatrist treated him, easing his distress but not completely alleviating the symptoms. The psychiatrist adhered to medical ethical codes and refused to medicate Maturana adequately enough to restore his sanity. According to the attorney general, Arizona law stated that the psychiatric institution must find a "willing" physician. No one came forward until they expanded to a nationwide search. Nelson Bennett, M. D., from Georgia, found Maturana competent and claimed that he did not require pharmacological intervention (Freedman, 2001).

The concept of forcing treatment in this population brings forth many questions to consider. Are psychiatric care providers who work with condemned prisoners obliged to respect choices, protect the individual, and support self-governance? Are detained criminals entitled to the same degree of autonomy as any other patient? Is the mental condition of the inmate after treatment relevant if he were ill at the time of the offense? What is the definition of "insane" and which mental conditions may result in the dismissal of criminal responsibility? Should the courts be able to order involuntary treatment on a case-by-case basis?

Conclusions

Unmistakably, ethical issues with psychiatric patients on death row are a web of morality, politics, and personal values. As with caring for any other patient, providers of mental health care to inmates must consider the best outcome for the client. The fate of the inmate is influenced by the legal system, the media, public opinion, and those who provide care in the system, all of whom may have conflicting views and values.

In terms of the mentally disabled inmate, it has recently been recognized that they should not be considered for the death penalty. Evolution in America's perspectives on capital punishment have progressed to a mere legal acknowledgment that execution of these individuals is cruel and unusual punishment. Nevertheless, in practice, this population of inmates continues to be sentenced with the ultimate punishment. Each case in each state may be viewed differently by the judge or jury; therefore, the sentencing of these individuals varies, regardless of the federal law. Thus, the late of the mentally handicapped within the court system continues to be a conflict between medical ethics, legal statues, and public beliefs.

Disclaimer

The views expressed in this article are those of the authors and do not reflect the official or position of the Department of the Navy Department of Defense, or the United States Government.

References

ACHSA. (1990). Code of ethics. Retrieved August 2, 2008, from: www.achsa.org.

Acker, J., Bohm, R., & Lanier, C. (2003). America's experiment with capital punishment: Reflections on the past, present and future of the ultimate penal sanction. National Commission on Correctional Health Care. Adapted from *CorrectiveCare.*

American Correctional Health Services Association. Retrieved August 2, 2008, from: www.achsa.org.

American Nurses Association: Press Release. (1996). Professional societies oppose health care professionals' participation in capital punishment. Silver Spring. MD: American Nurses Association.

Amnesty International. (2005). List of abolitionist and retentionist countries. Retrieved August 2, 2008, from: http://web.amnesty.org.

Broom, D. (2004). Prejudice and punishment. Retrieved August 2, 2008, from: http://umainetoday.umaine.edu/Issues/v4i2/prejudice.html.

Clark Prosecutor. (2005). Sammy Crystal Perkins. Retrieved August 2, 2008, from: www.clarkprosecutor.org.

Death Penalty Information Center. (2005). Mental illness and the death penalty. Retrieved August 2, 2008, from: www.deathpenaltyinfo.org.

Dieter, R. (2003). International influence on the death penalty in the United States. *Foreign Service Journal,* 80(10), 31–35.

Freedman, A. (2001). The doctor's dilemma: A conflict of loyalties. *Psychiatric Times,* 18(1). Retrieved August 2, 2008, from: www.psychiatrictimes.com.

Hausman, K. (2003). Court allows state to medicate death row inmate forcibly. *Psychiatric News,* 38(6), 2.

National Commission on Correctional Health Care. (2005). Standards of Care First Published in CorrectCare. Retrieved August 2, 2008, from: www.ncchc.org/resources/standards.html.

National Mental Health Association. (2005). Death penalty and people with mental illness. Retrieved August 2, 2008, from: www.nmha.org.

Prejean, H. (2005). *The death of innocents: An eyewitness account of wrongful executions.* New York, NY: Random House.

Wadley, J. (2004). Ellsworth to outline American attitudes toward death penalty. University of Maine. The University Record Online.

Zimring, F. (1999). Report from the Institute for Philosophy and Public Policy. *The executioners dissonant song: On capital punishment and American legal values.* Derived from *The killing state: Capital punishment in law politics and culture.* New York, NY: Oxford University Press.

Zimring, F. (2003). *The contradictions of American capital punishment.* New York, NY: Oxford University Press.

Critical Thinking

1. What are the ethical considerations of treating the mentally handicapped patient who is on death row?

From *Journal of Forensic Nursing,* vol. 4, September 2008, pp. 143–146. Copyright © 2008 by International Association of Forensic Nurses (IAFN). Reprinted by permission Wiley-Blackwell via Rightslink.

Covert Medication in Older Adults Who Lack Decision-Making Capacity

FRANCES TWEDDLE

I t is generally accepted legally, ethically, and in terms of good practice, that prior to administering treatment, conducting an investigation, or assisting with care, health-care professionals must obtain the patient's valid consent (Department of Health (DH), 2001a). The law holds bodily integrity in high regard (Griffith, 2004) and gaining consent recognizes and respects the individual's right to determine what happens to their body (DH, 2001a).

If a competent adult is touched without consent, they have the right to sue for trespass to their person (DH, 2001a). Should the health professional have gained valid consent, then they will normally be protected from a successful action for trespass (Dimond, 2008a). For a valid defence against a claim of trespass, consent must be full (the patient agrees to all of the proposed treatment); free (consent must be the free choice of the individual and cannot be obtained by undue influence); and informed (the patient needs to be aware of all relevant information in order to make an informed choice) (Griffith, 2004).

Rights of an Individual with Capacity

Any mentally competent adult has the right to give or refuse consent to treatment (Haxby and Shuldham, 2008). The underpinning ethical principle is respect for autonomy (Griffith, 2004). 'Autonomy' is usually associated with terms such as self-determination and being an autonomous person means running one's life according to a set of self-chosen rules or values (Hendricks, 2000). In health-care, one of the main ways that patients exercise autonomy is through the concept of consent (Hendricks, 2000).

Capacity is paramount to consent. The Mental Capacity Act 2005 and the The Mental Capacity Act Code of Practice (CoP), came into force in 2007. The Act 'provides the legal framework for acting and making decisions on behalf of individuals who lack the mental capacity to make particular decisions for themselves'.

If the patient has capacity, their decision with regard to consent and treatment is binding and the health-care professional must respect the patient's choice, even if it differs from their

professional recommendations (Dewing, 2002). This is outlined in the NMC Code of Conduct (NMC, 2008), thus a failure to respect the patient's right to accept or decline treatment would be a breech of professional duty.

Beneficence

The refusal of treatment can be difficult for health-care professionals as their duty to respect autonomy conflicts with their desire to act in a beneficent manner (Dunbar, 2003). Generally, the principle of beneficence is the nurse's responsibility to benefit others (Beauchamp and Childress, 2009). Beneficence is a major element of a nurse's professional duty and generates a moral obligation to undertake positive actions aimed at safeguarding the health and welfare of patients (Hendrick, 2000).

Beneficent acts are thought to encourage physical and psychological benefit, including actions taken to prevent disease, promote health, and reduce pain and suffering (Hendrick, 2000). These are all subjective terms and there is no way of determining objectively what counts as a beneficent act as people have individual attitudes to illness and pain (Hendrick, 2000).

Determining Decision-Making Capacity in Practice

Problems arise in relation to consent when the patient is deemed to be lacking capacity due, for example, to conditions such as dementia (DH, 2001b). It is important that patients are not automatically assumed to be lacking capacity simply because of their condition (DH, 2001b).

The Mental Capacity Act 2005 states that, 'a person must be assumed to have capacity unless it is established that he lacks capacity'. It is the responsibility of the individual proposing the treatment to assess the patient's capacity. If at the time a decision needs to be made an impairment or disturbance of the person's mental functioning renders them unable to comprehend and retain information relating to the intervention (particularly the consequences of accepting or refusing the treatment), use this information in the decision-making process and communicate their decision, then they fail the capacity assessment and are considered to lack capacity.

In practice, determining capacity is not always this clear-cut. A patient may have fluctuating competence, which can be a common consequence of dementia (Brazier and Cave, 2007). Prior to the The Mental Capacity Act 2005, patients with fluctuating capacity tended to be viewed by the courts as lacking capacity (Brazier and Cave, 2007). Although the The Mental Capacity Act still requires that the patient is able to understand and retain information, it indicates that they are considered able to make decisions even if they are only able to retain the information for a short period of time (The Mental Capacity Act, 2005).

Fluctuating capacity can be problematic. For example, if the patient appears to have capacity at the time the decision needs to be made and refuses to consent to treatment, the fact that they may have lacked capacity previously becomes irrelevant, and they resume the rights of an adult with capacity (Dimond, 2004). However, if they have had capacity yet at the time of the decision are assessed as lacking it, then the best interest provisions of the The Mental Capacity Act apply (which will be discussed later). This does not apply in the case of advanced directives in which the patient made specific instructions regarding treatment when they had capacity. An advanced directive would be binding upon the health-care professional.

There is further difficulty with regard to the passive accepting patient. Experience indicates that patients who would potentially fail the capacity assessment, such as those with dementia, are often encouraged to accept medication for staff convenience, without being fully informed about the drug. Despite the fact that the patients are often willing to consume the medication, it is questionable whether the professional has truly gained consent (Dimond, 2004). It seems that if the patient accepts treatment, i.e. does what the health-care professional wants them to do, their capacity is assumed to exist, even if they have a mentally debilitating condition such as dementia. Should they refuse, however, their capacity is questioned (Dimond, 2004).

Capacity has a relationship with dependency, with people mistakenly presuming that the greater the dependency the less capacity exists (Dewing, 2002). Vulnerable adults, such as those with dementia, are often regarded as being totally dependent and, therefore, completely lacking in decision-making capacity, usually without being given the opportunity to try (Dewing, 2002). Additionally, capacity is situation-specific and it is important in practice not to presume that because someone lacks the capacity to feed themselves they are also unable to make decisions regarding their treatment (DH, 2001b). Assessing capacity appropriately and not assuming a lack of capacity based on the presence of dementia is an area in which general clinical practice can improve.

The Rights of the Individual Who Lacks Decision-Making Capacity

The Mental Capacity Act 2005 states that if, and only if, an adult lacks capacity can decisions be made on his or her behalf. The Mental Capacity Act 2005 regulates decision-making for incapable persons (Griffith, 2009) and the Act states, 'an act done, or decision made, under this Act for or on behalf of a person who lacks capacity must be done, or made, in his best interests'. The Mental Capacity Act 2005 demands a holistic approach to determining best interests, setting out a statutory list of factors that decision-makers are required to consider (Griffith, 2009).

The checklist aims to assist decision-makers and ensure consistency in the way best interest decisions are made (Griffith, 2008), however, decision-makers are required to consider all factors relevant to the individual (Griffith, 2006). This holistic approach requires the decision-maker to consider the past and present wishes and feelings of the person concerned, their background, religious and cultural beliefs, and whether similar treatment to that proposed has been refused in the past (Griffith, 2008).

Additionally, the The Mental Capacity Act 2005 has made provisions for people other than health-care professionals to consent on behalf of another (Griffith, 2008). These decision-makers will have been granted authority to consent on behalf of an incapacitated person under the The Mental Capacity Act 2005. For example, a person appointed as an attorney under a Personal Welfare Lasting Power of Attorney can be given authority to consent to treatment (Griffith, 2008). If there is no designated decision-maker then the nurse will decide what intervention would be in the best interests of the individual lacking decision-making capacity.

To ensure that older adults who have been assessed as lacking capacity receive the appropriate medication to prevent pain or deterioration, the nurse might consider it to be in the patient's best interest to administer the medication covertly.

Covert Administration

In 2007, the NMC published a position statement on the covert administration of medication (Dimond, 2008). Covert administration refers to medication that is concealed, usually in food or drink, so that it is being provided to the patient unknowingly (Griffith, 2007). The statement highlights that it may be regarded as deceptive to disguise medication in the absence of informed consent.

Under the provisions of the Human Rights Act 1998, covertly administering medication to an incapable adult would not contravene their right to be free from torture, inhumane or degrading treatment if the substance could convincingly be shown to have therapeutic necessity (Griffith, 2007). The health-care professional must be accountable for this decision and must be certain that hiding medication is in the best interest of that individual, i.e. the medicine must be necessary in order to save life, prevent deterioration or ensure an improvement in the patient's physical or mental health (NMC, 2007).

The requirement that nurses are personally accountable for their practice is enshrined within the NMC's Code of Conduct (2008). Nurses must practise in an open and co-operative manner, show respect for the patient's independence and recognize and value their involvement in their care (NMC, 2008). The individual should not be ignored simply because they lack decision-making capacity. It is important that the decision that needs to be made is explained to the patient (Griffith, 2006).

The holistic approach to deciding best interests is further enhanced by the requirement that the decision-maker must make their decision with involvement and support from members of the multi-disciplinary team and the patient's carers/relatives; it is inadvisable to make this decision in isolation. The duty to consult with others must be balanced against the patient's right to confidentiality and nurses should only seek the views of people who it is appropriate to consult. Consulting family members may be problematic for the nurse as they may disagree about a person's best interests.

The Mental Capacity Act 2005 requires that the views of those consulted are given consideration but ultimate responsibility for deciding what would be in the best interests of the individual lacking capacity lies with the decision-maker (Griffith, 2008). Good record-keeping is vital when making decisions of this nature (NMC, 2008).

Covert administration should only be carried out as a last resort (Honkanen, 2001; NMC, 2007). A study by Treloar et al in 2000 investigated the incidence of covert administration in care settings including NHS hospitals and patients' homes and found that rather than being a last resort, covert administration was widespread, with incidences particularly high in clinical areas where patients lacked capacity. The overriding reason was to ensure that patients received the appropriate treatment (Treloar et al, 2000). While this would be condemned by some as being overly paternalistic, others would perceive it as a beneficent act (Dingwall, 2007).

Beneficence versus Non-Maleficence

The underlying question is whether the harm caused by giving the medication covertly is greater than the harm the patient may sustain if they do not receive the medication at all (Dingwall, 2007). The principle of non-maleficence refers to a nurse's duty not to harm their patients or subject them to risk (Beauchamp and Childress, 2009). Non-maleficence is considered to be less morally demanding than beneficence; it generates fewer obligations simply requiring nursing staff not to harm patients, rather than imposing an obligation to act positively to help (Hendrick, 2000). There are similar problems in practice to those of beneficence as what is viewed as harmful is subjective and dependent on the individual's outlook. Harm could be physical or psychological and be caused intentionally (abuse) or unintentionally (negligence) (Hendrick, 2000).

In practice, the benefits of ensuring the patient receives the appropriate treatment by administering medication covertly need to be balanced against the non-maleficent act of omitting important medications and the disrespect of patient autonomy (Lamnari, 2001). The principles of beneficence and non-maleficence are closely linked. In health-care, the moral objectives are to help those who are sick and to prevent harm. Nursing interventions aimed at benefiting patients (for example immunisations) may also cause an element of harm (pain at the injection site) (Hendrick, 2000).

Covert Administration in Practice

The way medicine is administered covertly may be harmful if the nurse does not have sound pharmacological knowledge and understanding of the possible consequences (Honkanen, 2001). Nurses need to be aware of the risks of opening capsules or crushing tablets. Firstly, the nurse risks potential legal consequences (Dingwall, 2007). If the way the medication is administered is outside the product licence and the patient is harmed as a result then the legal responsibility shifts from the manufacturer to the prescriber and administrator (Dingwall, 2007). Opening capsules, crushing tablets and mixing medicines into food or drink falls outside of the product licence of most drugs and effectively renders that medication unlicensed (Bending, 2001).

Tampering with medications can alter their therapeutic effectiveness. It is important that modified-release medications are swallowed whole, for example (Bending, 2001). If the tablet is damaged by crushing, then the entire dose will be released quickly, resulting initially in overdose followed by a period with no medication. While the nurse may think they are acting beneficently by ensuring the patient receives the medicine, in this instance their actions are in fact non-maleficent (Bending, 2001).

The same principles apply when considering the crushing of enteric-coated medicines. These coatings reduce stomach irritation and prolong drug release. Crushing exposes the patient to irritant active ingredients and potentially causes them harm (Bending, 2001). Additionally, crushing inevitably leads to a lower dose being administered as residue is left on the crushing device (Dingwall, 2007). Continually administering the drug at a lower dose than prescribed might harm the individual or eliminate the effectiveness of the drug.

Covert administration does not always impose these risks. If the nurse placed a dispersible paracetamol in a patient's drink for example, the medication is still being administered covertly, however, no risks apply and this does not fall outside the product licence. Some would argue that camouflaging medicines by hiding them in food or drink is beneficent as it is a 'kind' way of administering medication to distressed older adults (Lamnari, 2001). However, if the patient becomes suspicious that their nurse is disguising drugs they may lose the trust which is a fundamental principle underpinning the nurse-patient relationship.

Conclusion

If an adult is deemed to be lacking capacity in accordance with the The Mental Capacity Act 2005, then decisions about their treatment can be made on their behalf. These decisions must be based on the patient's best interests (according to The Mental Capacity Act, 2005).

Key Points

- Any mentally competent adult has the right to give or refuse consent to treatment or nursing intervention.
- The ethical principle underpinning this free choice is respect for autonomy, and the nurse's professional duty

to respect the decision of the patient is enshrined within the NMC Code of Conduct.

- Capacity is paramount to consent and problems arise in practice when the patient is deemed to lack capacity.
- The aim of this article is to present the issues that arise when patients lack decision-making capacity, especially in relation to older adults with dementia who lack the capacity to consent to medication, as well as the covert administration of medication.
- In relation to consent and covert administration, the nurse is required to balance respect for the patient's autonomy with their desire to act beneficently and in a non-maleficent manner.

References

Beauchamp, T, Childress J (2009) *Principles of Biomedical Ethics.* Oxford University Press, Oxford.

Bending A (2001) Hiding medicines. *Prim Health Care* **11(8):** 24–25.

Brazier ME, Cave E (2007) *Medicine, Patients and the Law.* Penguin, London.

Department of Health (2001a) *Reference Guide to Consent for Examination or Treatment.* Available at: www.dh.gov.uk/en/Publicationsandstatistics/Publications/PublicationsPolicyAndGuidance/DH_4006757 Accessed 2 March 2009.

Department of Health (2001b) *Seeking Consent:working with older people.* Available at: www.dh.gov.uk/en/Publicationsandstatistics/Publications/PublicationsPolicyAndGuidance/DH_4009325 Accessed 2 March 2009.

Dewing J (2002) Older people with confusion: capacity to consent and the administration of medicines. *Nurs Older People* **14(8):** 23–28.

Dimond B (2004) Medicinal products and consent to treatment by the older person. *Br J Nurs* **13(1):** 41–43.

Dimond B (2008) *Legal Aspects of Nursing.* Pearson Education Limited, Essex.

Dingwall L (2007) Medication issues for nursing older people (part 2). *Nurs Older People* **19(2):** 32–36.

Dunbar T (2003) Autonomy versus beneficence: An ethical dilemma. *Prim Health Care* **13(1):** 38–41.

Griffith R (2004) Consent to examination and treatment 1: the capable adult patient. *Nurse Presc* **2(4):** 177–79.

Griffith R (2006) Making decisions for incapable adults 1: Capacity and best interest. *Br J Comm Nurs* **11(3):** 119–25.

Griffith R (2007) Covert administration of medicines to adults must be the last resort. Nurse Presc 5 (2), pp. 79–81.

Griffith R (2008) Mental Capacity Act: determining best interests. *Br J Comm Nurs* **13(7):** 335–41.

Griffith R (2009) The Mental Capacity Act 2005 in practice: best interests. *Nurse Presc* **7(4):** 172–75.

Haxby E, Shuldham C (2008) Consent and lack of capacity. *Br J Cardiac Nurs* **3(12):** 559–63.

Hendrick J (2000) *Law and Ethics in Nursing and Health Care.* Stanley Thornes, London.

Honkanen L (2001) Point-counterpoint is it ethical to give drugs covertly to people with dementia? *West J Med* **174:** 229.

Lamnari A (2001) Point-counterpoint is it ethical to give drugs covertly to people with dementia? *West J Med* **174:** 228.

NMC (2007) Covert administration of medicines—disguising medicines in food or drink. Available at: www.nmc-uk.org/aDisplayDocument.aspx?documentID=4007 Accessed 2 March 2009.

NMC (2008) The Code: Standards of conduct, performance and ethics for nurses and midwives. Available at: www.nmc-uk.org/aArticle.aspx?ArticleID=3056 Accessed 2 March 2009.

Treloar A, Beats L, Philpot M (2000) A pill in the sandwich: convert medication in food and drink. *J R Soc Med* **93:** 408–11.

Critical Thinking

1. What is the nurse's role when treating a patient who lacks decision-making capability?

FRANCES TWEDDLE is a third-year student nurse studying towards an Advanced Diploma in Nursing at the University of Leeds.

UNIT 3

Drugs, Medications and Alternative Therapies

Unit Selections

Learning Outcomes

After reading this unit, you should be able to:

- Describe how extensively prescribing antibiotics may lead to drug-resistive organisms.

- Identify strategies to prevent harm from high-alert drugs.

- State the importance of ensuring that a patient is dynamically stable prior to accepting them onto the medical-surgical unit.

- Recognize the importance of knowing the side effects of medications and their follow-up labs.

- List the complications that may trigger nausea and vomiting.

Student Website

www.mhhe.com/cls

Internet References

The Center for Disease Control
www.cdc.gov

Institute for Healthcare Improvement: High-Alert Medications
www.ihi.org/IHI/Topics/PatientSafety/MedicationSystems/ImprovementStories/FSHighAlertMeds

The Institute for Safe Medication Practices
www.ismp.org/Tools/**highalertmedications**.pdf

The Joint Commission High Alert Medication and Patient Safety
www.jointcommission.org/sentinelevents/sentineleventalert/sea_11.htm

U.S. Pharmacopeia
www.usp.org

Changes in medical treatments, interventions and medications occur on a daily basis. Recently, the mind-body connection in the treatment of illness has become an important consideration in medical treatment. Because of these changes, nurses need up-to-date information on treatment modalities. Careful consideration should be used when there is an overuse of antibiotic therapies. The use of these drugs has become so widespread, that some bacteria have mutated to the level that they are resistant to available antibiotics. Hospital-acquired infections are of particular concern. Nurses and other health-care professionals must be aware of these drug-resistant organisms in order to prevent their spread. Also knowing who is medically at risk is an important consideration for everyone involved in health care. Nurses and other health care professionals must be aware of these drug-resistant organisms in order to prevent their spread.

Another consideration in the health care environment is the utilization of high alert drugs such as anticoagulants, sedatives, opioids, and insulin. These drugs require close monitoring during and after administration. It is crucial that procedures are in place to monitor and avoid incorrect administration of medications that fall into this category. Furthermore, to reduce fatal errors, nurses must be aware of these and other high alert medication and the potential effect they may have on patient. Additionally, recognizing and implementing procedures and highlighting drugs that look and sound alike can prevent confusion that can lead to fatal medication errors.

A different category of high alert medication that requires careful consideration prior to and during administration is the effective management of care for patients who are dynamically stable or unstable. While monitoring patients who are receiving vasoactive medication, it is critical that the nurse is aware of the facilities' policies and procedures. Also, utilizing safety precautions such as double checking dosage calculations and having a colleague verify those calculations can potentially decrease fatal errors. Possessing the knowledge of various effects of vasoactive drugs can reduce the potential adverse effect that these medications can have on patients.

Ineffective medication monitoring systems can lead to patient fatalities and drawn out lawsuits. It is imperative that nurses educate themselves on their facilities' policies, recognize medication contraindications, and always be aware of high alert medications.

Nausea and vomiting may be triggered by many factors such as surgery (post operatively), chemotherapy and radiation.

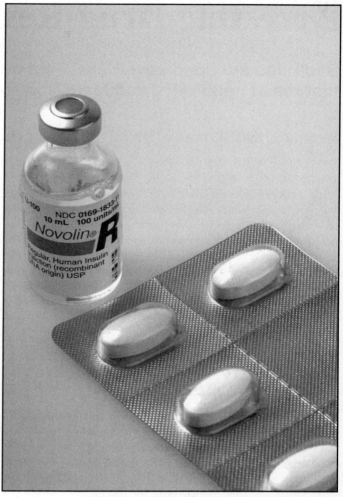

© The McGraw-Hill Companies, Inc./Jill Braaten, photographer

Nausea and vomiting may also have a tremendous impact on a patient's nutritional status and delay recuperation. Antiemetic drugs are used to treat the effects of nausea and vomiting. Additionally, antiemetic drugs may be used alone, or in combination with other drugs such as Antihistamines, Dopamine Antagonist, Anticholinergic Agents, and Corticosteroids. It is critical that nurses are aware of the potential side effects these medications can have on their patients.

Arresting Drug-Resistant Organisms

Bad bugs are up to new tricks. Find out about emerging pathogens that resist antibiotics and learn how to subdue them.

REBECCA KJONEGAARD, RN, CIC, BSN AND FRANK EDWARD MYERS III, CIC, CPHQ, MA

Anyone who's sick or otherwise compromised is especially vulnerable to emerging infectious diseases caused by organisms that are either new or manifesting themselves in new ways. In this article, we'll look at several emerging diseases characterized by antibiotic resistance: community-acquired methicillinresistant *Staphylococcus aureus* (MRSA), vancomycinresistant enterococci (VRE), vancomycin-resistant *S. aureus* (VRSA), and extended-spectrum betalactamase (ESBL)-producing organisms. Here's what you need to know to protect your patients and yourself.

Community-Acquired MRSA: Neighborhood Bully

You're probably familiar with hospital-acquired MRSA, which causes ventilator-associated pneumonia, bacteremia associated with central lines, and other nosocomial infections. In contrast, community-acquired MRSA most often causes soft-tissue infections.

Hospital-acquired MRSA typically affects the chronically ill and elderly, but community-acquired MRSA tends to infect younger people. The good news is that although it's resistant to methicillin, it's sensitive to other antibiotics.

Persons at higher risk include children in day care, amateur and professional athletes (who are in close contact and may share equipment), users of injected recreational drugs, men who have sex with men, and prisoners. It's transmitted primarily via direct physical contact, not through respiratory droplets. Transmission via indirect contact (for example, contact with contaminated door handles, eating utensils, and furniture) isn't likely. But in institutions or households, the disease may be transmitted via clothes and linens that are grossly contaminated with drainage from an infected wound. Athletes' sharing of soap and towels is another risk factor.

A soft-tissue MRSA infection often involves abscesses, furuncles, or cellulitis. Initially, the patient may assume he was bitten by a spider.

Although hospital-acquired MRSA may respond only to vancomycin, community-acquired MRSA is sensitive to many more antibiotics. After an infected wound is irrigated and debrided, the patient may be prescribed tetracycline, doxycycline, clindamycin, or sulfamethoxazole-trimethoprim. Besides methicillin, community-acquired MRSA is considered resistant to cephalexin, dicloxacillin, erythromycin, and quinolones.

To protect yourself and your patients, take contact precautions as recommended by the Centers for Disease Control and Prevention (CDC). (See *Taking steps to stay safe.*) Teach your patient not to use topical antibiotics if new abscesses or cellulitis develop; new culture specimens will need to be obtained first. Ask him about his direct contacts (anyone he's touched or who has touched him), to identify others with the same signs and symptoms who may need treatment.

To help him prevent further disease transmission in the household, teach him to clean household surfaces with regular soap (not antimicrobial) and water and to wash linens and clothing with detergent and dry them in a dryer.

VRE: First among Resistant Organisms

Enterococci were the first organisms to be found resistant to vancomycin. The morbidity and mortality rates for VRE are low. However, VRE occurs at higher rates in patients with MRSA. Researchers worry that treating MRSA with vancomycin has encouraged the development of vancomycin-resistant strains of *S. aureus,* which are rare.

In the hospital, VRE is dangerous to patients being treated with invasive devices, which increase infection risk, especially if their immune systems are compromised. However, although VRE can be found throughout an infected patient's room, the disease is spread by direct contact.

S. Aureus Resistance: This *VISA* Shouldn't Go Anywhere

As bacteria develop drug resistance, they evolve from sensitive to intermediate to resistant. Vancomycin intermediate *S. aureus* (VISA) and VRSA have recently been identified. Currently, vancomycin is the treatment of choice for strains of *S. aureus* that have developed resistance to methicillin (MRSA). This emerging resistance to vancomycin means that available antibiotics may become useless against these infections, which would be a serious development: Before antibiotics, *S. aureus* infections killed up to 40% of all patients undergoing major surgery.

Fortunately, neither VISA nor VRSA has been transmitted between patients to date. All patients known to have had VISA had a history of prolonged and recurrent vancomycin treatment, and all had the hospital strain of MRSA. Because VISA and VRSA evolved from MRSA, patients with VISA also are resistant to methicillin.

ESBLs: Upping the Antibiotic Ante

Microorganisms that produce ESBL are resistant to extended-spectrum (or third-generation) cephalosporins such as ceftazidime, cefotaxime, and ceftriaxone, and monobactams such as aztreonam. The most common ESBL-producing organisms are the Gram-negative bacteria, typically *Klebsiella* species and *Escherichia coli,* although ESBL production has also been found in some strains of the *Salmonella* species, *Proteus mirabilis,* and *Pseudomonas aeruginosa.* These organisms don't cause more serious disease or a different manifestation of disease than their non-ESBL-producing counterparts. But the infections they cause are more difficult to treat because the organisms resist first-line antibiotics used to treat Gram-negative bacterial infections.

Patients at risk include those with intravenous or indwelling urinary catheters or gastrostomies, patients on mechanical ventilation, patients who've had previous antibiotic therapy or emergency abdominal surgery, patients with gastrointestinal colonization or severe illness, and those with long stays in the hospital or intensive care unit.

Most ESBL infections respond to a carbapenem antibiotic, such as imipenem or meropenem. If your patient has a Gram-negative bacterial infection, culture specimens should be taken and tested for ESBL production and sensitivity so that he receives an effective antibiotic.

ESBL-producing organisms can spread by direct and indirect contact. If you're caring for a patient with this type of infection, don't touch your eyes, nose, or mouth during patient care. Perform appropriate hand hygiene before and after giving care and before drinking or eating.

Isolation Tactics

A patient with an antibiotic-resistant bacteria should be placed in a private room or in a room with another patient with the same infection. Follow isolation precautions recommended by the CDC.

- Don personal protective equipment before entering the room. Put on the gown first, then a mask or respirator (if needed because the patient has another illness such as

---

Taking Steps to Stay Safe

To reduce the risk of spreading antibiotic-resistant infections, follow these guidelines from the Centers for Disease Control and Prevention for contact precautions.

- Practice meticulous hand hygiene. Patients, their families, and health care providers should wash their hands before eating and before and after patient contact. Using warm water and soap, scrub for 15 seconds, rinse with water, and dry your hands with a paper towel. If your hands aren't visibly soiled, you can use an alcohol-based hand rub instead of soap and water.
- Use indwelling and I.V. catheters only when essential and remove them as soon as possible. Use proper insertion technique and follow catheter care protocols. If your patient has a central venous access device that hasn't been used in days, ask the health care provider if it can be removed. If your patient is incontinent, use a skin barrier cream and disposable incontinence products instead of an indwelling urinary catheter if possible.
- Implement and follow an antibiotic protocol. To ensure prudent use of vancomycin and extended-spectrum cephalosporins, your facility should have an antibiotic surveillance team and guidelines for the proper use of antibiotics. Know which antibiotics are being used to treat which organisms and treat infection, not colonization. Colonization doesn't harm the patient, and treating it often leads to more resistant organisms. Treatment is rarely indicated for patients with positive cultures but no signs of infection (such as fever, erythema, or increased white blood cell count).
- Place a patient with multidrug-resistant organisms in a private room or in a room with another patient who has the same organism.
- To reduce the risk of transmitting organisms to other patients or to staff, disinfect surfaces in the patient's room—especially high—touch surfaces such as countertops, bed rails, and door handles-daily.
- Follow standard precautions when handling and disposing of linen. No special precautions are needed for dishes, glasses, cups, or eating utensils; use disposable cups or plates only if ordered.

---

tuberculosis), then goggles or face shield (if splashing of body fluids is possible), and finally gloves.
- Change gloves if they tear or touch body fluids.
- Remove your gloves and goggles (or face shield) and gown and perform hand hygiene before leaving the patient's room.
- Remove the respirator outside the patient's room.
- Follow infection control protocols for bagging and disposing of protective gear.
- If possible, each patient should have dedicated equipment, such as a stethoscope, blood pressure cuff, and rectal

thermometer. If equipment must be shared, thoroughly clean and disinfect it before using it on another patient.

- Make sure that other staff members who enter the room understand and comply with these precautions. For more information, see "Getting the Most from your Personal Protective Gear," *Exposure Safety,* in the December issue of *Nursing2004.*

Supporting Your Patient

A patient who must be isolated is likely to be worried and depressed. Provide emotional support and encourage him to stay as engaged and active as his condition permits. Inform him about support services available from volunteers or the chaplain.

By understanding the newest trends in antibiotic resistance, you can help protect your patients and yourself from emerging diseases.

References

Centers for Disease Control and Prevention. Methicillin-resistant *Staphylococcus aureus* infections among competitive sports participants—Colorado, Indiana, Pennsylvania, and Los Angeles County, 2000–2003. *Morbidity and Mortality Weekly Report.* 52(33):793–795, August 22, 2003.

Centers for Disease Control and Prevention. Methicillin-resistant *Staphylococcus aureus* skin or soft tissue infections in a state prison—Mississippi, 2000. *Morbidity and Mortality Weekly Report.* 50(42):919–922, October 26, 2001.

Colodner R, Israel A. Extended-spectrum beta-lactamases: A challenge for clinical microbiologists and infection control specialists. *American Journal of Infection Control.* 33(2): 104–107, March 2005.

Lee MC, et al. Management and outcome of children with skin and soft tissue abscesses caused by community-acquired methicillin-resistant *Staphylococcus aureus. Pediatric Infectious Diseases Journal.* 23(2):123–127, February 2004.

Naimi T, et al. Epidemiology and clonality of community-acquired methicillin-resistant *Staphylococcus aureus* in Minnesota, 1996–1998. *Clinical Infectious Diseases.* 33(7):990–996, October 1, 2001.

Thomson KS. Controversies about extended-spectrum and AmpC beta lactamases. *Emerging Infectious Diseases.* 7(2):333–336, March–April 2001.

Critical Thinking

1. How can extensively prescribing antibiotics lead to drug-resistive organisms?

REBECCA KJONEGAARD is an infection control practitioner at Sharp Grossmont Hospital in La Mesa, Calif. **FRANK EDWARD MYERS III** is manager of clinical epidemiology and safety at Scripps Mercy Hospital in San Diego, Calif.

From *Nursing,* vol. 35, No. 6, June 2005, pp. 48–50. Copyright © 2005 by Lippincott, Williams & Wilkins/Wolters Kluwer Health. Reprinted by permission via Rightslink.

Reduce the Risk of High-Alert Drugs

Learn how adapting processes for prescribing, preparing, and administering can help reduce errors associated with certain high-alert medications.

HEDY COHEN

A ny Medication Error has the potential to harm your patient. When certain "high-alert" drugs and drug categories are involved, the threat of significant harm or death is even greater.[1]

The Institute for Healthcare Improvement (IHI) has included strategies to prevent harm from high-alert drugs as one of six new patient-safety interventions in its 5 Million Lives Campaign. Targeting anticoagulants, sedatives, opioids, and insulin in its high-alert drugs intervention, the IHI offers hospitals educational materials, research resources, suggestions for strategic change, and tools to monitor their success.[2] This article is the second in a five-part series focusing on each of this new campaigns practice interventions. (See *About the IHI* and *About this Series*.)

In this article, I'll highlight high-alert drugs that are especially dangerous if misused, and spell out practice changes you and your facility can make to minimize the chance of errors with these drugs.

Six Medications Stand Out

In 1995, the watchdog Institute for Safe Medication Practices (ISMP) conducted a landmark survey on more than 160 acute care hospitals to analyze which medications caused serious harm and death over a 1-year period. (*Living up to its name* provides details about ISMP.) At the study's end, six medications stood out as significantly risky in acute care settings. Coining the term "high-alert drugs," ISMP started spreading the word about the dangers associated with the following medications:[3]

- insulin
- heparin
- opioids
- injectable potassium chloride or potassium phosphate concentrate

- neuromuscular blocking agents
- chemotherapy drugs.

To continually refine and add to the list of high-alert drugs, ISMP reviews medication error reports and clinical and safety literature, confers with safety experts, and surveys health care practitioners and facilities. The comprehensive list appears at *www.ismp.org/Tools/highalert-medications.pdf.*

Organizations with Clout Set the Bar

Along with ISMP and the IHI, The Joint Commission recognizes the life-threatening risks associated with high-alert drugs and gives high priority to safeguard their use. In its hospital accreditation standards, The Joint Commission requires hospitals to "develop processes for managing high-alert drugs." After first identifying medications that pose great risk, a hospital should then develop safe processes for procuring, storing, ordering, transcribing, preparing, dispensing, administering, and monitoring them.[4]

The Joint Commission doesn't give specific suggestions on how to achieve these goals; each organization develops its own system and demonstrates the process and outcomes data to The Joint Commission. Despite the efforts of ISMP, the IHI, The Joint Commission, and other safety organizations errors with high-alert drugs are still too common.

Building a Culture of Safety

Having analyzed error reports for more than 30 years, ISMP recognizes that significantly reducing drug errors calls for the willingness of state and federal governments, regulatory agencies, the pharmaceutical industry, and everyone in health care organizations to change.

Living Up to Its Name

For over 30 years, ISMP has had a passionate dedication to patient safety. Through a confidential voluntary error-reporting program operated in cooperation with the United States Pharmacopeia, practitioners across the nation tell this independent, nonprofit organization about medication errors. The ISMP experts analyze lessons to be learned and share solutions with health care practitioners, government bodies, and the public and always adhere to a philosophy of fixing the system rather than blaming the individual.

www.ismp.org

Willingness to change clinical practices must start at the top: Facilities need to provide a culture where the staff feels safe reporting risks, errors, and near misses. The administrators need to survey everyone in the organization to set a baseline for improvement and resurvey them to gauge success. Getting "boards on board" to implement and support changes in practice is a key feature of the IHI's 5 Million Lives Campaign.[2]

Three formulas encompass all the steps to safer high-alert drug therapy and serve as a basis for change:

- Standardize error-prevention processes.
- Make errors apparent.
- Minimize the consequences of errors that reach the patient.

Standardize Processes to Prevent Errors

Following your facility's recommended processes and using equipment properly help prevent errors associated with high-alert drugs. Although not totally fail-safe, the following technology improves safety when used as recommended:

- *computerized prescriber order entry* (CPOE) systems, which identify certain drugs as high-alert during ordering
- *bar coding,* which can help ensure that the "five rights" are addressed
- *"smart" pumps* that have a drug library requiring confirmation of drug dose limits before they'll start infusing medication.

Besides adopting helpful technology, changing certain processes in the chain of medication administration also enhances patient safety. Here are some processes you and your facility can adopt to prevent errors. (*Zeroing in on safe guarding strategies* provides a concise list.)

Restrict high-alert drugs and administration routes. Removing these drugs or limiting their number, variety, and concentrations in patient-care units will reduce their involvement in errors. For example, hospitals that remove all concentrated potassium chloride stock from patient-care units eliminate errors associated with inadvertent administration of highly concentrated potassium. Removal of all neuromuscular blocking agents from units where patients aren't normally intubated helps prevent accidental administration to patients who aren't receiving mechanical ventilation.

Removing equipment that allows opportunities for mistakes is another effective way to eliminate errors. For example, buying and using epidural tubing that lacks injection ports makes injecting inappropriate medication into an epidural line much less likely.

Practice drug differentiation. Some high-alert drugs have look-alike and sound-alike names or look-alike packaging and labeling. To prevent potentially fatal mix-ups, we must highlight differences that help us distinguish among these drugs. For name pairs that are too close for comfort, such as heparin and Hespan, prevent confusion by using both the generic *and* brand name on prescriber order sheets, preprinted order forms, screens on all electronic devices used for medication therapy, storage shelves in patient-care areas, and medication administration records (MARs). Equally important, don't store these medications next to each other.

Using "tall-man" lettering for look-alike drug names is another safeguard. Research in a simulated dispensing pharmacy has shown that writing the unique parts of similar drug names in capital letters can reduce wrong drug selection by 35%.[5] The capitalization in hePARIn and heSPAn highlights their differences. (Go to *www.ismp. org/MSAarticles/drugnames.html* for additional look-alike drug names to watch out for.)

Use forcing functions. A system that eliminates or reduces the possibility of error by forcing conscious attention to a problem is called a hard or soft forcing function.

Hard forcing functions are built into equipment. These are "lock and key" devices that don't let you use something that isn't a perfect fit. A leaded gasoline nozzle that was manufactured larger than the opening of an unleaded gas tank won't let you put the wrong mix into your tank.

Medical equipment designed to prevent inadvertent drug infusion through the wrong route is another hard forcing function. For example, providing only oral syringes that don't fit into luer-lock intravenous (I.V.) tubing connections or ports prevents you from

Zeroing in on Safeguarding Strategies

Strategy	Examples
Fail-safes, forcing functions	• Use flow-control pumps for continuous I.V. infusions.
	• Use epidural tubing without side ports.
	• Use only oral syringes that can't be connected to I.V. tubing for administration of oral liquids.
Constraints	• Remove concentrated oral opioids from unit stock and dispensing cabinets.
	• If drugs such as paralytic agents must be stored in appropriate patient-care areas, sequester them from other drugs.
	• Separate look-alike products during storage and use.
	• Dispense I.V. and epidural infusions from pharmacy only.
	• Use smallest package, concentration, dose, and number of vials for unit stock.
	• Establish formulary drug restrictions (such as not keeping highly concentrated opioids in medical/surgical units) or criteria for selected opioids (such as their effects on opioid-naive patients).
	• Switch from I.V. to an oral or subcutaneous form as soon as possible.
Externalizing, centralizing error-prone processes	• Move I.V., epidural preparations from patient-care units to pharmacy.
	• Use commercially available premixed I.V. and epidural solutions or outsource production to companies that are inspected by the Food and Drug Administration.
Accessible information	• Post drug information charts in patient-care areas.
	• Make drug information available for simultaneous use by multiple staff (up-to-date reference books in nursing units or a database accessible online or via personal digital assistant).
	• Make the patient's medical record available to all appropriate caregivers.
	• Reconcile medications at every change in care setting.
Standardization, simplification	• Eliminate use of unapproved abbreviations (such as MSO4), drug name stems (such as "platins"), and acronyms.
	• Standardize drug and dose expressions.
	• Require use of CPOE, preprinted orders forms, or ordering protocols.
	• Require orders by metric weight, not by volume or ampule.
	• Include dose formula with calculated dose for pediatric patients.
	• Use drug preparation guideline in pharmacy.
	• Use dosing and infusion rate charts in patient-care units.
	• Standardize/limit drug concentrations and formulations.
	• Establish drug administration protocols for PCA, epidural, and I.V. infusions.
	• Dispense medications in unit-dose or unit-of-use packages.
Differentiation, warnings, reminders	• Apply warning lables or distinguish drug names and concentrations, such as with concentrated electrolytes and opioids.
	• Set dose limits on smart pumps and CPOE orders.
	• Differentiate drugs with different container sizes and outer wraps.
	• Provide checklists for complex processes such as PCA infusions.
	• Use tall-man lettering for look-alike drug names.
	• Build reminders into order sets, MARs, protocols, and dispensing cabinets.
Redundancies, independent double checks	• Recalculate the dose before administration to a neonate or child.
	• Verify the drug and drug preparation.
	• Establish double checks for unusual drugs, doses, regimens.
	• Verify patient diagnosis, indication for drug.
	• Establish double checks for pump rate, drug concentration, line attachments.
	• Use bar coding, smart pumps.
	• Educate patients and family and encourage their participation in care.
Patient monitoring	• Provide cardiac monitoring, pulse oximetry, and capnography for a patient receiving PCA.
	• Closely observe and monitor vital signs in patients receiving high-alert drugs.
	• Keep antidotes and resuscitation equipment close by.
Proactive risk evaluation	• Conduct a failure mode and effects analysis (FMEA) before a new process or device is put into place to look at what can go wrong and how it may lead to severe injury.

inadvertently administering an oral liquid into an I.V.line. Every health care facility should eliminate the possibility of this type of error by adopting these oral syringes for safety.

Medication Stand out text is not clear.

A CPOE system can be a technologic hard forcing function because it can be programmed to prohibit ordering medications until vital patient information such as allergies and weight are entered in the system.

Soft forcing functions are included in work processes but provide less protection. If your facility provides specific pathways for nurses to use and lessen error risk, you're using a soft forcing function each time you follow the steps. Preprinted standardized order forms and CPOE and automated dispensing cabinet (ADC) screens help reduce errors by offering limited lists of medications and dosages to choose from. Some ADC drawers are designed not to open until you correctly enter specific patient information and pharmacy has profiled the patient's medication to make sure it's appropriate.

Soft forcing functions aren't safe as hard forcing functions because you can override or work around their safety features.

Apply constraints. A constraint limits the performance of certain actions, such as limiting access to medications and related devices to lessen error risk. Examples include:

- limiting nurses' access to medications and devices in the pharmacy when it's closed
- stocking formulary drugs only in small quantities and in the smallest concentrations in the pharmacy night cabinet
- limiting access to high-alert medications in patient-care units, even when the pharmacy is open. For example, remove concentrated insulin products (U-500) to help prevent errors.
- limiting the use of certain error-prone products, such as intrathecal or epidural analgesics and toxic chemotherapy drugs, to specially educated or credentialed staff.
- requiring prescribers to consult specialists, such as a member of the pain management team, before ordering certain high-alert medications in high-risk patients: those with renal or hepatic dysfunction; neonates; older adults; patients with impaired immunity, cancer, HIV/AIDS, or other chronic illness; and pregnant and breast-feeding women.

Make patient information readily accessible. Members of the health care team need easy access to accurate and complete information about patients and drugs at each step of medication therapy: prescribing, transcribing,

preparing, dispensing, and administering. Electronic links between drug databases and patient information can provide immediate warnings about unsafe medications or doses at the various steps, but nurses and pharmacists must actively use these automated decision support systems to get the warnings.

The following scenario shows how a change in process could prevent an overdose or underdose of warfarin.

Problem: A prescriber orders warfarin for Mr. Lee at 0700, before the patient's international normalized ratio (INR) result is available, so he has no way of knowing whether the dose needs to be adjusted.

Solution: Develop a facility policy to administer warfarin at 2100, so all lab results would be available before dosing. This approach, which can apply to both electronic and "paper" systems, would let pharmacy verify the order before making the drug available. Always check and record your patient's INR result *before* you administer warfarin.

Another policy change might be in order if your facility uses wristbands to alert practitioners about patient allergies. Using certain colors to indicate specific drugs is risky because the meaning of colors isn't standardized. Interpreting color meaning can be especially difficult for nurses who work in more than one health care facility. A safer approach is to write "Allergy" on the wristband to alert health care providers to get complete allergy information from the patient or his chart.

Standardize and simplify. *Standardization* ensures that a process is completed the same way each time, reducing variability and opportunity for error. By reducing reliance on human memory and attention, standardization is useful to help new nurses to safely carry out unfamiliar processes. For example, checklists provide standard steps for complex processes.

A checklist is especially effective to prevent errors with high-alert drugs that are administered through patient-controlled analgesia (PCA) pumps. The items on a safety checklist include educating the patient about the proper use of PCA, warning family members and visitors about the dangers of PCA by proxy, and applying a warning label *For Patient Use Only* on the activation button. Points on a checklist for other medications might include identifying risk factors for respiratory depression and enhancing patient monitoring with pulse oximetry or capnography.

Using standard drug concentrations and volumes for high-alert drugs decreases the available options in patient-care areas. Errors are less likely because prescribers can order in only one concentration, such as, heparin, 25,000 units/500 mL. Once standard concentrations for infusions are established, dosing and infusion rate charts can be implemented to enhance safety, and smart pumps can be set to accommodate standard medication concentrations.

Simplification reduces what you need to remember and plan, and it helps with problem solving. For example, each additional step in the medication-use process adds to the cumulative risk of error. The probability of error increases with every additional step, so if each step is 99% reliable, a 5-step process is 95% reliable. A facility that buys standardized heparin concentrations reduces the chance of error by decreasing the need to calculate and mix each dose.

Apply reminders. Auxiliary labels on medication packages and highlights on MARs can remind nurses to take precautions for safe preparation and administration of a high-alert drug. These are a few sample labels:

> WARNING
> Highly Concentrated Drug—
> Must Be Diluted

> WARNING
> Paralyzing Agent—
> Causes Respiratory Arrest

> For ORAL Use ONLY

> For EXTERNAL Use ONLY

Another type of reminder is adding special precautions and patient-monitoring prompts to preprinted order sets and protocols for high-alert drugs such as heparin and insulin.

Use fail-safes. A fail-safe method minimizes the chance of harm or prevents error if there's a failure in the process. An example related to medication therapy is using a pump that doesn't allow free-flow: The fluid won't flow even if you forget to clamp the I.V. tubing.

The biomedical department or PCA pump vendor can program PCA pumps to be fail-safe in two ways:

- set the default drug concentration at zero to force an entry
- set the default drug concentration at the highest concentration of opioids used in the facility. Although this may seem dangerous, it actually helps reduce risk: If the default is 1 mg/mL and you load the pump with a 5 mg/mL cartridge but forget to change the setting, the patient would get 5 mg of opioid for each 1 mL infused. If the default setting is 5 mg/mL, the patient would get 0.2 mg for each 1 mL infused. A subtherapeutic

dose is much safer for the patient than risking oversedation and respiratory depression with an overdose.[6]

Include patients in therapy. An informed patient or patient representative may be the final stop for a harmful medication error. Patients and families (with the patient's permission) need clear, consistent information about his medications while he's in the hospital and at discharge.

For example, two patients getting the antineoplastic drug methotrexate need to have very different perspectives on the course of therapy. The patient who's getting it as cancer chemotherapy should understand that he'll probably get the drug over consecutive days each week and he'll require careful monitoring with blood counts; the patient getting methotrexate for a noncancerous condition such as psoriasis should understand that dosing is generally once a week.

Make Errors Apparent

When a medication error with a high-alert drug is about to happen, the following strategies make the error stand out and may prevent it from reaching the patient.

Perform independent double checks. Having another person review the steps you follow to prepare or administer a medication can uncover a mistake and possibly save a patient's life. For example, when a patient is receiving certain high-alert drugs (such as insulin or an opioid) by the I.V. route, two people should independently check the infusion pump setting against the order in each of these situations:

- just before starting the drug infusion
- every time the infusion rate is changed
- every time an empty infusion bag or cassette is replaced.

When you complete an independent double check, you verify that the drug, dose, and programming are correct according to the most recent order for this patient before the infusion starts. According to facility policy, you may have to add your signature to the MAR to indicate your verification.

Research shows that people checking the work of others will find about 95% of all mistakes,[7] although no studies have focused strictly on double checks in the medication administration process.

Doing the math shows how double checks, done before medications are administered, reduce risk. If the organizational error rate of administering medications after double checks is 3%, and only 5% of double checks fail to uncover an error, the chance of a medication error reaching the patient is low, as shown by multiplying the two:[8] $0.03 \times 0.05 = 0.0015$ or 0.15%.

Independent double checking is time-consuming when done correctly, so it should be used exclusively for high-alert drugs that pose the most problems in a specific patient-care area. Too much double checking may be counterproductive because it's burdensome. This can lead to double-check fatigue and result in complacency or work-arounds. Help identify the drugs that pose the most problems at your facility to establish realistic double-check policies.

Rely on redundancies. You can use other types of redundancies besides independent double checks to help reduce the risk of patient harm. They include matching high-alert drug orders to the patient's diagnosis, the drug's indication, and vital patient information to confirm that the drug and dose are appropriate. Verbal orders—especially for high-alert drugs—should be avoided if at all possible. Never take verbal orders for chemotherapy drugs. If you must take a verbal order for a high-alert drug, read back these details to the prescriber *after* you write the order in the patient's chart:

- patient name
- drug order as written
- spelling of the drug name, to ensure that you heard it right and transcribed it correctly.

Listen for bells and whistles. Technology such as well-designed ADCs and bar-code equipment can highlight errors with sights and sounds. For example, a bar-code system may sound an alarm and display a text warning if the correct drug, dose, and administration route aren't being used at the correct time for the correct patient.

To use bar-code technology correctly, observe these practices:

- Always scan drugs at the patient's bedside.
- Heed all alerts from the system.
- Retrain from the temptation to bypass or work around the alerts.

Minimize the Consequences of Errors

If, despite preventive measures, a medication error reaches the patient, these precautions can help minimize the harm.

Store smaller vials of risky drugs. Reducing the size of high-alert drug vials in patient-care areas is a relatively simple way to help minimize the effects of errors. For example, storing lidocaine in only 10 mL vials rather than 50 mL vials can prevent a fatal overdose if someone were to inadvertently administer the entire vial.

Closely monitor the patient. By closely monitoring patients who are receiving high-alert drugs, you can detect problems related to errors early and intervene promptly. At set intervals, evaluate your patient's neurologic status (including level of consciousness), vital signs, respiratory status, and lab results. Make sure reversal agents and resuscitation equipment are readily available and call the rapid response team at the first sign of trouble.

Take the Offensive against Risk

Among the many things you can do to reduce the risk of harm from errors with high-alert drugs, some of the most critical are watching for processes associated with their use that might fail and harm patients. You should feel free to inform your nurse-manager about your concerns and suggest techniques to make the medication-use process safer.

By monitoring external reports about errors in other settings, you can help develop error-prevention strategies in your practice. A medication error that occurs in one health care setting is likely to occur in other facilities where similar circumstances exist. The Institute for Safe Medication Practices details cases of errors and near misses, then adds tips on safe transcription, administration, and patient monitoring to reduce risk in several publications: *Nurse Advise-ERR* and *ISMP Medication Safety Alert!* (newsletters to which you can subscribe at *www.ismp.org*) and the monthly "Medication Errors" column in *Nursing 2007.*

Applying Safety

Medication safety experts at ISMP and organizations such as the IHI are striving to improve patient safety. By heeding their advice, you and your facility can apply many of the safety strategies they recommend to protect patients from errors with high-alert drugs.

References

1. Cohen MR (ed). *Medication Errors.* Washington, D.C., American Pharmacists Association, 2007.
2. Institute for Healthcare Improvement. Protecting 5 Million Lives from Harm. *www.ihi.org/IHI/Programs/Campaign/Campaign.htm.* Accessed June 8, 2007.
3. Cohen MR, et al. Survey of hospital systems and common serious medication errors. *Journal of Healthcare Risk Management.* 18(1):16–27, Winter 1998.
4. The Joint Commission. *Hospital Accreditation Standards,* standard MM.7.10. Oakbrook Terrace, Ill., The Joint Commission, 2006.
5. Grasha A, et al. A cognitive systems perspective on human performance in the pharmacy: Implications for accuracy, effectiveness, and job satisfaction. Technical report 062100-R, 1–170. Cincinnati, Ohio, Cognitive-Systems Performance Laboratory, 2000.

6. Institute for Safe Medication Practices. Misprogram a PCA pump? It's easy! *ISMP Medication Safety Alert.* 9(15):1–2, July 29, 2004.

7. Grasha AF, et al. Delayed verification errors in community pharmacy. Technical report 112212. Cincinnati, Ohio, Cognitive-Systems Performance Laboratory, 2000.

8. Institute for Safe Medication Practices. The virtues of independent double checks—They are really worth your time. *ISMP Medication Safety Alert.* 8(5):1, March 6, 2003.

Critical Thinking

1. What strategies can be used to prevent harm from high-alert drugs?

The author has disclosed that she has no significant relationship with or financial interest in any commercial companies that pertain to this educational activity.

Keeping Your Patient Hemo-Dynamically Stable

Follow this guide to use vasoactive medications safely in a medical/surgical unit.

JULIE MILLER, RN, CCRN, BSN

When I started as a critical care nurse 20 years ago, patients receiving dopamine, dobutamine, and other powerful vasoactive drugs were never sent to the medical/surgical unit. Today, transferring these patients to another unit isn't uncommon. If that unit is yours, are you prepared to manage their care? In this article, I'll discuss what you need to know about vasoactive drugs, including their intended effects and potential adverse reactions, so you can administer them safely and confidently.

Most facilities establish criteria for moving patients receiving these drugs out of the critical care setting; for example, patients must be stable, monitored with telemetry, and have their vital signs monitored at least every 4 hours. Many facilities don't let medical/surgical nurses to adjust intravenous (I.V.) infusions of vasoactive medications against vital signs parameters (also called titrating); learn and follow your facility's policy and procedure. (See *First, a word about safety* for other common precautions.)

Now let's look at some reasons why caring for these patients is so challenging.

What Are Vasoactive Drugs?

Besides dopamine and dobutamine, drugs categorized as vasoactive include vasodilators such as nitroglycerin and nitroprusside; vasoconstrictors such as epinephrine, norepinephrine, phenylephrine, and vasopressin; calcium channel blockers; beta-blockers; and milrinone. They're considered high-alert drugs because dosing errors carry an increased risk of patient harm.

Patients who need epinephrine, norepinephrine, nitroprusside, or phenylephrine infusions should be in an intensive care unit. Although you may not be administering these drugs, you should know how they work.

Administering vasoactive drugs is challenging because they can exert undesirable and potentially dangerous secondary effects in addition to their intended primary effects. For example, dobutamine is given to increase myocardial contractility (its intended effect), but it can also cause tachycardia and hypotension, which increase myocardial work-load and oxygen consumption.

First, a Word about Safety

Because of their potency and vesicant properties, vasoactive drugs should be administered through a central venous access device using an infusion pump.

Carefully monitor the patient's vascular access device to prevent infiltration and extravasation. Tell him to report a burning sensation or swelling at the I.V. site. If he experiences an infiltration of a vasoconstrictor such as dopamine, be prepared to administer an antidote subcutaneously to prevent tissue sloughing. Phentolamine, administered subcutaneously into and around the area of the infiltration up to 12 hours after infiltration, dilates local blood vessels and can stop tissue damage and potentially prevent limb loss.

Know your facility's policies and protocols for administering vasoactive drugs. Always double-check the math on the loading dose and have another nurse check the dose too. Be especially careful with drugs that are given based on the patient's weight so you don't inadvertently give a too-high dose to smaller patients.

If your patient is receiving multiple medications, label the I.V. lines at several points so you can quickly check or discontinue an infusion in an emergency.

To help sort out intended and undesirable effects, let's look at drugs according to their main actions: increasing or decreasing blood pressure (BP), increasing contractility, and decreasing heart rate. For more about the components of *cardiac output,* see *All about cardiac output.*

Drugs That Increase BP

Patients who are in shock or have an acute exacerbation of heart failure may need vasoactive medications to treat hypotension. Dopamine, epinephrine, and norepinephrine, for example,

---

All about Cardiac Output

Normal cardiac output is 4 to 8 liters/minute for an adult. The four components of cardiac output are:

- *preload,* which is determined by the stretch of the left ventricle during diastole, when it's filled with blood. Increased venous return to the heart increases preload. You can estimate right-sided heart preload noninvasively (you'll see jugular vein distension and peripheral edema) or invasively with central venous pressure, which is normally 0 to 8 mm Hg. Left-sided heart preload can be assessed noninvasively by crackles in the lung fields (indicating pulmonary edema) or invasively by pulmonary artery wedge pressure, which is normally 4 to 12 mm Hg.
- *afterload,* or the resistance the ventricles must overcome to eject blood into the aorta. If the patient has vasoconstriction, afterload and resistance to forward blood flow increase. This leads to decreased cardiac output and blood backing up in the heart. You'll see cool and pale or mottled extremities.
- *contractility,* or the force and strength of the heart's contraction, assessed by ejection fraction (normally 55% to 70%). Strong, bounding pulses reflect strong contractility; weak and thready pulses may indicate decreased contractility.
- *heart rate,* or the frequency of ventricular contractions. The ventricles fill and the heart is perfused during diastole. Heart rate determines diastolic filling time. Normally, a faster heart rate increases cardiac output and a slower heart rate decreases cardiac output, but if the heart rate is too fast, filing time is shorter, so cardiac output falls.

stimulate alpha receptors in the sympathetic nervous system, causing vasoconstriction and increasing venous return to the heart. This increases afterload and BP.

- *Dopamine* at less than 5 mcg/kg/min stimulates dopaminergic receptors, dilating renal vasculature. Titrate the dosage to achieve the desired hemodynamic or renal response, increasing by 1 to 4 mcg/kg/minute at 10 to 30-minute intervals.

At 5 to 10 mcg/kg/minute, dopamine is primarily a beta stimulator, increasing the rate and force of myocardial contraction and promoting smooth-muscle relaxation; at doses of 10 to 20 mcg/kg/minute, it's an alpha stimulator, causing vasoconstriction and increasing systemic vascular resistance (SVR). Dopamine is now given only rarely at 2 to 5 mcg/kg/minute to increase renal perfusion because studies have shown it's not as effective as once thought.

High-dose dopamine usually is restricted to critical care areas, so check your facility's policy to determine if you can administer this infusion in your unit. If so, check your patient's urine output often and monitor for tachycardia.

- *Norepinephrine* stimulates alpha$_1$ receptors to a greater degree than dopamine and stimulates beta receptors less strongly than epinephrine. Nor-epinephrine causes vasoconstriction and increases SVR. This drug also has been found to increase glomerular filtration rate and renal blood flow by its action on the efferent arterioles of the kidneys. Make sure you monitor for dysrhythmias, severe hypertension, and headache.
- *Epinephrine* is an alpha$_1$, beta$_1$, and beta$_2$ agonist that increases BP by increasing heart rate, cardiac output, and SVR. Watch for hypertension, respiratory distress, and dysrhythmias.
- *Phenylephrine* is a selective alpha$_1$ agonist. It's used to treat hypotension by increasing SVR without raising heart rate. This drug is a better option than norepinephrine, epinephrine, or dopamine if the patient's heart rate already is elevated. Watch for dysrhythmias, hypertension, and anaphylaxis.
- *Vasopressin* (antidiuretic hormone) is used for hypotension refractory to other medications. This drug also has been added to the advanced cardiac life support guidelines for managing cardiac arrest. Vasopressin binds to vasopressin receptor sites and causes vasoconstriction and increased BP. The drug also can increase the effect of other vasopressors given to patients in septic shock. Monitor your patient for dysrhythmias, coronary ischemia, bronchoconstriction, and anaphylaxis.

Drugs That Decrease BP

Complications of renal or cardiac disease can cause hypertension. Nitroglycerin, nitroprusside, and nicardipine are used to decrease BP through vasodilation. Monitor patients for signs and symptoms of hypotension such as dizziness, and palpitations. Watch especially for orthostatic hypotension.

- *Nitroglycerin* dilates coronary arteries and decreases myocardial oxygen consumption by decreasing venous return to the heart. At higher doses, it also can decrease afterload. This drug usually doesn't significantly lower a patient's BP.
- *Nitroprusside* has a more profound effect on BP and afterload reduction because it dilates peripheral arteries. Because of this, it's the drug of choice for hypertensive emergencies.
- *Nicardipine* is a slow-acting calcium channel blocker that's more selective to calcium channels in vascular smooth muscle than the heart. This drug causes vasodilation and decreases SVR. Because of its short half-life, it can be used to treat hypertensive emergencies. Nicardipine also has been added to the advanced cardiac life support guidelines as a choice in the ischemic stroke algorithm for treating hypertension. Monitor your patient for signs and symptoms such as peripheral edema, palpitations, flushing, and dizziness.

Drugs to Increase Contractility

You'll hear the term "inotropic" used frequently when contractility is discussed, and you can consider the terms interchangeable.

Positive inotropes such as digoxin, dobutamine, and dopamine are medications that improve contractility. *Negative inotropes* such as beta-blockers and calcium channel blockers reduce contractility.

Beta$_1$ receptors are located in just *one* organ, the heart. Beta$_2$ receptors affect the *two* lungs (causing bronchodilation), *two* kidneys, and the blood vessels (causing vasodilation). Cardioselective beta-blockers predominantly block the beta$_1$ receptors in the heart, but may affect beta$_2$ receptors to a small degree. Nonselective beta-blockers such as proptanolol block the beta$_1$ and beta$_2$ receptors, causing bronchoconstriction in addition to decreasing heart rate and contractility. This bronchoconstriction (a beta$_2$ effect) is why you must always be cautious when administering a selective beta-blocker to a patient with pulmonary disease and why you should never give a nonselective beta-blocker to a patient with pulmonary disease.

Beta$_1$ agonists, which stimulate the heart's beta$_1$ cells, increase heart rate, contractility, automaticity, and conduction velocity. Some examples are dobutamine and dopamine (at 5 to 10 mcg/kg/minute).

- *Dobutamine* is predominantly a beta$_1$ agonist, stimulating the beta$_1$ receptors of the heart and increasing contractility. To a lesser degree, it also increases heart rate and stimulates the body's alpha$_1$ receptors (causing vasoconstriction) and beta$_2$ receptors (causing vasodilation and smooth muscle relaxation). As the patient's cardiac output increases, his SVR will decrease. The patient's BP also may decrease, so watch for hypotension.

A patient with decompensated heart failure may he admitted to your medical/surgical unit to receive I.V. dobutamine to improve cardiac contractility. Dobutamine has the added benefit of reducing afterload in patients with cardiomyopathy and heart failure. With close monitoring, you can safely manage stable patients receiving dobutamine dosages up to 10 mcg/kg/minute.

- *Milrinone* is a phosphodiesterase inhibitor that relaxes smooth muscles; because of its vasodilating effects, it's a positive inotrope that improves contractility and decreases afterload. Infusions of milrinone are commonly used to treat dilated cardiomyopathy when dobutamine isn't effective for improving cardiac output.

The dosing range for milrinone is 0.375 to 0.5 mcg/kg/minute in the medical/surgical unit, and these patients are usually monitored with cardiac telemetry. Be alert for possible hypotension from vasodilation and possible rebound tachycardia.

Drugs to Decrease Heart Rate

Patients with heart failure have decreased cardiac output when their heart rate increases because of decreased diastolic filling time and poor myocardial contraction. Always monitor your patient's heart rate and BP when administering the following drugs to reduce heart rate:

- *Metoprolol,* a cardioselective beta$_1$-blocker, decreases heart rate, which reduces myocardial oxygen consumption. This is why beta-blockers are often ordered for patients with acute myocardial infarction. Monitor your patient for bradycardia and don't give this drug to a patient who has second- or third-degree heart block and doesn't have a pacemaker.
- *Alpha-beta blockers,* which have mixed blocking effects, include *labetalol* (used to treat hypertension) and *carvedilol* (used to treat heart failure). By blocking beta$_1$ receptors, these drugs decrease heart rate and contractility, so they're

also negative inotropes. By blocking alpha receptors and causing vasodilation, they reduce afterload. Monitor for bradycardia and hypotension. These drugs also shouldn't be given to patients with second- or third-degree heart block who don't have a pacemaker.

- *Diltiazem,* a calcium channel blocker, decreases heart rate and often is prescribed for patients with supraventricular tachycardias such as atrial fibrillation or flutter. Calcium channel blockers slow conduction through the atrioventricular node, decreasing heart rate. Monitor your patient for bradycardia, changes in pulse quality, and prolongation of the PR interval on electrocardiogram.

Diltiazem is commonly given as an infusion at 5 to 15 mg/hour. Initially, you'll administer a loading dose of 0.25 mg/kg over 2 to 3 minutes. Be sure to double-check the loading dose ordered for weight-based dosing and monitor the patient for bradycardia.

Diltiazem shouldn't be given to patients with second- or third-degree heart block who don't have a pacemaker. Use caution when administering diltiazem or any calcium channel blocker to a patient with heart failure or cardiomyopathy; remember, these drugs are negative inotropes that decrease cardiac contractility. If your patient receives an overdose, administer I.V. calcium (the antidote) as ordered.

Other Drugs You May Administer

In the medical/surgical unit, you may care for a patient with dilated cardiomyopathy or heart failure who has elevated afterload due to the compensatory vasoconstrictive actions of the sympathetic nervous system and the renin-angiotensin-aldosterone system. To reduce left ventricular afterload in these patients, you'll administer angiotensin-converting enzyme (ACE) inhibitors such as enalapril, or angiotensin receptor blockers such as candasartan, and positive inotropes such as dobutamine, milrinone, or dopamine.

In a patient with heart failure and volume overload, treatment is aimed at decreasing circulating volume. To reduce volume, the patient will be given a loop diuretic such as furosemide or bumetadine. Loop diuretics interfere with sodium and water reabsorption in the loop of Henle.

Aldosterone antagonists such as spironolactone inhibit the action of aldosterone on the kidneys. Aldosterone causes sodium and water retention. An aldosterone antagonist inhibits that action, letting the kidneys get rid of excess sodium and water. Aldosterone antagonists are potassium sparing, so you should use them cautiously in patients who are also on ACE inhibitors. Because ACE inhibitors are potassium sparing, patients taking both drugs are at risk for hyperkalemia.

Angiotensin-converting enzyme inhibitors and angiotensin receptor blockers inhibit the renin-angiotensin-aldosterone system, promoting vasodilation and sodium and water loss. (Vasodilation pools blood in the periphery, also contributing to a decrease in preload.)

Be sure to closely monitor your patient's vital signs when administering diuretics, ACE inhibitors, or angiotensin receptor blockers. These medications can cause symptomatic hypotension by decreasing preload, especially within the first hour of administration.

The newest drug for treating acute decompensated heart failure is *nesiritide,* a recombinant B-type natriuretic peptide that promotes diuresis and natriuresis, resulting in water and sodium loss and decreased preload. Nesiritide also reduces preload by acting as a vasodilator. Because it decreases SVR, nesiritide decreases systolic BP and afterload. This increases cardiac output without increasing the heart rate. The risk of hypotension usually occurs in the first 15 to 60 minutes after the infusion is started, so nesiritide should be administered only in a critical care or progressive care area, where the patients BP can be closely monitored.

Now let's look at agents that often are administered as adjuncts to vasoactive drugs.

A patient with a normal heart may have decreased preload because of blood loss (for example, bleeding during surgery or after traumatic injury). Decreased preload leads to decreased cardiac output and systemic hypotension. To increase preload via volume expansion, the patient will be given an infusion of an isotonic crystalloid solution such as 0.9% sodium chloride solution or lactated Ringer's solution. Colloid solutions, such as 5% or 25% albumin, hetastarch, or blood transfusions, also may be ordered to raise circulating vascular volume and increase preload.

Twenty-five percent albumin increases circulating blood volume by pulling fluid into the vascular space from the interstitial space. These solutions may be prescribed for a patient with excessive peripheral edema or a low total albumin or protein level.

Albumin increases circulating blood volume by pulling fluid into the vascular space from the interstitial space.

Hetastarch, a nonprotein colloid plasma volume expander, must be limited to 1,500 mL or less per 24 hours. Larger amounts of hetastarch can cause platelet and clotting factor VIII dysfunction, putting your patient at risk for bleeding and further reduction in blood volume.

Knowledge into Practice

Suppose you're caring for Nate Brandon, 68, who's receiving intermittent doses of I.V. furosemide and I.V. dobutamine at 7.5 mcg/kg/minute for heart failure caused by cardiomyopathy. He's awake, alert, and oriented, but over the last 4 hours, his BP has fallen to 78/50 and he's oliguric. The health care provider has ordered the dobutamine titrated to a mean arterial pressure (MAP) of 70 mm Hg. Mr. Brandon's current MAP is 60 mm Hg: his heart rate is 110, his skin is cool and pale, and his peripheral pulses are 1+ You auscultate a few bibasilar crackles in his lungs.

Mr. Brandon is stable, but his cool, pale skin and narrow pulse pressure indicate increased afterload, his bibasilar crackles mean he also has increased preload. An increase in the dobutamine infusion should reduce afterload and preload and increase contractility, with a net result of increasing Mr. Brandon's cardiac output. However, the drug also may increase his heart rate, which could *decrease* cardiac output.

You increase the dobutamine infusion to 10 mcg/kg/minute and reassess Mr. Brandon in 30 minutes. His urine output has increased, he has fewer bibasilar crackles, his skin is cool but pink, his heart rate is 108, his BP is 102/64, and his MAP is 76 mm Hg. His pulses are 2+, and he's awake, alert, and oriented. The slight increase in dobutamine has improved his clinical status.

Is Your Head Spinning?

By knowing each drug's effect on BP, contractility, and heart rate, you'll understand the various effects vasoactive drugs can have on patients like Mr. Brandon. By understanding and anticipating both intended and unintended drug effects, you can safely manage vasoactive drug administration in your unit.

Resources

Albert NM. et al. Evidence-based practice for acute decompensated heart failure. *Critical Care Nurse.* 26(4):14–29, December 2004.

Geisler GM, Delgado R. *Evaluation and Treatment of Patients with Acute Decompensated Heart Failure.* Houston, Tex., Texas Heart Institute, 2004.

Heart Failure Society of America Evaluation and management of patients with acute decompensated heart failure: HFSA 2006 comprehensive heart failure practice guideline. Rockville, Md., National Guidelines Clearing-house. www.guideline.gov/summary/summary.aspx?doc_id=9328. Accessed January 22, 2007.

Hyrniewiez K, et al. Comparative effects of carveddol and metoprolol on regional vascular responses to adrenergic stimuli in normal subjects and panents with chronic heart failure. *Circulation,* 108(8):971–976, August 26, 2003.

Prahash A. Lyoch T. B-type natriuretic peptide: A diagnostic, prognostic, and therapeutic tool in heart failure. *American Journal of Critical Care* 13(1):46–55, January 2004.

Urden LD, et al. (eds). *Thelan's Critical Care Nursing Diagnosts and Management,* 5th edition. St. Louis. Mo., Mosby, Inc. 2005.

Wiegand Lynn-Mettale DJ. Carlson KK (eds). *AACN Procedure Manual for Critical Care,* 5th edition. Philadelphia, Pa., W.B., Saunders Co. 2005.

Critical Thinking

1. Why is it important to ensure that a patient is dynamically stable prior to accepting them onto the medical-surgical unit?

JULIE MILLER is staff-development educator for critical care at Trinity Mother Frances Health System in Tyler, Tex.

The author has disclosed that she has no significant relationship with or financial interest in any commercial companies that pertain to this educational activity.

Medication-Monitoring Lawsuit: Case Study and Lessons Learned

Another Real-Life Case in Nursing Home Litigation

LINDA WILLIAMS, RN

The lack of a medication-monitoring system can be an expensive liability exposure. In addition to the mental anguish experienced by families over losing a loved one because of a medication error, courts have been quite harsh in ruling that nursing homes were derelict in their duty to safely administer and monitor the medications given to the residents. The following case study summarizes a lawsuit against a nursing home concerning this issue:

An 82-year-old woman was admitted to a hospital with left-sided weakness as a result of a cerebral vascular accident (CVA) or stroke. Within the prior week, she had been treated and released at the same hospital emergency room twice for uncontrolled nosebleeds. The woman remained at the hospital for stabilization of the CVA and was discharged six days later to a rehab hospital for aggressive physical and occupational therapies. She was unable to physically tolerate the aggressive therapies at the rehab hospital so, after 10 days, she was discharged to a nursing home for less aggressive therapies.

Within a month of her arrival at the nursing home, she fell from a commode and sustained a bloody nose, facial fractures, and a subdural hematoma. She was immediately sent to the hospital and treated for 11 days before returning to the nursing home in a much-deteriorated state. Within two months, she died at the nursing home from complications related to her injuries.

A year later, the woman's family filed a lawsuit against the nursing home for negligence in supervision of medication administration and monitoring the claimant for adverse consequences of medications.

When the resident was a patient in the first hospital, a neurologist wrote, "She is *not* a candidate for anticoagulation therapy due to her recent history of severely uncontrolled nosebleeds." However, when she was admitted to the rehab hospital, her medication orders included an anticoagulant medication, heparin 5000 units, administered subcutaneously, twice a day. She remained on the subcutaneous heparin and was discharged to the nursing home with the same order. No labs had been taken at either hospital regarding the therapeutic level of this medication.

When the patient arrived at the nursing home, she was alert and able to make her own decisions, requiring assistance for all transfers. She did not have a previous history of falling, but the facility determined that she was at risk because of her left-sided weakness, and kept a call-light within her reach at all times. Upon the initial skin examination, the nurse noted areas of mild bruising across her abdomen and arms, a common adverse effect of anticoagulant therapy. The use of heparin was mentioned on the resident's care plan, but there were no specific mentions of follow-up plans.

The following dates are significant in tracking and understanding the series of physician orders that eventually contributed to the resident's decline: Two weeks after the woman had arrived at the nursing home, a nurse noted that she had not undergone any lab work to monitor the effects of her heparin therapy; the nurse requested and received International Normalized Ratio (INR) lab orders from her physician. The following day, July 11, the woman's physician received the lab results and ordered another anticoagulant medication, Coumadin 5 mg, to be given orally once, and ordered a recheck of the INR levels the following day. This was done and the physician ordered the Coumadin to be continued with the heparin until a therapeutic blood level range was achieved.

By July 13, nursing noted that the resident's bruising began to escalate. The physician was notified of her INR level, which remained within normal range. The nurses did not administer any more Coumadin, but continued to give the subcutaneous heparin until they heard back from the physician, four days later. By then the INR effects from the Coumadin had diminished.

On July 17, the physician ordered 10 mg (i.e., doubled the dose) of Coumadin and asked the lab to draw another INR the next day. That night, the resident became concerned about the increased bruising on her abdomen from the heparin injections.

The anticoagulation effect of Coumadin persists beyond 24 hours, so the INR level that was phoned to the physician the next day, July 18, did not reflect the peak level from the 10 mg of Coumadin previously given. Despite this, the physician ordered 15 mg of Coumadin on July 18.

On July 19, the resident's INR level was climbing above the therapeutic range. Her physician lowered the Coumadin dosage to 2 mg daily, starting July 20, and discontinued the heparin.

The next day, the resident fell while leaning forward on her commode and sustained facial injuries and an uncontrolled nosebleed. A CNA was in the room with her but was unable to prevent the fall. When the resident arrived at the hospital, her INR level was nearly twice the normal range; the two large doses of Coumadin and her last heparin dose were still peaking in her system. A subdural hematoma developed that could not be surgically relieved immediately because her blood was so thin. As a result, the resident's health continued to deteriorate until her death, almost two months later. This case was settled out of court for a substantial amount.

Lessons Learned

Though the nurses followed the resident's physician orders, they overlooked several red flags along the way that might have brought about a less-devastating outcome. You may be able to protect your residents by heeding these red flags and taking the necessary precautions:

- **Obtain and read all history and physical information about the resident before, or at least during, the admit process.** Had the neurologist's statement from the first hospital been read, the nurses would have at least questioned the heparin orders. One nurse did finally catch the lack of lab values—two weeks later. *(Note: Although these problems originated with the hospitals, only the nursing home was sued.)*

- **Develop anticoagulant therapy care plan interventions.** The facility appropriately identified the lack of follow-up lab studies but did nothing to remedy it. Simple interventions, such as monitoring for and reporting signs of excess bleeding, would have heightened awareness by all nursing staff and possibly altered the physician's treatment plan, especially if he had known about the escalated bruising.

- **Develop policies and procedures for bleeding precautions.** The facility's nursing staff did not alter the resident's care, even though her labs, combined with her history of uncontrolled nosebleeds and escalated bruising, indicated that she was becoming a high risk. Sometimes, with critical INR levels, bed rest with a bedpan is the safest option until the resident is stabilized. Consult your medical director when developing bleeding precaution policies and procedures.

- **Obtain physician responses to labs and drug regimens in a timely manner.** The nurses allowed four days to go by during the most critical time of the resident's anticoagulant therapy conversion from subcutaneous heparin to oral Coumadin. This time lag, and subsequent diminished INR value, led the physician to order more medication for the resident than she needed. Even if the nurses have to phone the physician several times a day, critical responses such as this should not go unheeded for more than a day. The facility's medical director also can be used to hasten the response time or, possibly, write orders.

- **Provide periodic training on anticoagulation therapy for nursing staff members.** There was no evidence to indicate that the facility's nursing staff were aware of or questioned the increased doses of Coumadin, despite the fact that the resident's previous doses and heparin usage had not peaked. Since many elderly people require anticoagulant therapy, every nurse needs to be aware of the dosage, uses, contraindications, peak levels, precautions, and therapeutic lab values for anticoagulant drugs. When asked, most pharmacy consultants will provide this and similar training on other commonly used medications in the elderly.

By developing this knowledge and sensitivity in monitoring a difficult class of drugs, you should be able to protect your residents and facility from a fate similar to that reported here. A facility must plan to adequately monitor resident medication regimens and take appropriate action in keeping prescribing physicians informed. In geriatric facilities, this is especially necessary during anticoagulation therapy.

Critical Thinking

1. Why is it important to know the side effects of medications and their follow-up labs?

LINDA WILLIAMS, RN, is a long-term care risk manager for the Guide-One Center for Risk Management and its Senior Living Communities Division. The GuideOne Center for Risk Management is dedicated to helping churches, senior living communities and schools/colleges safeguard their communities by providing practical and timely training, and resources on safety, security and risk management issues.

From *Nursing Homes Magazine,* March 2003, pp. 42, 44–45. Copyright © 2003 by Vendome Press. Reprinted by permission via Copyright Clearance Center.

Antiemetic Drugs

LISA CRANWELL-BRUCE

This column features a review of the class of drugs known as the antiemetics, with particular focus on the two newest types: the serotonin (5-HT3) receptor antagonists and the neurokinin-1 receptor antagonists (NK).

Management of pain, nausea, and vomiting can be a challenge in nursing care and patient satisfaction in both ambulatory and hospital settings (Hache, Vallejo, Waters, & Williams, 2009). These common complications after surgical procedures also are known as postoperative nausea and vomiting (PONV). Nausea and vomiting (N&V) also may be experienced by patients receiving chemotherapy and radiation therapy treatments. Nausea and vomiting lead to patient distress and dissatisfaction, as well as prolonged hospital stays, increased costs, and increased nursing time (Bender et al., 2002; Tipton, McDaniel, Barbour, & Johnson, 2007). It places the patient at risk for malnutrition, dehydration, electrolyte imbalances, and aspiration pneumonia (Bender et al., 2002). For the patient undergoing chemotherapy or radiation therapy, the emotional factors linked to N&V can interrupt treatment and affect not only treatment adherence but also the patient's quality of life (Collins & Thomas, 2004). Nurses understand N&V affect the patient both physically and emotionally. As an adverse effect, it can be very difficult to manage.

Nausea is a subjective symptom characterized by the patient's report of an unpleasant sensation in the back of the throat. It usually precedes vomiting. The patient also may report symptoms of sweating, dizziness, and chills, and feelings of hot and cold or excessive salivation. Vomiting is the expulsion of stomach contents through the mouth due to the forceful contraction of the abdominal muscles. Retching, also called "dry heaves," is marked by rhythmic or spasmodic contractions of the diaphragm along with the muscles of the thorax and abdominal wall. However, retching does not cause expulsion of the stomach contents. The patient may have retching before or after vomiting (Bender et al., 2002; Garrett, Tsuruta, Walker, Jackson, & Sweat, 2003; Parkman, 2002).

Signals from the cerebral cortex, the sensory organs, and/or the vestibular apparatus of the inner ear can all activate the vomiting center. The chemoreceptor trigger zone (CTZ) located in the postrema surface of the brain also can stimulate the vomiting center indirectly. These signals can come from substances in the blood, the stomach, or the small intestine, or can result from direct action of emetogenic compounds in the blood. Emetogenic compounds include agents such as opioids and chemotherapy drugs or toxins (Garrett et al., 2003). The goal of antiemetic drug therapy is to interfere with the signal of any of the chemoreceptors and thereby prevent activation of the vomiting center. Antiemetic drugs are used to block one or more of these receptors between the CTZ and the vomiting center (Collins & Thomas, 2004; Garrett et al., 2003).

Many patients, especially those who previously experienced previous PONV or are receiving cancer therapy treatments, may experience *anticipatory nausea and vomiting*. This is a type of psychological nausea and vomiting. For these patients, certain signals from repeated associations can trigger nausea and vomiting. These signals or stimuli can include smells, sight, thoughts, and tastes (Bender et al., 2002). Treatment modalities for anticipatory nausea and vomiting can include behavioral interventions such as relaxation as well as benzodiazepines (e.g., lorazepam [Ativan®]) given the night or morning before chemotherapy (Garrett et al., 2003).

Chemotherapy and radiation treatment are known to increase a patient's risk of nausea and vomiting. N&V occur in almost 60% of patients receiving this treatment (Bender et al., 2002). Several factors increase the patient's risk of PONV including postoperative fasting time, type of anesthetic agent used, opioids used for postoperative pain management, anxiety, obesity, younger age, female gender, history of previous PONV or motion sickness, type of surgical procedure, and pre-existing gastroparesis (Flake, Scalley, & Bailey, 2004; Gundzik, 2008; Jolley, 2001).

Drugs commonly used to treat N&V include antihistamines, dopamine antagonists, and anticholinergics (see Table 1). Corticosteroids such as dexamethasone (Decadron®) and methylprednisolone (SoluMedrol®) are used as preventive agents, usually in combination with other medications. One of the newest class of antiemetics, known as the serotonin agents or as the 5-HT3antagonists, includes dolasetron (Anzemet®), granisetron (Kytril®), ondansetron (Zofran®), and palonosetron (Aloxi®). The single neurokinin (NK) receptor antagonist aprepitant (Emend®) is the newest agent (Flake et al., 2004; Flemm, 2004; Garrett et al., 2003).

Antihistamines

Antihistamine agents have been used historically to treat nausea and vomiting. They work by inhibiting the action of histamine at the H1 receptor, which in turn limits stimulation of the

Table 1 Drugs Commonly Used for Treatment of Nausea and Vomiting

Antihistamine Drugs	Dopamine Antagonist Drugs	Anticholinergic Agents
Dimenhydrinate (Dramamine®)	Promethazine* (Phenergan®)	Scopolamine (Transderm Scop®)
Meclizine (Antivert®)	Prochlorperazine (Compazine®)	Trimethobenzamide (Tigan®)
Diphenhydramine (Benadryl®)	Metoclopramide (Reglan®)	
	Droperidol (Inapsine®)	
	Chlorpromazine (Thorazine®)	

*The FDA recently issued a "black box" warning for promethazine regarding the risks of IV push administration. The preferred method of administration is muscular injection (Hill, 2009).

vomiting center in the vestibular system. They also act on the vestibular center of the middle ear (Flake et al., 2004). All antihistamines are indicated to treat motion sickness and nausea. However, only meclizine is indicated as an antiemetic (Adams & Koch, 2010). Antihistamines are inexpensive drugs available as generics; dimenhydrinate and diphenhydramine also are available as over-the-counter preparations. A significant adverse effect of this class is sedation (Gundzik, 2008).

Dopamine Antagonists

Dopamine antagonists are another older class of drugs, also available in inexpensive generic forms. This class of drugs works in the CTZ but is not effective for motion sickness. Adverse effects cause these drugs to be difficult to tolerate. These effects include Parkinson's-like side effects, which include bradykinesia, stiffness and lost of balance, and extrapyramidal symptoms (cogwheeling, dystonia, akathisia) (Gundzik, 2008).

Corticosteroids

Corticosteroids are used most frequently as the additional agent in an antiemetics combination protocol. They are useful for delayed nausea and vomiting; they also are prescribed frequently for patients receiving chemotherapy and reserved for severe cases (Bender et al., 2002; Hache et al., 2009). Adverse effects can include mood swings, delayed wound healing, insomnia, and facial flushing, and sodium and fluid retention (Adams & Koch, 2010).

Serotonin (5-HT3) Receptor Antagonists

The serotonin (5-HT3) receptor antagonists, one of the newer classes of antiemetics, now are considered first-line drugs for treatment of nausea and vomiting related to cancer treatment;

they are considered to be the gold standard in antiemetic therapy (Goodin & Cunningham, 2002). Serotonin, which is distributed widely in the gastrointestinal tract, is released from the enterochromaffins cells in the lumen of the gastrointestinal tract. Serotonin receptors are found in the postrema of the brain, where the CTZ is located (DiVall & Cersosimo, 2003). The 5-HT3 drugs inhibit the action of serotonin at the small bowel, the vagus nerve, and the CTZ (Flake et al., 2004).

An important area of difference among the four available 5-HT3 agents is their side effect profile. Awareness of this difference is critical for the nurse because typically the side effect profile of a class of drugs is similar to the prototype drug for the class (Goodin & Cunningham, 2002). The 5-HT3 antagonists differ in the chemical structure as well as dose response, half-life, and affinity to bind to the drug receptor. This in turn leads to different physiological responses to the drugs as well as a difference in their adverse effects (Gan, 2005).

Dolasetron (Anzemet®) is indicated for the prevention of postoperative nausea and vomiting as well as prevention of cancer chemotherapy-induced N&V. It can be given either orally or by intravenous injection, with both routes equally effective. The drug is given as a single dose before chemotherapy; the IV formulation is dosed by the patient's weight and the oral form given as a standard dose of 100 mg, with no adjustment needed for impaired renal or hepatic function. Both the oral and IV doses are based on weight in children ages 2–16. For prevention of postoperative N&V, the drug is administered as a single dose 15 minutes before the cessation of anesthesia. It also may be given as soon as the patient experiences postoperative N&V (Sanofi-aventis, 2008).

Adverse effects of dolasetron include headache, fatigue, fever, diarrhea, abdominal pain, and abnormal hepatic function, with headache reported as the most common adverse effect (Garrett et al., 2003). Of greater concern are asymptotic but reversible ECG interval prolongations for PR, QRS, QT, and QTc, with changes starting 2 hours after initiation of therapy (Goodin & Cunningham, 2002). This drug therefore should be used with caution in the patient with prolong conduction disorder and/or taking other medications that would prolong the QT interval (Garrett et al., 2003; Goodin & Cunningham, 2002).

Ondansetron (Zofran®) is the only 5-HT3 now available as a generic drug. It is available as an injectable solution for IM and IV administration, as well as a tablet, an orally disintegrating tablet, and an oral solution. It is indicated for nausea and vomiting from high and moderate emetogenic chemotherapy, prevention of nausea and vomiting from radiation therapy, and prevention of postoperative nausea and vomiting (GlaxoSmithKline, 2006). The oral route can be used for patients undergoing both chemotherapy and radiation therapy, while the injectable form of the drug is used only for patients receiving chemotherapy; it currently is not indicated for use before radiation therapy. Oral dosing is 24 mg before chemotherapy or radiation, and 8 mg IV 30 minutes before chemotherapy. For the prevention of PONV, ondansetron 4 mg IV is given 2–5 minutes before induction of anesthesia or 8 mg by mouth 1 hour before the induction or anesthesia. It also can be given as needed for PONV at a dose of 1-4 mg IV (Garrett et al., 2003).

GlaxoSmithKline (2006) offers limited information for pediatric use of the oral formulation for indication and specifically for ages under 4, but recommended dose is 4 mg orally for children over age 4. Limited information is available for injectable solutions in children over 1 month; dose is calculated by body weight, but the manufacturer warns of a prolonged half-life in children.

Adverse effects of ondansetron are mild-to-moderate headache, diarrhea, fever, abnormal vision, and dizziness. As with palonosetron, the drug also may cause asympathetic and reversible ECG prolongation of PR, QRS, QT, QTc; this is a concern for the patient with a prolonged conduction disorders as well as those receiving other drugs that may cause a prolonged QT interval. Unlike palonosetron, ondansetron also demonstrated a significant slowing of heart rate, which is noteworthy for the patient who has bradycardia or receives drugs that can cause bradycardia (Garrett et al., 2003; Goodin & Cunningham, 2002).

Granisetron (Kytril®) is the third 5-HT3 drug for review. It is available in tablets, oral solution, and an injectable solution for IV use. Tablets are indicated for nausea and vomiting with emetogenic chemotherapy and radiation therapy. The injectable solution is indicated for preventing nausea and vomiting with emetogenic chemotherapy, as well as preventing and treating PONV (Roche Exchange Oncology, 2005). Adult dosing for the oral formulation is 1–2 mg daily. Dosing of the injectable form for the patient receiving chemotherapy is 10 mcg/kg 30 minutes from the start of chemotherapy. For prevention of PONV, 1 mg of granisetron is given 30 seconds before the induction of anesthesia or immediately before the reversal of anesthesia (Garrett et al., 2003; Roche Exchange Oncology, 2005). No data on its use in the pediatric population exist at this time, and no dosing is available for the patient under age 18 (Roche Exchange Oncology, 2005).

The adverse effect profile of granisetron is similar in some aspects to the other 5-HT3 drugs. Possible effects include headache, constipation, diarrhea, abdominal pain, visual disturbance, dyspepsia, asthenia, and somnolence (Garrett et al., 2003; Goodin & Cunningham, 2002; Roche Exchange Oncology, 2005). However, granisetron rarely causes ECG changes, with fewer effects on ECG than intravenous ondansetron (Goodin & Cunningham, 2002; Roche Exchange Oncology, 2005). It thus may be a more appropriate choice for the patient who has conduction disorder and/or takes other drugs that can cause prolong QT conduction.

Palonosetron (Aloxi®) is the newest 5-HT3 agent. Palonosetron is different from the other 5-HT3 agents in that it has the highest affinity for the 5HT-3 receptor and a long half-life of 40 hours (DiVall & Cersosimo, 2003). The long half-life may help the patient undergoing chemotherapy who experiences delayed N&V (24 hours-6 days) after chemotherapy treatments (Bender et al., 2002; DiVall & Cersosimo, 2003). Palonosetron is indicated for nausea and vomiting from high and moderate emetogenic chemotherapy as well as prevention of PONV (Roche Exchange Oncology, 2005). It is only available as an injectable formulation dosed as 0.25 mg or 0.75 mg in one or two doses (Rittenberg, 2004). No data support its use in patients

under age 18 (MGI Pharma, 2007). Palonosetron is administered 30 minutes before the start of chemotherapy or immediately before the induction of anesthesia (DiVall & Cersosimo, 2003; MGI Pharma, 2007). The drug has shown better control of emesis than ondansetron and dolasetron, and higher reported quality of life scores with its use (Rittenberg, 2004). The longer half-life also allows it to be given as a single dose that is effective in preventing acute and delayed N&V (DiVall & Cersosimo, 2003); the single dose is more cost effective than other preparations. The 5-HT3 antagonists vary in cost, but still are considered expensive drugs for treatment when compared to some of the older antiemetic agents (Flake et al., 2004).

Adverse effects of palonosetron are rare and mild to moderate, including headache, constipation, dizziness, and abdominal pain. Visual disturbance and diarrhea have not been reported (DiVall & Cersosimo, 2003). Rare cases (1% occurrence) of QT prolongation, sinus bradycardia, or tachycardias have been reported with use of the drug (MGI Pharma, 2007).

Neurokinin-1 Receptor Antagonist

The newest antiemetic drug is aprepitant (Emend®), the first and only neurokinin-1 receptor antagonist available. It is used in combination therapy with a 5-HT3 antagonist and dexamethasone for prevention of chemotherapy induced N&V (Flemm, 2004; Merck & Co., 2008). Substance P is involved in the emetic reflex; aprepitant prevents substance P from binding to receptor sites in the brainstem, which inhibits emesis (Flemm, 2004).

Aprepitant comes in both oral and injectable formulations. The oral formulation is indicated in combination therapy (with dexamethasone and a 5-HT3 antagonist) for preventing chemotherapy-induced N&V (Flemm, 2004). It is administered for this indication as 125 mg in a single dose 30 minutes before chemotherapy on day 1 and then 80 mg oral on days 2 and 3 of chemotherapy. To prevent PONV, aprepitant is given as an oral dose of 40 mg 3 hours prior to the induction of anesthesia. The injectable formulation may be substituted for the first oral dose; it is given as a 115 mg dose. The injectable formulation does not have any other indications at this time (Merck & Co., 2008). Although aprepitant is indicated and has displayed excellent effectiveness for acute and delayed chemotherapy nausea and vomiting, it is only a preventive drug and is not effective for the patient already experiencing N&V (Flemm, 2004). It also is not recommended presently for the patient under age 18 (Merck & Co., 2008).

The adverse effects of treatment with aprepitant are asthenia, fatigue, constipation, diarrhea, nausea, anorexia, and hiccups (Flemm, 2004). Although nausea may seem to be an unusual adverse effect for an antiemetic, Hache and colleagues (2009) indicated dizziness or lightheadedness may have been reported as mild nausea by some patients. In their study of surgical patients receiving aprepitant in combination therapy for prevention of PONV, they also reported that the adverse effects of other drug therapies could be problematic while trying to predict or report adverse effects of an individual drug.

Conclusion

Prevention and treatment of N&V related to cancer therapy as well as PONV have been advanced greatly with the development of new classes of antiemetics. The nurse should be aware of the various indications for these medications as they can differ with the route of the drug. Reviewing the adverse effects of each drug also is important because the 5-HT3 antagonists have different adverse effect profiles from the prototype drug for the class. Nausea and vomiting will remain a challenge in nursing care and treatment. Understanding the classes of antiemetic drugs, as well as their individual and combined indications, doses, and adverse effects is an ongoing challenge as advances are made in this area of drug therapy.

Nausea and vomiting will remain a challenge in nursing care and treatment.

References

Adams, M.P., & Koch, R.W. (2010). *Pharmacology connections to nursing practice.* Upper Saddle River, NJ: Pearson.

Bender, C.M., McDaniel, R.W., Murphy-Ende, K., Pickett, M., Rittenberg, C.N., Rogers, M.P., et al. (2002). Chemotherapy-induced nausea and vomiting. *Clinical Journal of Oncology Nursing, 6*(2), 94–102.

Collins, K.B., & Thomas, D.J. (2004). Acupuncture and acupressure for the management of chemotherapy induced nausea and vomiting. *Journal of the American Academy of Nurse Practitioners, 16*(2), 76–80.

DiVall, M.V., & Cersosimo, R.J. (2003). Palonosetron. *Formulary, 7*(38), 414–430.

Flake, Z.A., Scalley, R.D., & Bailey, A.G. (2004). Practical selection of antiemetics. *American Family Physician, 69*(5), 1169–1174.

Flemm, L.A. (2004). Aprepitant for chemotherapy-induced nausea and vomiting. *Clinical Journal of Oncology Nursing, 8*(3), 303–306.

Gan, T. (2005). Selective serotonin 5-HT3 receptor antagonists for postoperative nausea and vomiting: Are they all the same? *CNS Drugs 2005, 19*(3), 225–238.

Garrett, K., Tsuruta, K., Walker, S., Jackson, S., & Sweat, M. (2003). Managing nausea and vomiting current strategies. *Critical Care Nursing, 23*(1), 31–50.

GlaxoSmithKline. (2006). *Zofran.* Retrieved June 1, 2009, from www.gsk.com/products/prescription_medicines/us/zofran.htm

Goodin, S., & Cunningham, R. (2002). 5-HT3-Receptor antagonists for the treatment of nausea and vomiting: A reappraisal of their side effect profile. *The Oncologist, 7*(5), 424–436.

Gundzik, K. (2008). Nausea and vomiting in the ambulatory surgical setting. *Orthopaedic Nursing, 27*(3), 182–188.

Hache, J.J., Vallejo, M.C., Waters, J.H., & Williams, B.A. (2009). Aprepitant in a multimodal approach for prevention of postoperative nausea and vomiting in high-risk patients: Is there such as thing as "too many modalities"? *TheScientificWorldJOURNAL, 34*(9), 291–299.

Hill, M. (2009, September 17). Nausea drug to bear warning. The Philadelphia Inquirer, C1.

Merck & Co. (2008). *Prescribing information: Emend®.* Retrieved June 1, 2009, from www.emend.com/aprepitant/emend/consumer/product_information/pi/index.jsp

MGI Pharma. (2007). *Aloxi prescribing information.* Retrieved June 1, 2009, from www.aloxi.com/common/downloads/pi.pdf

Parkman, H.R. (2002). New advances in the diagnosis and management of nausea and vomiting. *Case Manager, 13*(2), 83–87.

Rittenberg, C.N. (2004). The next generation of chemotherapy induced nausea and vomiting prevention and control: A new 5-HT3 antagonist arrives. *Clinical Journal of Oncology Nursing, 8*(3), 307–310.

Roche Exchange Oncology. (2005). *Kytril production information.* Retrieved June 1, 2009, from www.rocheexchange.com/oncology/productinformation/kytril

Sanofi-aventis. (2008). *Anzemet® production information.* Retrieved June 1, 2009, from http://products.sanofi-aventis.us/Anzemet_Injection/anzemetinj.pdf

Tipton, J.M., McDaniel, R.W., Barbour, L., & Johnson, M.P. (2007). Putting evidence into practice: Evidence-based interventions to prevent, manage and treat chemotherapy-induced nausea and vomiting. *Clinical Journal of Oncology Nursing, 11*(1), 69–78.

Critical Thinking

1. What are some of the complications that trigger nausea and vomiting?

LISA A. CRANWELL-BRUCE , MS, RN, FNPC, is a Clinical Instructor, Byridine F. Lewis School of Nursing, College of Health and Human Services, Georgia State University, Atlanta, GA.

From *MEDSURG Nursing*, September/October 2009, pp. 309–313. Copyright © 2009 by Academy of Medical-Surgical Nurses (AMSN). Reprinted by permission of Anthony J. Jannetti, Inc, East Holly Avenue, Box 56, Pitman, NJ 08071-0056; 856-256-2300; FAX 856-589-7463. www.medsurgnursing.net

UNIT 4

Disease and Disease Treatments

Unit Selections

Learning Outcomes

After reading this unit, you should be able to:

- Discuss the management of wound related pain and the essence upon which pharmacological treatment is based.

- Identify the three categories of abdominal pain.

- List the unique role the nurse plays in managing patient pain.

- Summarize how anxiety can impact a patient pre and post surgery.

- Explain the differences between type I and type II respiratory failure.

- Define the role of the Registered Dietician in the nutritional management of a diabetic patient.

- Describe the difference between hypovolaemic, cardiogenic and septic shock.

- Discuss the nurse's role in facilitating a successful transition from the hospital to homecare.

- Define the nurse's role in assessing the patient who is receiving insulin via an insulin pump.

- Identify the connection between stress and physical illness.

Student Website

www.mhhe.com/cls

Internet References

American Diabetic Association
www.diabetes.org/living-with-diabetes/?utm_source=WWW&utm_medium=GlobalNavLWD&utm_campaign=CON
Diabetes Monitor
www.diabetesmonitor.com
Health News Flash
www.healthnewsflash.com/conditions/respiratory_failure.php#8
National Heart Lung and Blood Institute
www.nhlbi.nih.gov/health/dci/Diseases/shock/shock_what.html

This unit focuses on a variety of new ideas and discoveries related to diseases and treatment modalities that occur within a variety of health care environments. Diseases, diagnosis, and treatment include topics such as the effective management of pain, anxiety and open heart surgery, respiratory failures, shock, facilitating a tracheostomy patient's transition from the clinical environment to the home environment, and caring for the patient's insulin pump and dealing with stress of an insulin pump.

Pain is a unique experience that can be felt and expressed only by the patient. As healthcare providers we utilize a variety of assessment tools to assist the patient in expressing the level of pain that she/he may be experiencing. As discussed in one of the upcoming articles, pain has been identified as one of the most common reasons patients seek treatment from a healthcare provider. However, as the nurse and an important part of the interdisciplinary team, it is crucial that the nurse understand the role she/he plays in assessing the patient's level of pain and work collaboratively with the interdisciplinary team to ensure that the patient's pain is effectively managed.

Anxiety is a symptom which is often experienced by patients who have to endure open heart surgery. The physiological manifestation that many patients exhibit as a result of fear of the unknown can be debilitating to the patient and can often exhibit itself in a variety of ways such as, chest pain, palpitations, shortness of breath, and dizziness. The nurse plays a vital role to ensure that the patient receives effective patient education, is able to administer pharmacological treatment to relieve the symptoms of anxiety, and allow for proper recuperation from the procedure.

A thorough respiratory assessment is one of the most important things that a nurse can and will do for patients. Although the clinical features for respiratory failure vary with each patient, it is essential that a base line physical assessment is obtained and all lab values are reviewed while beginning a shift. An important part of completing a physical assessment includes the ability to identify advantageous breath sounds and poor perfusions that may lead to respiratory complication. The nurse's role is essential during this assessment process in identifying the clinical features of early, intermediate, and late signs of type 1 and type II respiratory failure.

Medical nutrition therapy is an essential part of managing the care of a diabetic patient. The American Diabetic Association (ADA) 2009, clinical practice recommendation states that individuals who have pre-diabetes or diabetes should receive individualized medical nutritional therapy to improve treatment goals, preferably by a registered dietitian who is familiar with the components of diabetes. Since there is a direct correlation between an individual's diet and the manifestation of diabetes, it is important that proper medical nutritional therapy is received.

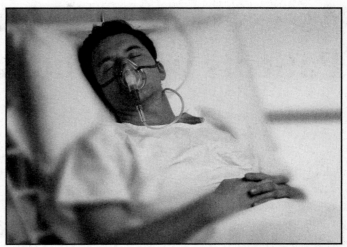

© Stockbyte/PunchStock

The registered dietician plays a vital role in assessing, diagnosing, intervening, and managing the diabetic patient.

Shock is described as a complex physiological syndrome (Garrestson & Malberti, 2007, p. 46) that can lead to death if the symptoms are not recognized and treatment is not provided promptly. The clinical manifestations of shock vary. It is necessary to be aware that hypovolaemic shock differs from cardiogenic and septic shock. Additionally, there is a variation in the treatment of each type of shock. The nurse plays an essential role in recognizing the symptoms of shock and the treatment modalities available for the patient.

The intricacy of discharging patients who require long-term tracheostomy care will require a collaborative working environment with the discharge team and homecare nursing team. Educating the patient prior to discharge and effective communication between the two teams will aid in facilitating a successful and safe transition from the hospital to the homecare environment.

Are you educated on how to use an insulin pump, or are you ready to care for the patient who is receiving insulin via an insulin pump? Over 375,000 Americans with diabetes are utilizing insulin pumps (Miller, 2009, p. 57). The role of the nurse shifts when a hospitalized patient is utilizing his/her insulin pump in the inpatient environment. The nurse plays a critical role in assessing the hospitalized patient's ability to manage their insulin pump and blood glucose level.

Reference

Garreston, S., & Malberti, S. (2007). Understanding hypovolaemic, cardiogenic and septic shock. *Nursing Standard* 50(21) 46–55.

Assessment and Management of Patients with Wound-Related Pain

ANDY RODEN AND ELAINE STURMAN

Patients' experiences of wound pain are subjective and multicausal in nature. However, optimal pain management is an essential part of the nurse's role and needs to be prioritised.

Acute pain occurs following injury to the body and generally dissipates when the injury heals. It is a short-term pain of less than 12 weeks duration (British Pain Society (BPS) 2008). Acute pain serves a useful purpose in warning healthcare professionals that deferring treatment may result in life-threatening consequences; for example, severe abdominal pain may signal appendicitis, which can lead to peritonitis. In acute wound-related pain, 'pain that occurs after tissue injury has a protective role, alerting the body to damage and inducing rest to allow tissue regeneration' (Wulf and Baron 2002).

Chronic pain serves no useful function, persisting beyond the time one would expect normal healing to occur. It is continuous long-term pain that may last anywhere from six weeks to three months after the onset of symptoms (BPS 2008). Chronic wound-related pain affects patients' wellbeing, level of function and quality of life (Goodridge et al 2005). Pain of this nature may also result in reduced social activities, increased family tensions for patients and care givers, limited employment and financial hardship.

Incidence of Acute and Chronic Pain

A UK survey conducted by Bruster et al (1994), of recently discharged patients (n = 3,162) from 36 NHS hospitals, showed that 33% (n = 1,042) of patients (medical and surgical) in moderate or severe pain experienced pain all or most of the time. Whelan et al (2004) carried out a prospective cohort study on 5,584 inpatients and found that 59% of patients had pain; 28% described severe pain, 19% described moderate pain and the remainder reported mild pain.

A study by Blyth et al (2001) investigated the incidence of chronic pain in a random selection of 17,543 adult patients. Pain was defined as 'pain experienced every day for three months in the six months prior to interview'. Chronic pain was reported in

17% of males and 20% of females. A similar study carried out by Elliott et al (1999), using the chronic pain definition of 'pain or discomfort, that persisted continuously or intermittently for longer than three months', found that 50.4% of respondents reported having chronic pain. It was also noted that reporting of chronic pain increased with age in both men and women from about one third in those aged 25–34 years to almost two thirds in those aged over 65 years (Elliott et al 1999).

Pain is frequently reported in patients with leg ulcers, with between 17% and 65% of individuals experiencing severe or continuous pain (Dallam et al 1995, Ebbeskog et al 1996). Nemeth et al (2003) suggested that the prevalence of pain in patients with pure or mixed venous ulcers is approximately 50%, with more than half of the patients using analgesia as part of their treatment regimen. Similarly, Puntillo et al (2002) examined 5,957 patients undergoing procedures such as drain removal, non-burn dressing and central venous pressure catheter insertion, and found that:

- Pain intensity increased at the time of the procedure.
- Around 63% of patients received no analgesia.
- Less than 20% of patients received opiates.
- Mean dose of opiate, if used, was 6.44mg morphine equivalent. No reliable recommendation for a morphine-equivalent dose for procedural pain is available, although a range of 1–10mg intravenous administration has been suggested (Puntillo et al 2002).
- Around 10% of patients received a combination of drugs, including local anaesthetics and/or anxiolytics and non-steroidal drugs as part of the treatment regimen.

Unrelieved pain remains a problem. Carr and Jacox (1992) found that analgesics failed to relieve pain in about half of post-operative patients. Pain that is poorly managed in the acute phase may lead to cardiovascular, pulmonary, thromboembolic or gastrointestinal complications (Middleton 2003).

These physical complications need to be considered as well as the potential emotional and psychological effects that may occur as a result of prolonged hospitalisation, medication anxiety and depression. The cost to the NHS of treating pain should also be considered (BPS 2008).

Nociceptive and Neuropathic Pain

Pain can be divided into nociceptive and neuropathic pain. It is important that nurses recognise the differences between the two to understand the nature of the pain experience and to determine how best to treat pain.

Nociceptive pain

This is mediated by receptors that are located in skin, bone, connective tissue, muscle and viscera (Anaesthesia UK 2009). The International Association for the Study of Pain (1994) defined a nociceptor as 'a receptor preferentially sensitive to a noxious stimulus or to a stimulus which would become noxious if prolonged'. A noxious stimulus is defined as one which is damaging to normal tissues.

Nociceptive pain can be described as either somatic, for example as a result of wound debridement, or visceral, for example deep pain that is difficult to isolate and locate. Acute nociceptive pain is usually time limited and generally dissipates when the injury heals (Gruener 2004).

Neuropathic pain

This can be described as 'burning', 'shocking' or 'shooting' in nature and is produced by damage to, or pressure on, nerves in the peripheral or central nervous systems (Johnson 2004). Neuropathic pain can manifest itself as excruciating pain in the wound bed (hyperalgesia) or in the surrounding skin (allodynia). Allodynia is defined as pain resulting from a stimulus that ordinarily does not elicit a painful response, for example delicate touch or a light draught of air on the open wound.

In relation to chronic wounds, pain of neuropathic origin may be found in people whose wounds have been open for some time. For these patients conventional analgesia aimed at tissue-based pain may not ease the pain and individuals may require different and often mixed agents; for example, drugs such as paracetamol, non-steroidals and opioids may be supplemented by drugs that have proven efficacy in relieving neuropathic pain, such as amitriptyline or gabapentin (McQuay and Moore 1998). However, the mechanism of action of antidepressant drugs in the treatment of neuropathic pain remains unclear (Saarto and Wiffen 2007).

It is important to note that some patients may experience both nociceptive and neuropathic pain, alternatively termed 'mixed pain'. The patient might also experience acute exacerbations of pain, for example during dressing changes (incident pain).

Causes of Wound-Related Pain

Wound pain is complex and can occur as a result of many factors. These include:

- Nociceptive pain, for example fractures, burns and inflammation.
- Neuropathic pain, for example peripheral neuropathy.
- An 'inflammatory soup' (Goodwin 1998) of macrophages and lymphocytes, histamine, serotonin, bradykinin, substance P, prostaglandins and cytokines in affected peripheral tissues, for example in a chronic wound, may result in patients experiencing increased sensitivity to pain; for example, pain may be triggered by a small stimuli, such as a light touch.
- Infection at the wound site.
- The presence of oedema.
- Iatrogenic causes, for example debridement, bandaging and wound cleansing.

The effect of these factors will be influenced by the patient's unique daily cycle, and pain episodes might be triggered throughout the day. This emphasises the importance of individualised patient assessment.

A 24-hour cycle is now outlined to highlight some of the problems encountered by a patient with a venous leg ulcer in need of bandaging.

On waking

There may be pain on movement during waking. Pain may result from bandage or dressing slippage during the night. Analgesia may also have worn off overnight and there may be stiffness in the joints as a result of immobility.

Mid morning

There may be anticipatory pain resulting from the patient's thoughts about the impending change of dressing. Pain may occur at dressing removal because of drying out of the wound bed and surrounding tissue and subsequent adherence of the dressing to the skin.

Midday

Venous disease may result in restricted mobility, causing the individual to sit in a chair with his or her legs in a dependant position. This may contribute to dependant oedema and swollen legs, resulting in pain.

Night-time

There may have been sleep disturbance because of pain. Pain can occur as a result of pressure on the wound bed during sleeping. Pain may also result from limb elevation during the night.

It is important that the nurse's role encompasses the identification and assessment of the origin of pain to implement the most appropriate treatment strategy and minimise patient discomfort.

The Effect of Pain

Pain can affect individuals physically, emotionally, psychologically and socially (Langemo 2005).

Arterial and venous leg ulcers can cause considerable pain. In a systematic review of the impact of leg ulcers on patients' daily life, Persoon *et al* (2004) found that leg ulcers pose a threat to physical functioning and have a negative effect on psychological and social functioning. Other problems included (Persoon *et al* 2004):

- Pain.
- Immobility.
- Sleep disturbance.
- Lack of energy.
- Limitations in work and leisure activities.
- Lack of self-esteem.

Pain control is essential in effective wound management and nurses need to have a good working knowledge of the causes and types of pain associated with wounds to provide timely, effective and individualised care.

Pain Assessment

One of the failures of modern medicine is the inadequate assessment and treatment of pain (Reddy *et al* 2003). The management of pain in patients with chronic wounds requires an individualised, patient-centred approach. This should be based on accurate patient assessment, which should include:

- Patients' reporting of pain, both numerical and descriptive.
- Non-verbal cues of pain.
- Location, duration, intensity and onset of pain.
- The effect of pain on the patient's quality of life.
- Efficacy of current analgesia.

Pain rating scales are available to help patients to identify and express the location, intensity and duration of pain. The visual analogue scale (VAS) uses numerical ratings from zero to ten, with verbal descriptors including 'no pain' and 'worst pain'. The VAS is easy to use and can provide precise scores in research where large patient numbers are involved. However, it does not allow for pain descriptors to be used, which might yield important information about the nature of the pain.

The verbal rating scale (VRS) uses a list of descriptors to help the patient describe increasing pain intensity. Words commonly used may include 'no pain', 'mild pain', 'moderate pain' and 'severe pain'. This tool is easy to understand, but lacks the level of precision of the zero to ten VAS scale, and it could be argued that it might be confusing for patients who might have used the VAS scale in the past.

Briggs and Closs (1999) demonstrated that the VRS proved most efficient in the reporting of acute pain, with only 0.5% of patients unable to score their pain as opposed to 14% of patients who were unable to use the VAS. In relation to chronic pain, Cork *et al* (2004) showed excellent correlation between the VRS and VAS, although the VRS showed a tendency to be higher than the VAS ($P = 0.068$). Cork *et al* (2004) proposed that the VRS provides a useful alternative to VAS scores in the assessment of chronic pain.

These tools can provide valuable information about the severity of patients' pain. However, assessment needs to continue at different times throughout the day to identify if pain is associated with interventions, such as dressing changes, which might otherwise give a false high score.

The McGill Pain Questionnaire (MPQ) (Melzack 1987) consists of 102 pain descriptors in groups that allow the patient

	None (0)	Mild (1)	Moderate (2)	Severe (3)
Throbbing				
Shooting				
Stabbing				
Sharp				
Cramping				
Gnawing				
Hot, burning				
Aching				
Heavy				
Tender				
Splitting				
Tiring—exhausting				
Sickening				
Punishing—cruel				
(Melzack 1987)				

FIGURE 1   The Short-Form McGill Pain Questionnaire

and practitioner to describe not only the nature, but also the severity of pain. The diagrams incorporated can also be used to mark areas of pain. The questionnaire, because of its complexity, may take some time to complete.

Although widely used, clinical practice has demonstrated variations between VAS and the MPQ scoring. VAS scores were significantly higher in patients with neuropathic pain compared to cancer pain. MPQ total score (pain rating index) related to neuropathic pain was significantly higher than scores reported in the other pain groups (Majani *et al* 2003).

The Short-Form McGill Pain Questionnaire (SF-MPQ) (Melzack 1987) consists of 14 descriptors of pain, which are rated on an intensity scale of zero = none, one = mild, two = moderate and three = severe (Figure 1). The SF-MPQ also includes the present pain intensity index of the standard MPQ and VAS. This allows the practitioner to assess the efficacy of current medication regimens.

Communication of pain severity using the traditional VAS or VRS may be problematic for some patients, such as those who are cognitively impaired. Patients with dementia may not be able to indicate accurately the location of pain (Kerr *et al* 2006), as dementia can affect motor, cognitive, language and social and emotional abilities of individuals. Smith (2005) noted that patients with dementia are likely to have one or more chronic health conditions that cause pain. In patients with cognitive impairments, observation of non-verbal cues, such as alteration in posture, facial expressions, bracing, rubbing, guarding, facial contortion and agitation are vital.

Use of the above assessment tools can help the practitioner to identify the types of pain and then plan treatment accordingly.

Treatment

Pharmacological treatment of wound-related pain is based on the World Health Organization's (2009) three-step pain relief ladder developed for cancer pain relief (Figure 2). The patient's

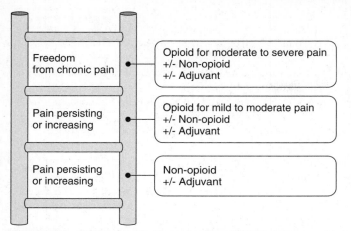

FIGURE 2 World Health Organization (WHO) cancer pain relief ladder adapted for wound care
(Adapted from WHO 2009)

pain is managed by either stepping up or down the ladder as appropriate. At any time other supportive non-drug therapies may be used to help relieve pain, for example transcutaneous electric nerve stimulation. The aim of giving mixed agents, or multimodal analgesia, is to provide superior dynamic pain relief with reduced analgesic-related side effects (Joshi 2005).

The evidence-based pharmacological treatment of neuropathic pain includes anticonvulsants and antidepressants, for example gabapentin and amitriptyline (McQuay and Moore 1998). The use of opioids in neuropathic pain remains controversial although use over the short to intermediate term may be of benefit in some patients (Eisenberg *et al* 2005). Sedation can be a troublesome side effect of drugs such as gabapentin and amitriptyline and patient education is vital to ensure that individuals remain vigilant for such complications.

Case studies and small trials (Twillman *et al* 1999, Grocott 2000) support the use of topical opioids such as diamorphine in a gel preparation for treating wound-related pain, but without larger controlled studies the evidence is not compelling (Ashfield 2005). Initial findings of similar dressings that contain a nonsteroidal anti-inflammatory drug, for example Biatain Ibu®, are promising (Gottrup *et al* 2007, Jørgensen *et al* 2008), but these are relatively new products and evidence is at present limited.

Both Argoff *et al* (2004) and Lin *et al* (2008) found that using topical local anaesthetic (lidocaine) patches proved effective in the treatment of pain associated with post-herpetic neuralgia. Saber *et al* (2009) also reported promising results from the use of local anaesthetic patches used in the treatment of acute post-operative pain following ventral hernia surgery.

De Jong *et al* (2007) conducted a systematic review of the use of non-pharmaceutical interventions in the treatment of pain for dressing changes in patients with burns. Relaxation, imagery, distraction, music therapy and rapid induction hypnosis were found to have positive outcomes for patients. There were, however, methodological shortcomings in the reviewed literature and further research is needed.

Multimodal analgesia may be more effective in providing pain relief than relying on one drug. As different analgesics have different modes of action, side effects may be reduced (Melzack and Wall 2003). Timing of administration is vital to achieve optimum analgesia. For example, medicines should be given 'by the clock' and potentially painful interventions such as dressing changes planned accordingly; for example, if Oramorph® is used, this needs to be given at least 30 minutes before the procedure.

It is important to note that the sight of the wound, previous painful experience, lack of pain assessment, trauma to the wound bed if there has been dressing adherence, cold cleansing solutions, nursing technique and wounds with an ischaemic element may contribute to patients' pain.

Conclusion

Pain is a multi-faceted unique experience requiring individualised assessment to ensure the correct treatment options are implemented. Pharmacological treatments should be multimodal and the importance of debilitating side effects such as constipation, nausea and sedation should not be overlooked.

The psychological affect of wound-related pain on patient wellbeing and quality of life also needs to be considered in the patient's overall plan of care.

References

AnaesthesiaUK (2009) *Management of chronic pain*. www.frca .co.uk/article.aspx?articleid=100512 (Last accessed: July 1 2009.)

Argoff CE, Galer BS, Jensen MP, Oleka N, Gammaitoni AR (2004) Effectiveness of the lidocaine patch 5% on pain qualities in three chronic pain states: assessment with the Neuropathic Pain Scale. *Current Medical Research and Opinion*. 20, Suppl 2, S21-8.

Ashfield T (2005) The use of topical opioids to relieve pressure ulcer pain. *Nursing Standard*. 19, 45, 90–92.

Blyth FM, March LM, Brnabic AJ, Jorm LR, Williamson M, Cousins MJ (2001) Chronic pain in Australia: a prevalence study. *Pain*. 89, 2–3, 127–134.

Briggs M, Closs JS (1999) A descriptive study of the use of visual analogue scales and verbal rating scales for the assessment of postoperative pain in orthopedic patients. *Journal of Pain and Symptom Management*. 18, 6, 438–446.

British Pain Society (2008) *FAQs*. www.britishpainsociety.org/ mediafaq.htm (Last accessed: July 1 2009.)

Bruster S, Jarman B, Bosanquet N, Weston D, Erens R, Delbanco TL (1994) National Survey of Hospital Patients. *British Medical Journal*. 309, 6968, 1542–1546.

Carr DB, Jacox AK (1992) *Acute Pain Management: Operative or Medical Procedures and Trauma. Clinical Practice Guideline*. Rockville, MD: Agency for Health Care Policy and Research, Public Health Service, US Department of Health & Human Services.

Cork RC, Isaac I, Elsharydah A, Saleemi S, Zavisca F, Alexander L (2004) *A comparison of the Verbal Rating Scale and the Visual Analog Scale for Pain Assessment*. www.ispub.com/journal/the_ internet_journal_of_anesthesiology/volume_8_number_1_12/ article/a_comparison_of_the_verbal_rating_scale_and_the_ visual_analog_ scale_for_pain_assessment.html (Last accessed: July 1 2009.)

Dallam L, Smyth C, Jackson BS *et al* (1995) Pressure ulcer pain: assessment and quantification. *Journal of Wound, Ostomy, and Continence Nursing*. 22, 5, 211–218.

de Jong AE, Middelkoop E, Faber AW, Van Loey NE (2007) Non-pharmacological nursing interventions for procedural pain relief in adults with burns: a systematic literature review. *Burns*. 33, 7, 811–827.

Ebbeskog B, Lindholm C, Öhman S (1996) Leg and foot ulcer patients. Epidemiology and nursing care in an urban population in south Stockholm, Sweden. *Scandinavian Journal of Primary Health Care*. 14, 4, 238–243.

Eisenberg E, McNicol ED, Carr DB (2005) Efficacy and safety of opioid agonists in the treatment of neuropathic pain of nonmalignant origin: systematic review and meta-analysis of randomized controlled trials. *Journal of the American Medical Association*. 293, 24, 3043–3052.

Elliott AM, Smith BH, Penny KI, Smith WC, Chambers WA (1999) The epidemiology of chronic pain in the community. *The Lancet*. 354, 9186, 1248–1252.

Goodridge D, Trepman E, Embil JM (2005) Health-related quality of life in diabetic patients with foot ulcers: literature review. *Journal of Wound, Ostomy, and Continence Nursing*. 32, 6, 368–377.

Goodwin SA (1998) A review of preemptive analgesia. *Journal of Perianesthesia Nursing*. 13, 2, 109–114.

Gottrup F, Jørgensen B, Karlsmark T *et al* (2007) Less pain with Biatain-Ibu: initial findings from a randomised, controlled, double-blind clinical investigation on painful venous leg ulcers. *International Wound Journal*. 4, 1, 24–34.

Grocott P (2000) The management of malignant wounds. *European Journal of Palliative Care*. 7, 4, 126–129.

Gruener DM (2004) *New Strategies for Managing Acute Pain Episodes in Patients With Chronic Pain*. www.medscape.com/viewarticle/484371 (Last accessed: July 1 2009.)

International Association for the Study of Pain (1994) Part III pain terms: a current list with definitions and notes on usage. In Merskey H, Bogduk N (Eds) *Classification of Chronic Pain*. Second Edition. IASP Press, Seattle. 209–214.

Johnson L (2004) The nursing role in recognizing and assessing neuropathic pain. *British Journal of Nursing*. 13, 18, 1092–1097.

Jørgensen B, Gottrup F, Karlsmark T, Bech-Thomsen N, Sibbald RG (2008) Combined use of an ibuprofen-releasing foam dressing and silver dressing on infected leg ulcers. *Journal of Wound Care*. 17, 5, 210–214.

Joshi GP (2005) Multimodal analgesia techniques and postoperative rehabilitation. *Anesthesiology Clinics of North America*. 23, 1, 185–202.

Kerr D, Cunningham C, Wilkinson H (2006) *Pain Management for Older People with Learning Difficulties and Dementia*. Joseph Rowntree Foundation, York.

Langemo DK (2005) Quality of life and pressure ulcers: what is the impact? *Wounds*. 17, 1, 3–7.

Lin PL, Fan SZ, Huang CH *et al* (2008) Analgesic effect of lidocaine patch 5% in the treatment of acute herpes zoster: a double-blind and vehicle-controlled study. *Regional Anesthesia and Pain Medicine*. 33, 4, 320–325.

Majani G, Tiengo M, Giardini A, Calori G, De Micheli P, Battaglia A (2003) Relationship between MPQ and VAS in 962 patients. A rationale for their use. *Minerva Anestesiologica*. 69, 1–2, 67–73.

McQuay HJ, Moore RA (1998) *An Evidence-Based Resource for Pain Relief*. Oxford University Press, Oxford.

Melzack R (1987) The short-form McGill Pain Questionnaire. *Pain*. 30, 2, 191–197.

Melzack R, Wall PD (2003) *Handbook of Pain Management*. Churchill Livingstone, Oxford.

Middleton C (2003) Understanding the physiological effects of unrelieved pain. *Nursing Times*. 99, 37, 28.

Nemeth KA, Harrison MB, Graham ID, Burke S (2003) Pain in pure and mixed aetiology venous leg ulcers: a three-phase point prevalance study. *Journal of Wound Care*. 12, 9, 336–340.

Persoon A, Heinen MM, van der Vleuten CJ, de Rooij MJ, van de Kerkhof PC, van Achterberg T (2004) Leg ulcers: a review of their impact on daily life. *Journal of Clinical Nursing*. 13, 3, 341–354.

Puntillo KA, Wild LR, Morris AB, Stanik-Hutt J, Thompson CL, White C (2002) Practices and predictors of analgesic interventions for adults undergoing painful procedures. *American Journal of Critical Care*. 11, 5, 415–429.

Reddy M, Keast D, Fowler E, Sibbald RG (2003) Pain in pressure ulcers. *Ostomy/ Wound Management*. 49, Suppl 4, 30–35.

Saarto T, Wiffen PJ (2007) Antidepressants for neuropathic pain. *Cochrane Database of Systematic Reviews*. Issue 4.

Saber AA, Elgamal MH, Rao AJ, Itawi EA, Martinez RL (2009) Early experience with lidocaine patch for postoperative pain control after laparoscopic ventral hernia repair. *International Journal of Surgery*. 7, 1, 36–38.

Smith M (2005) Pain assessment in nonverbal older adults with advanced dementia. *Perspectives in Psychiatric Care*. 41, 3, 99–113.

Twillman RK, Long TD, Cathers TA, Mueller DW (1999) Treatment of painful skin ulcers with topical opioids. *Journal of Pain and Symptom Management*. 17, 4, 288–292.

Whelan CT, Jin L, Meltzer D (2004) Pain and satisfaction with pain control in hospitalized medical patients: no such thing as low risk. *Archives of Internal Medicine*. 164, 2, 175–180.

World Health Organization (2009) *WHO's Pain Relief Ladder*. www.who.int/ cancer/palliative/painladder/en/ (Last accessed: July 1 2009.)

Wulf H, Baron R (2002) The theory of pain. In Calne S (Ed) *Pain at Wound Dressing Changes*. European Wound Management Association Position Document. Medical Education Partnership, London, 8–11.

Critical Thinking

1. How is wound-related pain managed with regard to pharmacological treatment?

ANDY RODEN is lecturer, School of Healthcare Sciences, Bangor University and Archimedes Centre, Wrexham and **ELAINE STURMAN** is directorate lead pharmacist, Wrexham Maelor Hospital, North Wales NHS Trust (East), Wrexham.

Acute Abdomen: What a Pain!

So many things–some life-threatening–can cause abdominal pain. Here's how to capture the clues quickly and accurately.

Susan Simmons Holcomb, ARNP, BC, PhD

Determining the cause of abdominal pain is often tricky and time-consuming. Because pain can be nonspecific and the abdomen has many organs and structures, numerous potential causes have to be ruled out as clinicians try to pin down the source of pain.

Reasons for abdominal pain fall into three broad categories: inflammation, organ distension, and ischemia. In some cases, the underlying cause is life-threatening, so a fast, accurate assessment is essential. In this article, I'll describe how to assess a patient with abdominal pain and intervene appropriately. Let's start by looking at a couple of hypothetical situations.

Laurie Greene, 42, comes to your ED complaining of intermittent abdominal pain and bloating that she's had for a month. She reports no change in bowel habits and no other significant medical history.

When you examine Ms. Greene, you find a protrusion in the umbilical, hypogastric, and left iliac regions. This area is dull to percussion; when you palpate it, you note a firm mass. A complete blood cell (CBC) count reveals severe anemia. A stat computed tomography (CT) scan of the abdomen and pelvis reveals a benign, totally encapsulated left ovarian tumor. The tumor is removed, her anemia is resolved, and she's doing well.

Patrick Leeson, 45, comes in complaining of dull, achy periumbilical pain that migrated to his right lower quadrant. He says it started about 24 hours ago. He has no nausea or vomiting or changes in bowel habits. He says he was treated for testicular cancer 10 years ago and it began with abdominal pain similar to what he's experiencing now.

Mr. Leeson has an elevated white blood cell count with a left shift, possibly indicating a bacterial infection or inflammation. A CT scan of the abdomen and pelvis shows an inflamed appendix. Mr. Leeson is admitted to the hospital for an appendectomy and recovers completely.

Narrowing Things Down

So where do you start when a patient has abdominal pain? Anatomically the abdomen is divided into four quadrants and nine regions. You can use these divisions to narrow down the area of complaint and document your findings (see *Where does it hurt?*). Remember, however, that abdominal pain can be referred to many locations, including the shoulders, cardiac area (substernal and left chest), low and mid back, and groin.

Besides pain location, the kind of pain provides clues to its cause. Type A delta nerve fibers innervate cutaneous tissues and the parietal peritoneum; stimulation from an irritant such as pus, blood, bile, or urine often leads to localized pain. Type C fibers innervate visceral tissue, so visceral pain is more generalized and deeper.

Visceral pain can be divided into three subtypes:

- **Tension pain** is caused by organ distension, as from bowel obstruction or constipation. Blood accumulation from trauma and pus or fluid accumulation from infection or other causes also can cause this pain. Tension pain that's described as colicky may be caused by increased peristaltic contractile force as the bowel tries to eliminate irritating substances. Patients with tension pain may have trouble getting comfortable and squirm a lot trying to find a comfortable position.
- **Inflammatory pain** may arise from inflammation of either the visceral or parietal peritoneum, as in acute appendicitis. This pain may be described as deep and boring. Initially, if the visceral peritoneum is involved, the pain may be poorly localized; as the parietal peritoneum becomes involved, the pain may become localized. Most patients with inflammatory abdominal pain want to lie still.
- **Ischemic pain** is the most serious type of visceral pain—and, fortunately, the least common because the affected area will become necrotic if blood flow isn't promptly restored. Sudden in onset, this pain is extremely intense, progressive in severity, and not relieved by analgesics. Like patients with inflammatory pain, patients with ischemic pain won't want to move or change positions. The most common cause of ischemic abdominal pain is strangulated bowel.[1]

Where Does It Hurt?

You can use the four abdominal regions shown here and the nine abdominal regions shown below to help you determine what's causing your patient's abdominal pain.

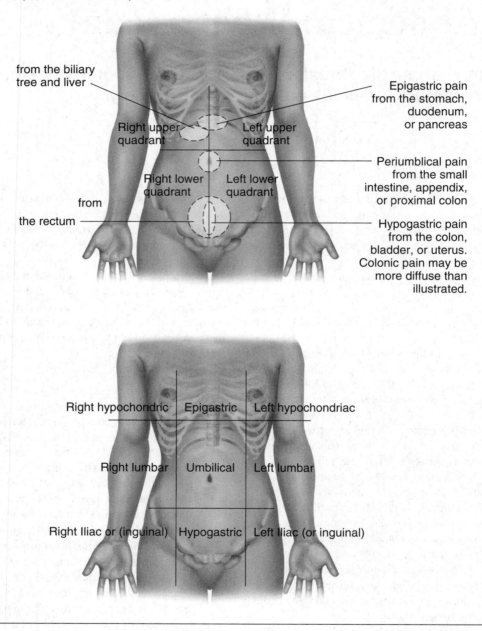

from the biliary tree and liver

Right upper quadrant

Left upper quadrant

Right lower quadrant

Left lower quadrant

from the rectum

Epigastric pain from the stomach, duodenum, or pancreas

Periumblical pain from the small intestine, appendix, or proximal colon

Hypogastric pain from the colon, bladder, or uterus. Colonic pain may be more diffuse than illustrated.

Right hypochondric   Epigastric   Left hypochondriac

Right lumbar   Umbilical   Left lumbar

Right Iliac or (inguinal)   Hypogastric   Left Iliac (or inguinal)

Assessment Pointers

Now let's look at how to begin your assessment of a patient with abdominal pain. Take a health history, gynecologic history for women, and family history of abdominal conditions such as gastroesophageal reflux disease (GERD), gallbladder disease, renal calculi, colon cancer, or inflammatory bowel disease.

Ask the patient when the pain began, where it's located, and how he'd describe its quality and intensity. Ask if the pain is constant or intermittent, if it wakes him at night, and if anything

aggravates or relieves it. If he says that food worsens or relieves abdominal pain, ask him what kind of food. Assess and document whether he has associated signs and symptoms such as fever, nausea or vomiting, change in bowel habits, weight loss, heartburn, or rectal bleeding.

If he reports nausea and vomiting and a change in bowel habits, ask if he's recently traveled, eaten food that was recalled, drunk water that might have been contaminated, or gone swimming in lakes or public pools. Ask about frequency of bowel movements. If he reports diarrhea, ask if the diarrhea is liquid, loose, or a combination and whether he's noticed blood in the stool.

If he's had a change in bowel habits without diarrhea, ask about the color and consistency of the stool, whether it floats or sinks, and if it's associated with mucus or change in odor.

Vomiting that precedes abdominal pain, or is associated with the onset of abdominal pain, may suggest infection as a possible cause of pain. Abdominal pain that began before vomiting may indicate appendicitis or, more rarely, cholecystitis. Suspect cholecystitis in patients with right upper quadrant abdominal pain and a family history of early gallbladder disease.[2] Other risk factors for cholecystitis include being female, age 40 or older, and overweight. Associated signs and symptoms may include vomiting and fever.[3]

As you continue your assessment, ask if the patient is taking new medications that might cause abdominal pain (such as non-steroidal anti-inflammatory drugs) or has recently been diagnosed with a condition that might be associated with abdominal symptoms, such as GERD.

In children, abdominal migraine can cause abdominal pain. Look for a pattern of symptoms, especially if the abdominal pain is associated with vomiting and is cyclical. Ask about frequency, duration, and associated symptoms such as vomiting. Also ask about a personal or family history of migraines.[1]

Your physical assessment should include inspection, auscultation, percussion, and palpation.

- *Inspect* the abdomen for movement, such as fluid waves or increased peristalsis. Look for scars from past surgeries; the patient may have adhesions that could lead to bowel obstruction. Note the contour of the abdomen; generalized distension may indicate increased gas, but local bulges may indicate a distended bladder or a hernia.
- *Auscultate* for bowel sounds or additional sounds such as bruits. Normal bowel sounds consist of peristaltic clicks and gurgles occurring at a rate of 5 to 34 per minute. Hypoactive bowel sounds may indicate an ileus. Hyperactive bowel sounds may indicate early intestinal obstruction. Arterial bruits with both systolic and diastolic components are abnormal sounds made by blood traveling through narrowed arteries such as the aorta or renal, iliac, or femoral arteries.
- *Percussion* can help you identify the borders of organs such as the liver, as well as determine the presence of air or solid masses such as tumors. Normally you'll hear tympany (a drumlike sound) over the stomach and intestines, areas that normally are air filled. You'll hear dullness over solid areas such as the liver, spleen, tumors, or other masses.
- *Palpation* lets you assess local versus generalized areas of tenderness, as well as check for masses and enlarged organs. Palpation can go from light to deep, but keep in mind that a patient with abdominal pain may not tolerate abdominal palpation at all. He may guard (tighten his abdominal muscles), preventing you from assessing

the abdomen adequately via palpation. If this happens, flexing his knees may relax the abdomen so you can palpate it. If he's very ticklish, you can circumvent the tickle response by placing his hand below yours and palpating the area first with his hand, then switching hands so you can palpate.[4] If the presence of a bruit leads you to suspect that the patient has an aortic aneurysm, palpation may be contraindicated or best left to the physician.

If you suspect an aortic aneurysm, palpation maybe contraindicated or best left to the physician.

To assess for specific areas of tenderness, use specific palpation techniques. *Murphy's sign* evaluates gallbladder tenderness and inflammation. Hook your fingers under the patient's right lower ribs or press them under the ribs, then ask the patient to take a deep breath. A sharp increase in tenderness with a sudden stop in inspiratory effort constitutes a positive Murphy's sign, indicating acute cholecystitis.

Leave checking for rebound tenderness for last because it may elicit enough pain that the patient won't let you touch his abdomen again. Push your fingers into the area of tenderness slowly and firmly, then quickly lift them away. Rebound tenderness is present if the pain worsens when you withdraw your fingers. Rebound tenderness suggests peritoneal inflammation; for example, from appendicitis.

If you suspect that your patient has appendicitis, check for *Rovsing's sign* and for referred rebound tenderness. Press deeply and evenly in the patient's left lower quadrant, then quickly withdraw your fingers. Pain in the right lower quadrant during left-sided pressure (a positive Rovsing's sign) suggests appendicitis, as does right lower quadrant pain on quick withdrawal (referred rebound tenderness).

You can also assess for appendicitis by looking for a *psoas* or *obturator sign*. Place your hand just above the patient's right knee and ask him to raise his thigh against your resistance. Alternatively, ask him to turn onto his left side; then extend his right leg at the hip. Flexing the leg at the hip makes the psoas muscle contract; extension stretches it. Increased abdominal pain on either maneuver (a positive psoas sign) suggests that the psoas muscle is irritated by an inflamed appendix.

To elicit the obturator sign, ask the patient to bend his right knee, then flex his right thigh at the hip and rotate the leg internally at the hip to stretch the internal obturator muscle. Right hypogastric pain (a positive obturator sign) suggests irritation of the obturator muscle by an inflamed appendix.

If you think the patient's abdominal pain may be related to pyelonephritis or renal calculi, assess for costovertebral angle tenderness. Place the palm of one hand in the right costovertebral angle and strike it with the ulnar surface of your fist. Repeat in the left costovertebral angle. Pain with percussion suggests pyelonephritis.

Some Causes of Acute Abdominal Pain

Cause	Signs and symptoms
Abdominal aortic aneurysm	• Usually asymptomatic, but may cause back and abdominal pain • Pulsatile mass may be palpable.
Appendicitis	• Abdominal pain over umbilicus, moving to right lower quadrant • Often associated with fever. Clinical exam may show rebound tenderness and positive obturator, psoas, and Rovsing's signs. • Complete blood cell count will show an increase in white blood cells with a shift to the left and increased neutrophils.
Cholecystitis	• Pain in the right upper quadrant (toward the epigastric area) can radiate to the shoulder or back. • Nausea and vomiting may occur. • Biliary colic (pain that increases over 2 to 3 minutes and is sustained for 20 minutes or more) • Positive Murphy's sign
Constipation	• Possible colicky to sharp pain that can mimic appendicitis • Patient may have diffuse tenderness on palpation as well as palpable stool.
Diverticultis	• Left lower quadrant pain, often worse after eating and improved after defecation • Possible fever • Possible diarrhea or constipation • Abdomen may be distended and tympanic and tender to palpation over the left lower quadrant.
Ectopic pregnancy, ruptured	• Sudden onset of lower left or right quadrant pain • Possible vaginal bleeding
Gastroenteritis	• Diffuse abdominal cramping, possibly with nausea, vomiting, diarrhea, and fever • Possible hyperactive bowel sounds, abdominal distension, and diffuse tenderness on palpation
Ileus or bowel obstruction	• Diffuse pain that comes in cramping waves lasting 5 to 15 minutes • Nausea, followed by vomiting when the bowel obstructs • Stool may be passed distal to the obstruction and may also involve diarrhea • Abdomen may be distended with high-pitched bowel sounds. • Diffuse tenderness and guarding
Pancreatitis	• Pain in the right upper quadrant to epigastric area, possibly radiating to the back; can be associated with nausea and vomiting as well as fever • Possible ileus • In severe cases, shock, jaundice, and pleural effusion • Rare signs include Grey Turner and Cullen's signs.
Peptic ulcer disease	• Usually epigastric pain 1 to 3 hours after meals and often associated with nighttime awakenings • Sudden and severe pain with radiation to the right shoulder, along with peritoneal signs, may indicate perforation. • Hematemesis or melena suggests hemorrhage.
Peritonitis	• Acute diffuse abdominal pain that can be associated with fever, nausea, and vomiting. Pain increases with any motion. • Abdominal distension and rigidity. Rebound tenderness present but, unlike in appendicitis, is diffuse rather than localized. Guarding may be present. • Possible signs and symptoms of shock

Diagnostic Testing and Treatment

The following lab studies may help narrow down causes of abdominal pain:

- CBC count, for signs of infection, cancer, and inflammation
- complete metabolic profile, for blood glucose level, renal or hepatic dysfunction, electrolyte imbalances, or problems related to low albumin level
- stool sample to look for infection or parasites
- urinalysis to look for infection or evidence of renal calculi
- amylase and lipase levels, which will be elevated in a patient with pancreatic problems
- *Helicobacter pylori* level to check for peptic ulcer disease
- pregnancy test and microscopic examination of vaginal secretions in women, to rule out ectopic pregnancy and infections such as bacterial vaginosis or vulvovaginal candidiasis.

- sexually transmitted disease testing in sexually active men and women.

The following imaging studies may be done:

- A CT scan may be the first imaging test performed because it's more sensitive, specific, and accurate than a plain radiographic abdominal series.[2]
- An abdominal/pelvic ultrasound is more diagnostic than a plain X-ray and can help clinicians identify renal stones, gallstones, appendicitis, and gynecologic problems.
- Abdominal plain film radiography may reveal stones, bowel dilation, fluid levels indicating bowel obstruction, and stool and gas patterns.

For details on signs and symptoms specific to common abdominal problems, see *Some causes of acute abdominal pain.*

Because the causes of abdominal pain are so varied, so are the treatments. Generally, surgery is indicated for bowel obstruction, acute appendicitis, a ruptured ovarian cyst, and aortic aneurysm. Antibiotics will be prescribed if the cause of pain is an infection such as pyelonephritis or a lower urinary tract infection. However, if the infection is due to an abscess, surgical drainage may also be performed. Abdominal pain due to viral gastroenteritis will be treated with fluids, bowel rest, and antiemetics if the patient is over age 12.

Your Role

Because some causes of abdominal pain are life-threatening, triaging patients quickly and accurately is crucial. Other nursing interventions include ongoing assessments, managing the patient's pain, providing emotional support, restoring fluid and electrolyte balance, and specific interventions to treat the pain's underlying cause. If pain is associated with infection, for example, you'll also take steps to regulate your patient's body temperature and administer antibiotics as prescribed.

Because some causes of abdominal pain are life-threatening, triaging patients quickly and accurately is crucial.

Manage your patient's pain with medications as ordered and nonpharmacologic interventions, including positioning, back rubs, and heating pads (if not contraindicated).

To protect your patient against complications such as cardiac dysrhythmias and seizures, maintain electrolyte balance. Patients with diarrhea, vomiting, or fever are the most prone to electrolyte imbalances. Make sure electrolyte levels are evaluated before electrolyte replacement begins and periodically reassessed during replacement.

If your patient's abdominal pain was caused by GERD, hiatal hernia, peptic ulcer disease, or diverticulitis, teach him about foods to avoid as well as how to time meals in relation to activities and bedtime. He should avoid overeating in general and stay away from fats, fried foods, spices, coffee, tea, tomato products, and alcohol. (Some patients should avoid certain other foods as well.) Tell him not to eat within 2 to 3 hours before bedtime and not to lie down or exercise immediately after eating.

Advise him to try to maintain a normal weight and to lose weight if he's overweight or obese; the risk of GERD and gallbladder disease increases with weight. He should reduce stress, quit smoking, decrease or eliminate alcohol consumption, and reduce his use of medications that can damage the esophagus, such as corticosteroids and nonsteroidal anti-inflammatory drugs, including aspirin.

Divide and Conquer

Although abdominal pain can be tricky to diagnose and treat, remembering which structures lie in which section and understanding the different types of pain can help you net clues to the source of the patient's pain so he gets the help he needs—and fast.

References

Miller SK, Alpert PT. Assessment and differential diagnosis of abdominal pain. *The Nurse Practitioner.* 31(7):39–47, July 2006.

MacKersie AB, et al. Nontraumatic acute abdominal pain: Unenhanced helical CT compared with three-view acute abdominal series. *Radiology.* 237(1):114–122, October 2005.

Trowbridge RL, et al. Does this patient have acute cholecystitis? *JAMA.* 289(1):80–86, January 1, 2003.

Smeltzer SC, et al. *Brunner & Suddarth's Textbook of Medical-Surgical Nursing,* 11th edition. Lippincott Williams & Wilkins, 2008.

Critical Thinking

1. What are the three categories of abdominal pain?

SUSAN SIMMONS HOLCOMB is a nurse practitioner at Olathe (Kan.) Medical Services, Inc., and a consultant in continuing nursing education at Kansas City (Kan.) Community College.

Pain Management: The Role of the Nurse

FAUSTNA OWARE-GYEKYE, BA (NURSING), MPHIL, FWACN

Introduction

Pain is a universal experience. Everyone has known pain to some degree. Pain is often a useful protective signal because it is a warning of potential health problems, e.g. the dysuria caused by a bladder infection or earache of otitis media. Pain is also a consequence of some normal bodily functions, e.g. mild dysmenorrhoea or the pain of childbirth. Pain can also warn of emotional or stress related problems. (e.g. headache, gastritis or low-back pain caused by tension, anxiety or the stress of daily living).

It disables and distresses more people than any single disease. However, pain is a totally unpleasant, subjective, personal and multidimensional experience. No one can fully appreciate the pain of another. No two persons feel the same degree of pain in the same way. Pain is a complex and perceptual phenomenon. One of the earliest fears of any client with a diagnosed illness is the concern over possible pain.

Relief from pain has probably been a concern of mankind since the beginning of time, but health professionals have not until recently devoted much clinical effort or research to pain relief. Yet pain control must be recognized as a priority. Both palliative and curative care must be used. Health professionals often hold prejudices against clients in pain. When no obvious source of pain can be found, nurses may stereotype clients as complainers or difficult clients. Meanwhile, the client who bears his/her pains silently is looked upon as a good or co-operative patient. This action of labelling patients', stops them from calling for assistance. Many nurses distrust patients' self-report of their pain which suggests that they have their own bench mark which is acceptable as to when and how patients should express their pain. Very often a nurse who has personally experienced pain is better able to provide support.

The nurse plays a key role in assuring good pain control for people with unrelieved pain. She is with the patient more than any other health team member and is in a position to constantly assess and evaluate the effectiveness of the pain treatments. Nurses care for clients in many settings and situations. However, the care of patients with pain demands skills in both the science and the art of nursing. The nurse's responsibility is to make the patient comfortable as possible, physically and emotionally. The nurse, client, family and members of the health care team must collaborate to find the most effective approach to pain control. Effective pain management not only reduces physical discomfort but also promotes early mobilization, shortened hospital stays and reduced health care costs. The nurse plays a unique role in managing pain as an assessor, planner, implementer, educator, advocate and supporter. The nurse performs these roles by employing variety of approaches.

Numerous factors influence the pain experience of the individual thus the nurse must consider all factors affecting the client in pain to accurately assess the client's pain and to select appropriate pain therapies. These factors are age, sex, culture, meaning of pain, attention, anxiety, previous experience, coping style, family and social support, upbringing, or socialization, religion, part of the body involved and beliefs about health and illness—to name a few.

Children and adults express pain differently. Children are at times unable to express their pain accurately because it is unfamiliar, while older people may have learnt to cope with pain so that they either accept it and may be quiet or may complain about it. Thus, studies have shown that children are grossly under medicated for pain, especially comparing children with adults having the same medical diagnoses; children received fewer medication doses[1,2]. Generally, men and women do not differ significantly in responses to pain[3]. However, cultural influences on gender may produce different expressions of pain. For example, making it acceptable for a little boy to be brave and not cry whereas a little girl in the same situation will cry. In Ghana this idea is expressed in most of our languages (men should not cry). Yet, from experience, men are unable to bear or cope with pain. Culture influences how people learn to react to and express pain. Understanding cultural background, socio economic status and personal characteristics help the nurse more accurately to assess pain and its meaning for the clients.

Clients perceive pain differently if it suggests a threat, loss, punishment or challenge. Thus, the meaning a client associates with pain affects the pain experience.

The degree to which a client focuses on pain can influence pain perception. For example, there is more reaction to pain during the night and early morning hours, when the person's physiological processes are at low ebb and there is little distracting activity. The patient's thoughts may easily turn to self concern and anxiety may increase the reaction to pain. Very often, terminally-ill patients with chronic pain are put in side-wards, isolated and with no one to talk to or sympathize with them. This increases fear, loneliness and anxiety thereby exacerbating the pain. On the other hand, there is a relationship between pain and anxiety. Elevated anxiety levels cause an increase in pain perception.[4] In reverse, pain may also cause anxiety. However, fatigue heightens pain perception. This intensifies pain and decreases coping abilities. Pain is often experienced less after restful sleep than at the end of a tiring day.

The fact that previous pain experience of clients will let him/her accept pain more easily in future is a bit controversial and may not be true. Very often, frequent episodes of pain without relief may produce anxiety or fear. In contrast, experiences with the same type of pain that have successfully been relieved makes it easier for the client to interpret the pain.

A client often depends on the support of spouse, family or friends when coping with pain. Although pain still persists, the presence of a loved one can minimize loneliness and fear. The question is 'how does the nurse manage pain of her/his clients?' Some of the processes will be discussed.

Assessment
History Taking
Accurate and factual pain assessment is necessary for judging clients progress and response, arriving at proper nursing diagnoses and selecting appropriate therapies. Pain assessment is one of the most common and yet one of the difficult activities a nurse performs. Pain is rarely static therefore its assessment is not a one-time process but is ongoing. The assessment of pain is the beginning step in understanding and working towards the patient's goals. Unfortunately, this crucial step may be neglected.

Information is collected about the type of pain (burning, aching, dull, tingling, shooting and electric shock-like), location, intensity, quality, onset and duration of pain (acute or chronic pain), effects of pain, way and manner of expressing pain and pain pattern i.e. what relieves the pain and what causes or increases the pain.

Observation
Close observation of the patient often gives clues to the intensity of pain. The nurse should learn verbal or non-verbal ways or cues that the client communicates discomfort. It may vary from one person to the other. Vocalizations such as moaning, crying, screaming and gasping may be ways of expressing pain. Facial expression such as grimace, clenched teeth, lip biting, wrinkled brows or forehead, tightened fists, tightly closed or opened eyes or mouth, pursed mouth, curled up and tapping with toes or drumming with fingers will also be body movements to indicate pain. These signs may be accompanied by restlessness, immobilization, muscle tension, pacing activities, sleeplessness and tossing about in bed. Clients may react to pain by avoiding conversation or social contact and may exhibit reduced attention span. Diaphoresis and rapid pulse also are valuable clues. In all these, the nurse must exhibit patience, tolerance, gentleness and the power of observation. In assessing the severity of pain, people are advocating for use of a pain rating scale which will also evaluate the changes in clients. Presently, the faces rating scale for pain designed by Wong and Baker[5] to assess pain in children is also being used for adults.

Physical Assessment
The nurse assesses the vital signs to determine the effect of the pain on the client. At times, when clients find it difficult to describe the pain, the nurse examines physically to locate the pain as well as the severity.

Interventions to Relieve Pain
After the nurse has assessed the patient's pain, nursing intervention is planned. The nurse is supposed to be the advocate of the client thus, the nurse should report to the doctor if both medical and the nursing interventions do not work efficiently.

Explaining the Problem
The nurse explains in simple terms what is causing the pain, if and when it is to be expected and how long it may last. This explanation should continue with each experience.

Decision-Making
Allowing patients to regain some control over their daily activities may allow them to exert better control over their daily activities. A plan of care can be developed in which patients and families' decisions are respected, the day to day activities are put in order and daily routines can be revised to reduce anxiety and frustration about constant changes. Assigning the same nursing staff to care for the patients results in more consistent approach and plan of care.

Distraction
Many individuals can be distracted from constant preoccupation with discomfort. The nurse can employ various measures such as playing games, i.e. chess, ludo, music, watching television or simply talking with someone. This could be used during diagnostic procedures. The nurse can also assist in breathing exercises (eg. Lamaze breathing during labour)

The nurse can determine the patient's usual rest and sleep patterns, decide if they are adequate, determine why the patient is not getting sufficient rest and develop a plan to improve the situation. The plan must include decreasing the number of interruptions during the night with nursing activities. It is the responsibility of the nurse to ensure that the environment is quiet, putting the patient in a comfortable position, providing a night cup if necessary, or administering sedatives or analgesics to ensure sound rest.

Reducing Social Isolation
In the hospital, careful selection of roommates may provide mutual support for patients. The need for actual isolation cannot be ignored but perhaps more frequent visits by health care team members would help. Allowing the significant others to visit patient will also help.

Reducing Painful Stimuli
Many mechanical devices are used to minimize the pain of client, for example, hoisters, body casts, corsets in binders and braces. The nurse caring for patients in pain must use the proper technique and skill in handling patients with generalized

pain, especially changing or turning the patient. Special beds (eg stryker frame, foster bed, Bradford frame) allow movement with minimal handling of the body and thereby lessen pain.

Reducing Noise and Visual Stimulation

Some patients suffer from sensory overload and therefore may need a quiet environment and lights turned off in the night. The patient may be moved from a busy nurse's station to a quiet place. Minimize the number of persons who enter the client's room. The nurse, of all members of the health care team is in the best position to give attention to the patient. In the attempt to assist the patient, the nurse should not neglect her/his client.

Counterirritants and Cutaneous Stimulation

For some individuals, a change in the type of stimulation at the site of pain may result in pain relief. The nurse with the help of the patient may be able to find a satisfactory and relatively simple stimulus modification to ease the patient's discomfort. Various forms of touch such as back rub, application of heat or cold or simply holding the patient's hand may be helpful in diminishing pain. In addition, transcutaneous electrical nerve stimulator (TENS) may be used by the patient. The nurse may be valuable in (TENS) encouraging and assisting the patient to make these small manipulations.

Various forms of medication and routes of administration are used to render relief to patients in pain. The nurse needs to know the precise effect of the medication on the body. The time curve of beginning effect, the height of effectiveness and the time of declining effect must be understood. The nurse should be aware of the side effects of the various drugs and monitor the patient's responses constantly. Nurses are always worried about addiction but a responsible nurse who is working with the patient will continually monitor client's response and adjust the medication schedule accordingly. It is necessary to administer analgesic medication on a time basis, rather than on the basis of the patient's complaint or report of pain which prevents the serum drug level from falling to sub-therapeutic level. At times, nurses become helpless when they are unable to control situations but nursing activities involving touch, explanation and listening enhances the effect of the drug. Withholding medication or encouraging the patient to wait a little longer only increases the client's level of discomfort. Ill timing and inadequate dose of drugs will eventually worsen the patient's plight. There is a growing consensus that the use of placebos to assess and manage pain should be avoided especially without consent. It involves deception. The nurse must therefore respect the patient's rights to be informed of treatment[6]. Studies suggest that placebos tend to be used for patients who are disliked or who have failed to respond to standard agreement[7].

Patient Teaching Discomfort

Teaching has been mentioned frequently in relation to helping the person who has pain. Every nurse is a teacher and each patient's plan of care must be individualized based on his/her needs. If possible, the family should be included. The nurse should educate the patient and family on the causes of pain, the various ways of pain relief and alternate ways to do daily activities. In case of surgery or diagnostic procedures, the nurse should honestly inform the patient what is expected in terms of duration and intensity of pain and what measures will be available to assist with the discomfort. Teaching of antenatal and postnatal exercises must be done to control labour pains as well as after pains. However, the nurse should observe the patient during the exercises to make sure they are not having difficulty or perhaps performing them wrongly.

Providing Spiritual Assistance

Many patients who have prolonged pain with no hope of relief can and do derive benefit from their religious faith. This may help them to consider pain in a more positive way and thus make it more bearable for them. The nurse can arrange for the appropriate religious adviser to be available to the patient who so desires. The nurse should respect the religious faith of his/her clients without being judgmental or imposing their own beliefs on their clients, assuming an authoritarian role rather than the advocacy role. When the patient's spiritual comfort is provided, it boosts the patient's confidence and general morale.

Psychiatric Assistance

In some situations, either a psychologist or psychiatrist may be asked to evaluate the patient's condition. The nurse should be supportive and help the patient understand why these types of therapy may be helpful. The nurse's role also includes the careful recording and reporting of objective data about the patient's behaviour and interactions with family and others.

Behaviour Modification

Behaviour modification can be used frequently with the person in pain. The nurse uses positive reinforcement such as praises, congratulations and encouragement when the patient does her/his postoperative exercises well or any teachings. In recent work with patients in pain, the focus has been on perception of pain. In certain settings, hypnosis has been employed with success. Others have tried various forms of distraction, suggestion or operant conditioning with success. This mechanism is known as biofeedback.

Nutrition

Patients in pain usually experience loss of appetite. Thus, the patient's appetite may be improved by small, attractive servings and by a sincere interest in the patient's reactions to food. Care must be taken to ensure that foods the patient likes are prepared appropriately. Foods that the patient does not like should not be offered. Feeding of patients should be the sole responsibility of nurses and not the family members. Unless, it is on request, then they should be involved after necessary counselling has been done. When certain nursing cares are assigned to the family these tend to increase the anxiety level of the members. The nurse should also encourage intake of nourishing fluids and patient's appetite should be monitored satisfactorily.

Suicide Prevention

When caring for the patient who is experiencing severe continuous or intractable pain, the nurse must keep in mind the possibility of suicide. Pain is demoralizing especially when it is difficult to control. The patients dread being dependent or becoming a burden to the family. Thus, the nurse must anticipate some of these reactions and plan individualized counselling sessions for the clients.

Challenges

The challenges of caring for people who are terminally ill and in constant pain can be immense, however the rewards are fulfilling and worthwhile. The experience can evoke a spectrum of human emotions ranging from confusion, grief, helplessness, fear, anger and pity. Providing care can offer nurses new perspectives on their own lives and what it means to be human. But the challenges involved in caring for people in pain may also result in exhaustion, burnout and even career changes. In addition to the above challenges discussed, other challenges are;

- Misconceptions about pain
- Attitude of nurses towards patients in pain
- Gender differences in response to pain
- Inadequate knowledge about the pain process
- Addiction and dependence of patient
- Choice of easy options
- Resource constraints
- Lack of culturally appropriate pain assessment scales
- Alienating the contribution of the experiencing person
- Effects of patient's experience on the nurse as care giver
- Use of outdated strategies

Conclusion

In conclusion, the nurse helps in relieving pain by administering pain-relieving interventions (both pharmacologic and non-pharmacologic approaches) assessing the effectiveness of those interventions, monitoring for adverse effects, and serving as an advocate for the patient when the prescribed intervention is ineffective. In addition, the nurse serves as an educator to the patient and family to enable them manage the prescribed interventions themselves when appropriate. Innovation is what we need in the 21st Century nursing. Innovation depends on knowledge. However, knowledge is the bridge to achievement while education is the bridge to knowledge. There is the need for continuing education to re-educate ourselves on pain management.

Recommendation

I hope I have been able to discuss the role of the nurse in pain management. The following recommendations are made:

- Education of nurses about pain to clarify misconceptions and promote positive attitude
- Reorientation to cultural values that promote or prevent expression of pain. Providing resources for managing the patient in pain
- Developing tools for assessing pain
- Collaborating care for patients in pain
- Conducting research into the variations in pain experience and testing the existing pain management strategies
- Developing pain management protocols that are culturally sensitive. Organizing pain clinics
- Providing counselling and support for nurses who are constantly working with patients with chronic pain

From the above discussion, it is very necessary to address the following issues;

- How do we delineate our role as advocates and inflictors of pain?
- To what extent are we involving our patients in making decisions about managing pain?
- What happens if we disagree with other team members on the kind of treatment being used?
- Do we have to use placebos to manage pain?
- Is it our business to care for those we perceive as inflicting pain on themselves?

References

1. Gil K. Psychologic aspects of acute pain. Anaesthesial Report, 1990; 2:246.
2. Eland J M and Anderson J.E. The experience of pain in children. In Jacox. A editor pain: a source book for nurses and other health professionals. Boston, 1977, Little Brown & Co.
3. Beyer et al. Patterns of post operative analgesic use with adults and children following cardiac surgery. Pain. 1983; 17: 71–81.
4. Seers K. Perceptions of pain. Nursing Times, 1987; 83: 37–39.
5. Wong D and Baker C. Pain in children, comparison of assessment skills, Paediatric Nursing. 1988; 14:9–17.
6. Kozier B, Erb E and Blais K. Professional Nursing Practice. Concepts and Perspectives. 3rd Ed. California, USA, 1997.
7. Rushton C H. Placebo pain medication: Ethical and legal issues, Paediatric Nursing. 1995; 21: 2: 166–168.

Critical Thinking

1. What unique role does the nurse play in managing patient's pain?

From *West African Journal of Nursing*, May 2008, pp. 50–54. Copyright © 2008 by West African College of Nursing. Reprinted by permission.

Anxiety and Open Heart Surgery

JOY VIARS

Mrs. Miller, age 56, is noticeably upset because results of a cardiac catheterization led to a recommendation for open heart surgery due to coronary artery disease. Her mind races with multiple questions for her cardiologist and surgeon, and anxiety begins to mount. "How serious is this?" "Is it an emergency?" "How could this happen to me?" "What did I do wrong?" "Could I die during the operation?" The nurse must determine how to decrease Mrs. Miller's anxiety after the physicians explain her current heart condition and future surgery.

Anxiety

Anxiety is an unpleasant feeling that affects patients emotionally, psychologically, and physically (Smith, Kemp, Larson, Jaffe, & Segal, 2006). It is characterized by a feeling of foreboding, dread, or threat that could be real or imaginary (Moser et al., 2007). Anxiety in the medical setting is somewhat expected but can become a clinical issue (Cassem, Fricchione, Jellinek, Rosenbaum, & Stern, 2004) that can result in significant lifestyle impairment if left untreated. Anxiety can alter the way a person thinks, cause behavioral problems, and stimulate cognitive changes as well (Vaughn, Wichowski, & Bosworth, 2007). Anxiety after a major cardiac event can slow the patient's recovery and increase morbidity and mortality. During this time, the patient is confronted with awareness of his or her mortality, and has concerns regarding how open heart surgery will impact life, work, and relationship with others. The patient facing open heart surgery may have some normal anxiety-related behavior, but health care providers should monitor the patient's anxiety to preclude development an anxiety disorder (McCann, Fauerbach, & Thombs, 2005). Symptoms of anxiety can be physical or cognitive.

Anxiety has a negative effect on patients undergoing cardiac surgery. Preoperative and postoperative complications, pharmacologic and holistic treatment, and nursing interventions for patients undergoing open heart surgery will be discussed.

Physical Symptoms of Anxiety

Physical symptoms alone can be debilitating. Physical symptoms of anxiety reflect autonomic arousal, including headache, flushing, dry mouth, and nausea, or more intense symptoms such as chest pain, shortness of breathe, and heart palpitations (Cassem et al., 2004).

Chest Pain

Chest pain, a common symptom of anxiety, often is described as localized, fleeting, and sharp. The chest pain can be described as visceral tightness that lasts greater than 30 minutes (Braunwald et al., 2005). Chest pain and anxiety in the patient who is preparing for open heart surgery are a reminder of the severity of the heart problem and the possibility that chest pain could indicate a fatal heart attack (Fitzsimons, Parahoo, Richardson, & Stringer, 2003).

Palpitations

Palpitations, common in the patient dealing with anxiety, can be a frightening symptom for a patient diagnosed with a heart condition requiring open heart surgery. Palpitations are defined as an awareness of the heart's beating, and often are described as pounding, fluttering, skipping beats, or a sensation the heart is stopping. Although palpitations can be caused by a cardiac arrhythmia, they are a common symptom of depression and panic disorders. The patient may wonder if the heart condition is worsening and whether emergency surgery may be needed. Palpitations can occur when the heart just is working hard or if a patient is becoming anxious. They are diagnosed by electrocardiogram (EKG) or a Holter monitor (Braunwald et al., 2005).

Shortness of Breath

In a cyclical manner, anxiety can cause shortness of breath; the patient becomes more anxious, and the shortness of breath worsens. It is characterized by an abnormally uncomfortable awareness of breathing (Kahn & Smith, 2004). It is one of the cardinal symptoms of a pulmonary or cardiac disease but can result from other abnormalities which include anxiety. The health care provider's awareness of this symptom and its cause is critical to helping the patient avoid chronic shortness of breath.

Dizziness

Anxiety can be a predisposing factor to dizziness. A study done on chronic subjective dizziness assumed that balance functioning and emotion share a common neurologic pathway. The patient who complains of chronic imbalance frequently suffers from anxiety (Odman, 2008). Although this symptom has not been deemed serious, the patient who is preparing for open heart surgery may believe it is part of the heart problem and thus experience increased anxiety.

Psychological Symptoms

Fear and worry. Fear and worry are the primary psychological symptoms of anxiety (Smith et al., 2006). A patient facing open heart surgery may be consumed by fear of the unknown and worry regarding the recovery process, the uncertainty of the surgery's timing, and the physical incapacity after the surgery (Fitzsimons et al., 2003). This anxiety tends to cause other changes in the body due to the increase of adrenaline (Vaughn et al., 2007). Powerful catecholamines are released from the body when a person experiences anxiety. They cause increased heart rate, blood pressure, and cardiac output, which could cause myocardial ischemia. In a patient with known heart disease, catecholamines may cause a change in the EKG (Fitzsimons et al., 2003).

Preoperative Anxiety in Open Heart Surgery

A patient preparing for open heart surgery may have many anxieties. Most of them result from the concern about pain, poor or uncertain surgical outcomes, and surgery itself (Gallagher & McKinley, 2007). In one study, patients discussed their anxiety prior to surgery (Fitzsimons et al., 2003). Their descriptions included questions such as, "Will I even live to have the surgery?" and "Will I make it off the table?" Some even compared waiting for open heart surgery to being on death row waiting for execution.

Psychological concerns can manifest as fatigue, shortness of breath, chest pain, anxiety, and depression (Fitzsimons et al., 2003). A patient preparing for surgery goes through a crisis (Ivarsson, Larsson, & Sjoberg, 2004). During this time, the patient's ability to cope with the surgery depends on how much support he or she receives from family, friends, and health care professionals. The patient welcomes the support. As the patient prepares for surgery, the goal is a speedy recovery. For this to happen, the patient needs to be educated fully before surgery to enhance the opportunity for a complication-free postoperative period that is manageable for the patient, family, and health care professionals.

Postoperative Complications from Anxiety

Preoperative anxiety has been associated with slower wound healing, decreased immune response, fluid and electrolyte imbalance, and increased rate of infection. It also has been linked to increased postoperative pain (Scott, 2004). Anxiety can alter a patient's vital signs by releasing epinephrine in the blood stream, causing blood vessel constriction and increased heart rate (Vaughn et al., 2007). Abnormal vital signs can contribute to the development of other postoperative problems. If the deep somatic, cutaneous, and visceral pain of heart surgery is left untreated, the recovery process for the patient can be compromised (Kshettry, Carole, Henly, Sendeibach, & Kummer, 2006).

After surgery, the patient may face continued anxiety because of financial stress, uncertain health, and decreased ability to work (Karlsson, Johansson, & Lidell, 2005). In one study, 37% of patients experienced psychological distress and cognitive impairment after cardiac surgery (Tolmie, Lindsay, & Belcher, 2006). Even though these effects resolved some time after the operation, the patients had periods when psychological distress and cognitive impairment had detrimental effects on their rehabilitation process.

Nursing Interventions

Assess vital signs. Assessing a patient's vital signs can reveal a great deal about his or her condition, and may be the only option to make a quick diagnosis. Anxiety can alter a patient's vital signs by causing a negative impact on the body, such as an increase in blood pressure and heart rate. When the patient preparing for open heart surgery develops

anxiety, epinephrine released in the blood causes vasoconstriction that results in chest pain. Anxiety can increase the patient's heart rate and cardiac contractility, leading to increased blood pressure and temperature, flushing, and sweating (Vaughn et al., 2007).

Preoperative education

Ku, Ku, and Ma (2002) found 76% of patients who underwent open heart surgery indicated their anxieties were relieved as a result of a pre-admission education program regarding the open heart procedure. The researchers noted that extensive preparation prior to cardiac surgery had a positive effect on patients' physical well-being and anxiety. Another study discussed surgical patients' anxiety from loss of control, uncertainty, and unpredictable postoperative stays in the ICU (Berg, Fleischer, Koller, & Neubert, 2006). Anxiety was found to contribute to postoperative conditions, such as delayed wound healing and an increased use of analgesics. Only 16% of patients indicated they were informed accurately and prepared for their postoperative hospitalization.

Relaxation Techniques

Relaxation techniques can be offered to patients without a physician's order. Allowing the patient to express his or her feelings, and providing a quiet, serene setting, and a listening ear are strategies the nurse should use with the anxious patient. These techniques can assist the patient in tolerating anxiety and minimize escalating anxiety that can result from autonomic dyscontrol (Gorrol & Mulley, 2006).

Complementary and Alternative Therapies

Complementary and alternative medicine includes the use of massage, guided imagery, and music. These therapies evoke the relaxation response through stimulation of the parasympathetic nervous system, which is believed to decrease pain and anxiety. A recent study by Kshettry and colleagues (2006) found patients had a decreased heart rate, diastolic blood pressure, pretreatment and post-treatment pain, and tension scores as a result of complementary therapies such as touch and music.

Music can decrease anxiety and pain after surgery. Music intervention is used to change a specific behavior or feeling such as anxiety. Music therapy has positive physical and psychological effects prior to and after surgery. Music also has a positive effect on a patient's pain intensity and anxiety (Sendelbach et al., 2006).

Pharmacologic Treatment of Anxiety

Pharmacologic treatment of anxiety preoperatively and postoperatively may differ. Three classes of drugs are dispensed for anxiety: anxiolytics, antidepressants, and beta blockers. Although medications can help with anxiety, they carry risks of adverse effects, dependence, and withdrawal. For example, alprazolam (Xanax®) is an anxiolytic (benzodiazepine) commonly used for anxiety. A usual starting dose of alprazolam is 0.25–0.5 mg by mouth three times daily, increased as needed to a maximum daily dose of 4 mg. Dependence on the drug is possible, along with serious side effects, such as somnolence, confusion, lightheadedness, and dry mouth.

Another commonly used anxiolytic is buspirone (Buspar®). This drug is not related pharmacologically to alprazolam and does not appear to cause physical dependence. Buspirone is given initially in 5 mg doses three times daily, but dosing can be increased to as much as 20–30 mg daily; administration should not exceed 60 mg per day. This drug also can have side effects, such as lightheadedness, dizziness, and dry mouth. Prescribers must consider potential side effects compared to the potential harm of uncontrolled anxiety. Patients may need medications such as antidepressants that have a more long-term effect until they are able to resume their normal daily activities (DiPiro, 2003).

Loss of Control

Patients' loss of control over daily function can cause anxiety. In a recent study, patients who perceived they had more control had less anxiety; the patients who perceived they had less control developed higher levels of anxiety and experienced more complications (Moser et al., 2007). Patients' involvement in their treatment plans is important at all times but appears to be more significant when patients are dealing with open heart surgery due to the severity of the operation.

Conclusion

An estimated 7 million people underwent cardiovascular surgeries or other procedures in the United States in 2006 (American Heart Association, 2009). As this number continues to grow, nurses need to understand the impact of anxiety preoperatively and postoperatively. This anxiety can cause slower wound healing in patients, increased rate of infection, increased postoperative pain, and abnormal vital signs (Vaughn et al., 2007). As patients prepare for open heart surgery, they depend on health care professionals to explain the procedure and the perioperative expectations. If health care professionals consistently educate these patients, their anxiety will decrease and the postoperative recovery rate will improve.

References

American Heart Association. (2009). *Heart disease and stroke statistics-2009 update*. Dallas: Author.

Berg, A., Fleischer, S., Koller, M., & Neubert, T. (2006). Preoperative information for ICU patients to reduce anxiety during and after the ICU-stay: Protocol of a randomized controlled trial. *BMC Nursing, 5*(4). doi: 10.1186/1472-6955-5-4.

Braunwald, E., Fauci, A., Hauser, S., Jameson, J., Kasper, D., & Longo, D. (2005). *Harrison's principles of internal medicine* (16th ed.). New York: McGraw-Hill.

Cassem, N., Fricchione, G., Jellinek, M., Rosenbaum, J., & Stern, T. (2004). *Massachusetts General Hospital handbook of general hospital psychiatry* (5th ed.). Philadelphia: Mosby.

DiPiro, J., (2003). *AHFS drug handbook* (2nd ed.). Springhouse, PA: Lippincott Williams & Wilkins.

Fitzsimons, D., Parahoo, K., Richardson, S.G., & Stringer, M. (2003). Patient anxiety while on a waiting list for coronary artery bypass surgery: A qualitative and quantitative analysis. *Heart & Lung, 32*(1), 23–31.

Gallagher, R., & McKinley, S. (2007). Stressors and anxiety in patients undergoing coronary artery bypass surgery. *American Journal of Critical Care, 16*(3), 248–265.

Gorrol, A., & Mulley Jr., A. (2006). *Primary care medicine office evaluation and management of adult patients* (5th ed.). Philadelphia: Lippincott Williams & Wilkins.

Ivarsson, B., Larsson, S., & Sjoberg, T. (2004). Patients' experiences of support while waiting for cardiac surgery. A critical incident technique analysis. *European Journal of Cardiovascular Nursing, 3,* 183–191.

Kahn, S., & Smith, E. (2004). *In a page: Signs and symptoms*. Philadelphia : Lippincott Williams & Wilkins.

Karlsson, A., Johansson, M., & Lidell, E. (2005). Fragility-the price of renewed life. Patients experiences of open heart surgery. *European Journal of Cardiovascular Nursing, 4,* 290–297.

Kshettry, V., Carole, L., Henly, S., Sendelbach, S., & Kummer, B. (2006). Complementary alternative medical therapies for heart surgery patients: Feasibility, safety, and impact. *The Society of Thoracic Surgeons, 81,* 201–206.

Ku, S., Ku, C., & Ma, F. (2002). Effects of phase I cardiac rehabilitation on anxiety of patients hospitalized for coronary artery bypass graft in Taiwan. *Heart & Lung, 31*(2), 133–140.

McCann, U., Fauerbach, J., & Thombs, B. (2005). Anxiety and cardiac disease. *Primary Psychiatry, 12*(3), 47–50.

Moser, D., Riegel, B., McKinley, S., Doering, L., An, K., & Sheahan, S. (2007). Impact of anxiety and perceived control on in-hospital complications after acute myocardial infarction. *Psychosomatic Medicine, 69*(1), 10–16.

Odman, M., (2008). Chronic subjective dizziness. *Acta Oto-Laryngologica, 128*(10), 1085–1088.

Scott, A. (2004). Managing anxiety in ICU patients: The role of preoperative information provision. *Nursing in Critical Care, 9*(2), 72–79.

Sendelbach, S., Halm, M., Doran, K., Miller, E., Hogan, E., & Gaillard, P. (2006). Effects of music therapy on physiological and psychological outcomes for patients undergoing cardiac surgery. *Journal of Cardiovascular Nursing, 21*(3), 194–200.

Smith, M., Kemp, G., Larson, H., Jaffe, J., & Segal, J. (2006). Anxiety attacks and disorders: Symptoms, types, and treatment. Retrieved October 1, 2007, from www.helpguide.org/mental/anxiety_types_symptoms_treatment.htm

Tolmie, E., Lindsay, G., & Belcher, P. (2006). Coronary artery bypass graft operation: Patients' experience of health and well-being over time. European Journal of Cardiovascular Nursing, 5, 228–236.

Vaughn, E, Wichowski, H., & Bosworth, G. (2007). Does preoperative anxiety level predict postoperative pain? (Clinical report). AORN Journal Retrieved September 29, 2007, from www.encyclopedia.com/printable.aspx?id = 1G1:161011252

Critical Thinking

1. How can anxiety impact a patient's pre- and post surgery?

JOY VIARS, MSN, NP-C, is a Nurse Practitioner, Bay Heart Group, Tampa, FL.

From *MEDSURG Nursing,* September/October 2009. Copyright © 2009 by Academy of Medical-Surgical Nurses (AMSN). Reprinted by permission of Anthony J. Jannetti, Inc. www.medsurgnursing.net

Respiratory Assessment in Adults

Tina Moore

Aim and Intended Learning Outcomes

The aim of this article is to enable readers to understand respiratory physiology and assessment, focusing on the signs and symptoms of respiratory dysfunction and appropriate interventions.

After reading this article you should be able to:

- Understand respiratory physiology.
- Assess respiratory status appropriately and correctly.
- Identify signs of hypoxaemia and hypercapnia.
- Demonstrate knowledge and understanding of type I and II respiratory failure.
- Initiate appropriate nursing interventions for a patient experiencing respiratory difficulties.

Introduction

Deterioration of respiratory function is one of the major causes of critical illness in the UK (Department of Health 2000). The primary purpose of respiratory assessment is to determine the adequacy of gas exchange, that is, oxygenation of the tissues and excretion of carbon dioxide. By undertaking a full and systematic assessment of the patient's respiratory status, nursing staff are in a prime position to act on findings and ensure that appropriate medical and/or nursing interventions are initiated. A glossary is provided in Box 1.

Respiratory Physiology

The main function of the respiratory system is to provide life-sustaining oxygen to all cells in the body and to remove carbon dioxide, a byproduct of cellular metabolism. The respiratory system consists of the upper airway, including the nasal passages, sinuses, pharynx and larynx, and the lower airway includes the trachea, bronchi, lung, bronchioles and alveoli.

The control of ventilation occurs through voluntary and involuntary mechanisms. Voluntary control of the muscles of respiration is

Time Out 1

Consulting an anatomy and physiology book, draw and label a diagram of the upper and lower respiratory tract. Explain the process of gaseous exchange in the lungs.

regulated through the central nervous system (CNS). The CNS enables individuals to maintain conscious control over their breathing rate. Involuntary ventilation is dependent on the respiratory centre, comprising the medulla oblongata and pons.

The respiratory centre transmits impulses to the respiratory muscles, causing them to contract and relax. Normally, carbon dioxide levels influence the respiratory centre. When $PaCO_2$ levels, that is, partial pressure of carbon dioxide in arterial blood, in the blood rise, the respiratory centre is stimulated to increase the rate and depth of breathing, resulting in increased excretion of carbon dioxide. Low $PaCO_2$ levels in the blood eventually inhibit stimulation of the respiratory centre. This results in an initial increase In the respiratory rate, which then becomes slow and shallow to retain carbon dioxide in an attempt to achieve homeostasis.

Patients with chronic obstructive pulmonary disease (COPD) have experienced long-standing lung damage resulting in an alteration in gas exchange. Here, the central chemoreceptors become tolerant of high levels of carbon dioxide, resulting in reliance on hypoxia to stimulate the respiratory (hypoxic) drive. If patients with COPD are given too much oxygen, the hypoxic drive will be lost causing respiratory failure and possibly respiratory arrest. The transfer of oxygen from the atmosphere to the tissues is a four-stage process.

Diffusion of Oxygen into the Alveoli

Diffusion of oxygen is dependent on a normal airway diameter, adequate respiratory rate and depth and a functioning nervous supply. Airway passages can narrow in the presence of sputum, vomit, trauma, pulmonary oedema and irritants such as smoke. Chemoreceptors located in the circulatory system and brain stem sense the effectiveness of ventilation by monitoring the pH status of the cerebrospinal fluid, PaO_2, that is, partial pressure of oxygen in arterial blood and $PaCO_2$. Chemoreceptors respond to hypercapnia, acidaemia and hypoxaemia by sending impulses to the medulla oblongata to alter the rate of ventilation. There are two main types of chemoreceptor:

- Central chemoreceptors located in the medulla oblongata.
- Peripheral chemoreceptors located in the carotid and aortic bodies, which are more sensitive to decreases in oxygen levels in the blood.

Stretch receptors are located in the bronchial smooth muscle. They are stimulated by lung hyperinflation. Impulses are sent to the respiratory centre to limit further inflation, avoiding over distension of the lung, and to increase expiratory time. When the patient hypoventilates he or she should be encouraged to take deep breaths because a small increase in lung size may stimulate the stretch receptors to cause further inspiration, thus increasing lung expansion.

Transfer of Oxygen across the Alveolar Capillary Membrane

Gas exchange in the lungs occurs across the alveolar capillary membrane, which has a vast surface area, a thin membrane and a constant supply of both air and blood. These are ideal conditions for oxygen diffusion and transfer. Gases move from an area of high pressure in the alveoli, to an area of low pressure in the capillaries, until equilibrium is achieved. Surfactant is secreted by the alveolar cells and maintains its integrity by covering the inner surface of the alveolus and lowering alveolar surface tension at the end of expiration, thus preventing atelectasis and enabling greater transfer of oxygen. The rate at which oxygen diffuses across the alveolar capillary membrane is dependent on conditions in the alveoli, partial pressure of oxygen molecules and the adequacy of pulmonary circulation.

Transport of Oxygen Via Haemoglobin

Oxygen is transported within the circulation in two interrelated ways. Approximately 3% of oxygen is dissolved in plasma and the remaining 97% is transported by binding with haemoglobin. As oxygen diffuses across the alveolar capillary membrane, it dissolves in the plasma where it exerts pressure. As the partial pressure of oxygen increases in the plasma, oxygen moves into the erythrocytes and binds with haemoglobin until saturated. Measurement of haemoglobin concentration is important when assessing individuals with respiratory dysfunction. This is because a decrease in haemoglobin concentration below the normal value of blood reduces oxygen content. Increases in haemoglobin concentration may increase oxygen content, minimising the effect of impaired gas exchange. An adequate plasma level of PaO_2 is essential for the remaining oxygen to bind with haemoglobin to aid tissue perfusion; respiratory dysfunction impairs this process.

Movement of Oxygen from the Haemoglobin to the Tissues

Oxygen enters the tissues by diffusing down the concentration gradient from high concentrations in the alveoli to lower concentrations in the capillaries. This process is influenced by haemoglobin level, oedema, fibroses and destruction of the alveoli (Pierce 2007). Inadequate alveolar ventilation may cause a decrease in the normal pH level. $PaCO_2$

Box 1
Glossary

Acidaemia: state in which the pH of the blood falls below 7.35 (normal = 7.35–7.45)

Atelectasis: collapse of lung tissue with consequent reduction in gas exchange.

Hypercapnia: increased amount of carbon dioxide in arterial blood.

Hypoxaemia: insufficient oxygen content in arterial blood.

Hypoxia: diminished amount of oxygen in the tissues.

Orthopnoea: difficulty in breathing unless in an upright position.

Platypnoea: shortness of breath when sitting upright.

Time Out 2

1. After you have assessed a patient's respiratory status, write down what data you have collected.
2. On your next shift count how many patients have been diagnosed as having an identified respiratory problem or classified as being acutely or critically ill. How many of these patients have had a comprehensive respiratory assessment?
3. Describe the type of data you have observed during respiratory assessment of such patients.

is increased, CO_2 diffuses across the blood-brain barrier until $PaCO_2$ in the blood and cerebrospinal fluid (CSF) reach equilibrium. As the central chemoreceptors sense the resulting decrease in pH they stimulate the respiratory centre to increase the depth and rate of ventilation. Increased ventilation causes $PaCO_2$ of arterial blood to decrease below that of the CSF. As a result, $PaCO_2$ diffuses out of the CSF, returning its pH to normal (Huether and McCance 2006). In patients with COPD, these receptors become insensitive to small changes in $PaCO_2$ and as a result regulate ventilation poorly.

Hypoxaemia

The respiratory assessment may indicate that the patient has hypoxaemia. Hypoxaemia is the reduced oxygenation of arterial blood cells. Air and blood both arrive at the alveoli, the aim is that all the circulatory blood volume should be available for gas exchange. Adequate gas exchange requires ventilation and perfusion of blood flow to be matched. The relationship between ventilation and perfusion in the lungs is measured by calculating the difference between the alveolar and arterial partial pressure of oxygen (Huether and McCance 2006). At rest alveolar ventilation equals 4L/minute and perfusion equals 5L/minute. The ventilation to perfusion ratio is 4:5 = 0.8. In a 'perfect lung' gas exchange will be evenly distributed or perfectly matched. In other words, all alveoli receive an equal share of alveolar ventilation and the pulmonary capillaries receive an equal share of cardiac output. Abnormal ventilation to perfusion ratios are the most common cause of hypoxaemia (Huether and McCance 2006). These can be caused by either inadequate ventilation of well-perfused areas of the alveoli or good ventilation with poor perfusion, as occurs in pulmonary embolism.

Causes of hypoxaemia include:

- Reduced oxygen content of inspired gas, most commonly associated with a drop in atmospheric pressure, for example, high altitudes.
- Hypoventilation of the alveoli resulting in hypercapnia, which can occur in unconscious patients or those with COPD. A reduced amount of oxygen enters the alveoli, for example, when a patient takes a shallow breath.
- Thickened alveolar capillary membrane or decreased surface area for diffusion resulting in impaired diffusion of oxygen.
- Low cardiac output or complete vessel occlusion.
- Histotoxic or cytotoxic: histotoxic relates to substances that cause tissue poisoning and cytotoxins are substances that are toxic and hazardous to the cells.
- Atelectasis resulting in partial or complete collapse of the alveoli.

Nursing staff may use arterial blood gas (ABG) measurement to assess for hypoxaemia. This involves obtaining a sample of arterial blood either through the 'stab' method, usually from the radial or femoral artery, or through an established indwelling arterial catheter. The latter should he used for frequent sampling. The use of arterial catheters is not recommended in general ward settings because of the complications of disconnection and accidental intra-arterial injection, which can be life threatening.

Respiratory Assessment

The purpose of respiratory assessment is to determine the adequacy of gas exchange, that is, oxygenation of the tissues and excretion of carbon dioxide. Wherever possible the same nurse should be involved in the assessment and/or monitoring of the patient's respiratory status for the duration of the shift. This should enable consistency of assessment and the identification of subtle as well as overt changes in respiratory function. Depending on the severity of respiratory impairment, history taking may be limited and observational skills may need to be used (Moore 2004).

Factors that may influence the patient's respiratory function include:

- Pregnancy—fluid retention is caused by increasing oestrogen levels resulting in oedema. Progesterone levels rise six-fold during pregnancy (Lumb 2005) and have a significant effect on the control of respiratory function and ABGs. Enlargement of the uterus in the third trimester may cause the diaphragm to become misplaced, affecting lung expansion.
- Obesity—the poor positioning of obese patients in bed may impede lung expansion.
- Circulatory problems—pulmonary oedema and anaemia may impede gas exchange.
- Environmental influences—such as exposure to the cold, may cause shivering, thus the nurse will not be able to conduct the assessment properly, distorting findings.
- Trauma—particularly of the chest. A patient with chest pain will be unwilling to take deep breaths. If be or she has fractured ribs, the lung may be deflated and cause hypoinflation of the alveoli.
- Known allergies—may cause anaphylaxis, which could cause swelling of the upper airways and subsequent difficulty in breathing.
- Pathophysiological problems—in particular, those which can cause abdominal distension, for example, bowel obstruction and ascites. The lungs are unable to inflate fully as a result of distortion of the diaphragm.

When conducting respiratory assessment, the patient should he positioned upright, if possible. This position not only makes lung expansion easier, but also enables access to the anterior and posterior thorax. Alternative positions may distort findings and should be acknowledged, if unavoidable, when interpreting data. If appropriate, the patient's clothing should be removed because this may act as a barrier to visible and auscultation assessment, again distorting findings. Some patients may be aware that their respiratory function is being assessed and this may lead to a subconscious response that influences their breathing rate. Closed questions should be used to minimise any distress in the acutely breathless patient. Generally, respiratory assessment can be broken down into four areas: inspection, palpation, percussion and auscultation. Nurses do not perform percussion as a mode of respiratory assessment unless additional training has been undertaken. Nurses should identify and determine the meaning of different sounds over different parts of the thorax. This is an advanced and complex skill.

Time Out 3

With reference to Time Out 2, reflect on the following:

1. How many of these patients have their respiratory status recorded regularly on an observation chart?
2. List any other respiratory observations made by nursing staff.

Inspection

Inspection involves a direct, critical, purposeful observation, which includes vision, hearing and smell. The purpose of inspection is to observe for normal patient data and deviations, paying attention to obvious and subtle changes which will require further investigation.

Rate

The ratio of respiration to pulse rate in the healthy adult is 1:4 (Moore 2004). The respiratory rate should be counted for one full minute and categorised into one of the following:

- Eupnoea or 'normal' rate. Opinion as to what the normal rate should be varies, but parameters are between 10–17 breaths per minute.
- Tachypnoea, greater than 18 breaths per minute, is usually the first indication of respiratory distress. Possible causes include anxiety, pain, left ventricular failure and circulatory problems such as anaemia.
- Bradypnoea, less than 10 breaths per minute, may be an indication of increased intracranial pressure, depression of the respiratory centre, narcotic overdose and severe deterioration in the patient's condition.
- Hypopnoea or abnormally shallow respirations may vary with age. Shallow breathing is considered part of the normal ageing process.

Rhythm

The normal respiratory rhythm has regular cycles, with the expiratory phase slightly longer than the inspiratory phase. A short pause is normal between expiration and the next inspiration. Chest movement should be equal, bilateral and symmetrical (Ahern and Philpot 2002). Generally, respiratory rhythm varies between men and women. In men, the respiratory rhythm appears to originate from the abdomen or diaphragm whereas women have a tendency to breathe via their thorax or costal muscle. Patients who are sleeping are also inclined to use their abdominal muscles when breathing. There is an assumption that the use of abdominal muscles relates to an increase in respiratory effort (Moore 2004). It is important that nurses are aware of the different circumstances in which patients appear to use their abdominal muscles because this will prevent incorrect diagnosis. Altered rhythms may indicate underlying disorders, for example, Kussmaul respirations or rapid deep breathing resulting from the stimulation of the respiratory centre in the brain is caused by metabolic acidosis and occurs in diabetic ketoacidosis. Cheyne-Stokes respirations, periods of apnoea alternating with periods of hypoxia, may indicate left ventricular failure or cerebral injury and are sometimes present at the end stage of life.

Quality of Breathing

Normally, there is symmetry in chest movement. Failure of the chest wall to rise adequately may indicate fibrosis, collapse of upper lobes or bronchial obstruction. It may also indicate severe pleural thickening, which may cause flattening of the anterior chest wall and diminished respiratory effort. Sudden, sharp chest pain, for example, caused by pneumothorax, can inhibit the patient from taking deep breaths, resulting in hypoventilation of the alveoli.

Degree of Effort

The use of accessory muscles such as the sternocleidomastoid muscle, which passes obliquely across the side of the neck, the scalenus muscles at the side of the neck and the trapezius muscle spanning from the neck, shoulders and vertebrae, may suggest that the patient has difficulty in breathing. The patient may also have orthopnoea or even platypnoea. Patients who have difficulty in expiration may have abnormalities of lung recoil and/or airway resistance, such as emphysema, pulmonary oedema or asthma. Increased inspiratory effort can indicate upper airway obstruction, for example, anaphylaxis and epiglottitis. Tracheal deviation may indicate pneumothorax. The influence of the severity of breathlessness on restricted activity such as walking or talking should be noted. Other physical symptoms indicating difficulty in breathing may include breathing through 'pursed lips' on expiration as patients try to force air out of the overdistended alveoli. Nasal flaring can indicate respiratory distress in adults, although this is more common in children.

Skin Colour

Cyanosis, a bluish colour of the skin and mucous membranes, may occur when large amounts of unsaturated haemoglobin are present, and may be detectable when oxygen saturation of arterial blood drops below 85% (Moyle 2002). Cyanosis is usually considered a late sign of respiratory dysfunction, however, this is subject to considerable variability. Cyanosis is often difficult to appreciate in artificial lighting, unless quite defined and is best seen on the lips and under the tongue. There are two types of cyanosis. Peripheral cyanosis, usually indicating poor circulation, is observed in the skin and nail beds and is most noticeable around the lips, ear lobes and fingertips. Central cyanosis, usually indicating circulatory or ventilatory problems, is indicated by a bluish colour of the tongue and lips. Cyanosis can easily be overlooked and requires diligent observation. In the absence of central cyanosis, peripheral cyanosis normally indicates circulatory problems rather than respiratory disease (Casey 2001).

Prolonged hypoxaemia can lead to erythrocyrosis and produces a ruddy appearance of the skin. Particular caution needs to be taken when assessing skin colour on patients with dark pigmentation because colour changes, particularly cyanosis, are not easily detectable. It is important to note that anaemic patients may have insufficient haemoglobin to produce the blue colour of the mucous membrane that characterises cyanosis.

Deformities

Clubbing of the finger digits occurs as a result of a chronic condition forming over a long period of time. This may be indicative of hypoxaemia from chronic pulmonary or cardiovascular disease. Deformities of the posterior thorax can affect the quality of breathing. The diameter of the anterior and posterior chest should he compared with the side-to-side diameter. If the anterior and posterior diameter is approximately double the measurement of the side-to-side diameter, this indicates a 'barrel chest' caused by emphysema. Spinal deformities such as kyphosis also influence lung expansion.

Mental Status

A reduction in the patient's level of consciousness and/or altered mental status may indicate hypoxaemia. Symptoms may include inappropriate behaviour, drowsiness and confusion. Any change in mental status should be reported immediately because this may signal that the brain is being deprived of oxygen. If appropriate and immediate action is not taken the patient could deteriorate into unconsciousness, which may result in irreversible brain damage. Assessment of the patient's mental status should be conducted with care because he or she may demonstrate fear and anxiety, but may not he hypoxic. Language barriers and cultural approaches to disorders should also be considered during the assessment process because some patients may not understand certain instructions or questions.

Cough

Assessment of the patient's cough is important because it can indicate if a patient has difficulties in clearing the lungs of sputum or fluid.

The assessment of a patient's cough should include a number of important observations (Box 2). Sputum is a useful indicator of lung pathology (Box 3).

Palpation

Palpation is used to assess bilateral movements of the chest and diaphragm. It is also used to assess surgical emphysema. The palm of the hand, which should be warm, is placed on an area of the patient's chest where vibrations are felt for.

Auscultation

Assessment of breath sounds, with or without a stethoscope, should form part of nursing assessment. Knowledge of the different types of breath sounds aids description and diagnosis. Without a stethoscope, normal breathing should be quiet. Normal breath sounds are categorised as vesicular, bronchovesicular and bronchial:

- Normally, vesicular sounds, which are low pitched, low intensity and often described as 'soft and breezy', can be heard over most of the lung fields.
- Bronchovesicular sounds should he heard in the anterior region, near the main stem bronchi and posterior chest wall only between the scapulae. Bronchovesicular sounds are usually more moderate in pitch and intensity.
- Bronchial sounds are high pitched, loud and hollow. These sounds are usually heard over the larger airways and the trachea. If bronchial sounds are heard in other areas this could indicate consolidation of lung tissue, for example, in pneumonia.

Abnormal breath sounds, known as adventitious sounds, including crackles, as heard in pulmonary oedema, wheezing, usually associated with obstruction of the airways by bronchospasm or swelling, and rubbing or pleural friction, should be listened for. Stridor, a high-pitched sound, usually occurs on inspiration and is caused by laryngeal or tracheal obstruction. This requires immediate attention because it can be potentially life threatening. Crackles are discontinuous, non-musical, brief sounds heard more commonly on inspiration—small airways open during inspiration and collapse during expiration causing the crackling sounds. They can be classified as fine, high pitched, soft and very brief, or coarse, low pitched, louder and less brief. When listening to crackles, special attention should be paid to their loudness, pitch, duration, number, timing in the respiratory cycle, location, pattern from breath-to-breath and change after a cough or shift in position. Fine crackles are high pitched and are heard at the base of the lungs near the end of inspiration and usually represent the opening of the alveoli. Medium crackles are lower in pitch and are heard during the middle or latter part of inspiration. Course crackles heard on both inspiration and expiration are usually associated with mucus, which may clear after the patient has coughed.

Pulse Oximetry

The main function of pulse oximetry is to detect hypoxaemia before obvious symptoms are displayed (Moyle 2002). The pulse oximeter provides continuous, non-invasive monitoring of the oxygen saturation from haemoglobin in arterial blood. A pulse oximeter is a clip-like device that measures the amount of haemoglobin saturation in the tissue capillaries. The device transmits a beam of light through the tissue to a receiver. The wavelengths of the transmitted light are altered by the amount of saturated haemoglobin. Light is translated by the receiver into a percentage of oxygen saturation of the blood. Changes can be detected immediately. It is important to remember that pulse oximetry does not provide comprehensive information on the patient's ventilatory status, but can calculate oxygen saturation status and detect hypoxaemia. Events that may interfere with the reading include:

- Nail polish—particularly dark colours, for example, black, dark blue (Wahr and Tremper 1996) and green.
- Poor peripheral perfusion—possibly resulting from hypotension, may lead to poor readings. It may help to rotate or transfer the probe to different sites frequently because peripheral perfusion may be better in different parts of the body. Probes that are applied too tight will cause vasoconstriction and interfere with readings. A dampened waveform could indicate a reduction in arterial flow or a misaligned sensor. In the case of misalignment, the probe will need to be repositioned. The probe should be checked regularly for tightness and misalignment. If this occurs the tape should he loosened or the position of the probe should be changed.
- Recording blood pressure—the pulse oximeter sensor needs to he placed on a finger of the opposite side of the arm where the blood pressure is being taken because inflation of the cuff will cause the readings to be inaccurate.
- Carbon monoxide poisoning—patients with, or suspected of, carbon monoxide poisoning should not be monitored using pulse oximetry. Carbon monoxide poisoning causes abnormal haemoglobins in thecase of carboxyhaemoglobin, which can occur in patients with carbon monoxide poisoning resulting from smoke inhalation. The pulse oximetry sensor cannot differentiate between oxyhaemoglobin and carboxyhaemoglobin (Moyle 2002), and will therefore provide a falsely evaluated oxygen saturation reading. It is considered dangerous practice to rely on pulse oximeter readings in this situation. Instead, ABG analysis should be undertaken (Moore 2004).
- Movement—sudden movements and restlessness may cause the pulse oximetry sensor to partially dislodge, or cause motion artefact (distortion of the wave form caused by movement). This affects the ability of light to travel from the light emitting diode to the photo detector in the pulse oximeter probe. It may also be difficult to determine the pulse in patients who have rhythmic movement, for example, seizures and shivering. 'The importance of keeping still should be explained to the patient. If the patient is unable to limit his or her movement, nursing staff should consider moving the probe to the ear lobe because movement here least affects the equipment. However, it is important to consult the pulse oximetry manufacturer's guidance for alternative sites. To minimise potential problems, it may be useful for nursing staff to test the equipment on themselves before placing it on the patient. The pulse reading should be correlated with the patient's heart rate. Variation between pulse and heart rate may indicate that not all pulsations are being detected. In this case a replacement monitor may be required.

Respiratory Failure

Respiratory failure is a syndrome in which the respiratory system fails in one or both of its gas exchange functions: oxygenation and/or carbon dioxide elimination (Sharma 2006). The condition can be acute or chronic. Chronic respiratory failure develops over several days or longer, allowing time for metabolic compensation and an increase in bicarbonate concentration. Therefore, the pH of arterial blood usually only decreases slightly. Acute respiratory failure is characterised by life-threatening derangements in ABGs and acid-base status. The manifestations of chronic respiratory failure are less dramatic and may not be as readily apparent. Bloodgas disturbances occur as a result of ventilation to perfusion inequality, inadequate alveolar ventilation or a combination of both. Unventilated alveoli result in vasoconstriction, which then diverts the blood flow to ventilated alveoli, resulting in atelectasis. There are two types of respiratory failure depending on the cause:

- Type I respiratory failure: oxygenation.
- Type II respiratory failure: ventilation.

Type I Respiratory Failure: Oxygenation

Oxygenation respiratory failure occurs in the presence of hypoxia without hypercapnia (Box 4), It is typically caused by a reduction in inspired oxygen, a ventilation to perfusion mismatch and alveolar hypoventilation (Priestley and Huh 2006). Most pulmonary and cardiac conditions can result in respiratory failure with inadequate oxygenation, pulmonary oedema and COPD being the more common causes. As a result of alveolar hypoventilation, the $PaCO_2$ rises resulting in a fall in $PaCO_2$. Respiratory failure that develops slowly allows renal compensation with retention of bicarbonate, often resulting in near normal pH levels. A change in the pH of the blood, together with an increase in carbon dioxide, affects the saturation of haemoglobin (Moyle 2002). This is the most common form of respiratory failure, and can be associated with virtually all acute diseases of the lung, which generally involve fluid filling or collapse of the alveoli. Examples of type I respiratory failure include cardiogenic, noncardiogenic, pulmonary oedema, pneumonia and pulmonary haemorrhage (Sharma 2006).

The distinction between acute and chronic hypoxic respiratory failure cannot readily be made on the basis of ABGs.

Clinical Features

Type I respiratory failure may have a variety of clinical manifestations (Box 5). However, these are non-specific and respiratory failure may be present in the absence of dramatic signs or symptoms. This emphasises the importance of ABG measurements in all patients who are acutely or critically ill or in those where respiratory failure is suspected.

Pulmonary arteries respond to hypoxia by vasoconstriction, producing vascular resistance and pulmonary hypertension. Right ventricular enlargement or right-sided heart failure develops later. Nursing care should be directed at preventing the patient from developing late clinical features, through early identification of increased respiratory rate, reduced oxygen saturation and neurological changes.

Type II Respiratory Failure: Ventilation

Type II respiratory failure can he caused by increased airway resistance and reduced lung compliance, as indicated in severe asthma and pulmonary oedema (Sharma 2006). Both oxygen and carbon dioxide blood levels are affected (Box 4). As the alveoli are microscopic and prone to collapse, the secretion of surfactant via the alveolar cells facilitates its expansion during inspiration. However, surfactant production is inhibited by hypoxia, acidosis, poor perfusion, smoking and dry gas, for example, unhumidified oxygen.

Clinical Features

In addition to the signs of hypoxaemia the patient may show clinical signs of hypercapnia, including: irritability; aggression; confusion; coma; headaches and papilloedema (oedema of the optic disc); and warm flushed skin and a bounding pulse (the effect of carbon dioxide on the peripheral vascular smooth muscles may also produce vasoconstriction by sympathetic stimulation).

Management

Identifying the type of respiratory failure is important as it determines the intervention. Underlying causes of respiratory failure, such as chest infections or trauma, should always be treated. The aim of managing respiratory failure is to enable adequate oxygen delivery to the tissues with an adequate PaO_2. This can be achieved through supplementary oxygen via nasal cannula or a face mask. In the case of severe hypoxaemia, intubation and mechanical ventilation may be warranted. Generally, type I respiratory failure may require supplementary oxygen. However, some local policies advocate non-invasive ventilation therapy. Type II respiratory failure requires additional intervention, for example, bi-level noninvasive ventilation, continuous positive airway pressure or full ventilation. Treating hypoxaemia will not improve the $PaCO_2$ and may make it worse (Lumb 2005). It is therefore essential to ensure that palliative relief of hypoxia does not result in hypercapnia, and arterial $PaCO_2$ should be monitored closely. Hypercapnia unaccompanied by hypoxemia is well tolerated and is not likely to threaten organ function unless accompanied by severe acidosis (Sharma 2006). Many experts believe that hypercapnia should he tolerated until the arterial blood pH falls below 7.2 (Sharma 2006). Appropriate management of the underlying disease is an important component in the management of patients with respiratory failure.

Oxygen Therapy

The need for oxygen therapy should be assessed in patients with cyanosis, oxygen saturations less than or equal to 92% without additional oxygen support and all patients with severe air flow obstruction (National Institute for Clinical Excellence 2004). With the exception of resuscitation, oxygen should always he prescribed by a doctor, with clear guidance regarding the flow rate, delivery system, duration and monitoring of treatment. When oxygen is being administered, the patient should be positioned upright if possible, to maximise lung expansion. If using nasal cannula, the flow rate of oxygen must not exceed four litres per minute, to prevent discomfort and damage to the nasal mucosa. A full respiratory assessment should he undertaken and the patient should be closely monitored throughout treatment.

Masks

Fixed performance masks provide a steady concentration of inspired oxygen. Such masks should always be used in patients who have COPD unless the patient's $PaCO_2$ is known to he normal. The flow of oxygen delivered by variable performance masks varies with changes in the breathing pattern. Masks can be used when there is no danger of carbon dioxide retention. If the patient is severely hypoxic a non-rebreathing mask with a reservoir bag attached can be used. A reservoir bag fills up with oxygen during the patient's expiratory phase and this oxygen is breathed in during inspiration. The use of a reservoir bag enables the delivery of high concentrations of oxygen to the patient.

Nasal Cannulae

Patients who are expectorating copious amounts of sputum, as in the case of gross pulmonary oedema, may be required to receive oxygen via nasal cannula. Nasal cannulae are simple, unobtrusive and allow eating, talking and washing to continue relatively unimpeded. As an approximate guide 21 minutes produces an inspired oxygen concentration of 25–30% (Sharma 2006). Nasal cannulae can cause drying and nasal crusting, which in turn can result in obstruction, therefore, maintaining nasal hygiene is important.

Conclusion

A comprehensive assessment of respiratory status should be performed on all patients who have an identified respiratory disorder and those who are classified as acutely or critically ill. Respiratory assessment should be performed by a competent nurse and used to identify potential respiratory problems. Early intervention is essential to improve the prognosis of patients **NS**

Box 4
Respiratory Failure

Type I respiratory failure is defined as:

- PaO_2 < 8KPa
- $PaCO_2$ <6KPa

Type II respiratory failure is defined as:

- PaO_2 >8KPa
- $PaCO_2$ >6KPa

PaO_2 = partial pressure of oxygen in arterial blood.
$PaCO_2$ = partial pressure of carbon dioxide in arterial blood.
KPa = kilopascals (a type of unit used to measure pressure).
(British Thoracic Society Standards of Care Committee 2002)

Box 5
Clinical Features of Type I Respiratory Failure

Early clinical signs include:

- Irritability, altered level of consciousness, confusion.
- Restlessness, anxiety, fatigue.
- Cool and dry skin.
- Increased cardiac output, tachycardia and headache as a resuit of stimulations of ventilation via the carotid chemoreceptors.

Intermediate clinical signs include:

- Confusion.
- Aggression.
- Lethargy.
- Tachypnoea.
- Dyspnoea can cause an uncomfortable sensation of breathing.
- Hypotension.
- Tachycardia, bradycardia and a variety of arrhythmias may result from hypoxaemia and acidosis.

Late clinical signs include:

- Cyanosis.
- Oxygen saturations of less than 75%.
- Diaphoresis or sweating.
- Coma and convulsions.
- Cardiac arrhythmias.
- Respiratory arrest.

Suggested answer to Time Out 5

1. Type 1 respiratory failure.
2. The most important intervention is to commence oxygen therapy and to improve gas exchange. This can be achieved by:

Box 6
Scenario

John, 55 years old, has been transferred to the respiratory care unit from the ward after developing respiratory problems. He underwent extensive abdominal surgery three days ago. He looks unwell and is experiencing severe abdominal pain. He is a known smoker. Ward documentation suggests that John's abdominal pain was never under control and he always lay in a semi-recumbent position.

Assessment Data

John is breathing spontaneously but is dyspnoeic with a respiratory rate of 30 breaths per minute. The pattern is regular but shallow and be is using his accessory muscles. He has an unproductive cough. He looks pale, but no central cyanosis is present Auscultation of the lungs indicates reduced air entry at both bases with some coarse crackles in the right mid zone and widespread mild expiratory wheeze, which is also heard without a stethoscope.

John's heart rate is variable at approximately 120 beats per minute. He looks pale and clammy. His blood pressure is 160/11OmmHg, and oxygen saturation levels are 87% on 60 % oxygen. Blood gas analysis shows that John's PaO_2 is 6.4 and $PaCO_2$ is 4.9. John responds to verbal instructions but appears drowsy. He feels cold and slightly clammy.

- Close monitoring of John's respiratory status, including ABG analysis and pulse oximetry. Hypoxic patients can deteriorate rapidly and require more advanced respiratory intervention.
- Pain control—if the patient's abdominal pain is not controlled he will be unable to expand his lungs fully.
- Position upright (blood pressure is nor compromised)— this position will also facilitate lung expansion and enable ventilation of the alveoli within the bases of the lungs.

Time Out 5

Read the scenario in Box 6, and answer the following questions:

1. Identify the type of respiratory failure.
2. What type of nursing and/or medical intervention will John require?

A suggested answer is provided below.

Time Out 6

Now that you have completed the article, you might like to write a practice profile.

- Administration of prescribed oxygen—this may be more effective via a non-rebreathing mask and reservoir bag.
- Monitoring temperature—for a possible underlying chest infection.
- Psychological care—explain all procedures to the patient.

References

Ahern J, Philpot P (2002) Assessing acutely ill patients on general wards *Nursing Standard,* 16. 47, 47–54.

British Thoracic Society Standards of Care Committee (2002) Non-invasive ventilation in acute respiratory failure. *Thorax.* 57, 3, 192–211.

Casey G (2001) Oxygen transport and the use of pulse oximetry. *Nursing Standard.* 15, 47, 46–53.

Department of Health (2000) *Comprehensive Critical Care: a Review of Adult Critical Care Services.* The Stationery Office, London.

Huether SE, McCance KL (2006) *Understanding Pathophysiology.* Third edition, Mosby, St Louis MO.

Lumb AB (2005) *Nunn's Applied Respiratory Physiology.* Sixth edition. Butterworth Heinemann, Oxford.

Moore T (2004) Respiratory assessment. In Moore T, Woodrow P (Eds) *High Dependency Nursing Care: Observation, Intervention and Support.* Routledge, London, 124–134.

Moyle J (2002) *Pulse Oximetry.* British Medical Journal Books, London.

National Institute for Clinical Excellence (2004) *Chronic Obstructive Pulmonary Disease. Clinical Guideline 12. NICE,* London.

Pierce L (2007) *Management of the Mechanically Ventilated Patient,* Second edition. Saunders/Elsevier, St Louis MO.

Priestley MA, Huh J (2006) *Respiratory Failure.* www.emedicine. com/ped/topic1994.htm (Last accessed: July 24 2007).

Sharma S (2006) *Respiratory Failure.* www.emedicine.com/med/ topic2011.htm (Last accessed: July 24 2007).

Wahr JA, Tremper KK (1996) Oxygen measurement and monitoring techniques. In Prys Roberts C, Brown BR (Eds) *International Practice of Anaesthesia.* Butterworth-Heinemann, Oxford.

Critical Thinking

1. What are the differences between type I and type II respiratory failure?

TINA MOORE is senior lecturer. School of Health and Social Sciences, Middlesex University, Middlesex. Email: t.moore@mdx.ac.uk

From *Nursing Standard,* vol. 21, no. 49, August 15, 2007, pp. 48–56. Copyright © 2007 by RCN Publishing Company Ltd. Reprinted by permission.

Medical Nutrition Therapy: A Key to Diabetes Management and Prevention

SARA F. MORRIS, MAT, MPH, RD AND JUDITH WYLIE -ROSETT, EdD, RD

The link between diabetes and diet has been well documented, as has the importance of diet in conjunction with medical interventions for diabetes. Patients often look to their primary care physicians for advice about general diabetes care, including diet, but survey studies have revealed that doctors feel uncomfortable advising patients on the sensitive issues of weight loss and diet.[1] Research is increasingly demonstrating that medical nutrition therapy (MNT), administered by a registered dietitian (RD) or nutrition professional, is a key component of diabetes management and a complement to treatment of diabetes by physicians.

The American Diabetes Association (ADA) 2009 clinical practice recommendations state that "individuals who have pre-diabetes or diabetes should receive individualized MNT as needed to achieve treatment goals, preferably provided by a registered dietitian familiar with the components of diabetes MNT."[2] Furthermore, the ADA's position statement titled "Nutrition Recommendations and Interventions for Diabetes" emphasizes the importance of MNT in preventing diabetes, managing existing diabetes, and preventing and slowing the onset of diabetes-related complications.[3] The integration of MNT into diabetes care has the potential to improve patients' diabetes management and to lessen the burden on physicians to provide nutrition information.

Defining MNT

MNT is defined as "nutritional diagnostic, therapy, and counseling services for the purpose of disease management, which are furnished by a registered dietitian or nutrition professional."[4] The American Dietetic Association, the professional organization of RDs, defines the nutrition counseling component of MNT as "a supportive process to set priorities, establish goals, and create individualized action plans which acknowledge and foster responsibility for self-care."[5]

In general, MNT consists of multiple, one-on-one sessions between an RD and a patient, in which the RD performs the nutrition assessment, diagnosis, counseling, and other therapy services according to the American Dietetic Association's "MNT Evidence-Based Guide for Practice/Nutrition Protocol" or according to the best available current evidence in the nutrition community.[5] After an RD receives a referral from a physician, the framework of counseling is standardized and documented as part of the American Dietetic Association's Nutrition Care Process (nutrition assessment, nutrition diagnosis, nutrition intervention, nutrition monitoring, and evaluation).[6]

The components of MNT provided by an RD, according to the Nutrition Care Process for any nutrition-related disease (not just diabetes), are outlined in Table 1.[7] Initially, the dietitian performs a nutrition assessment of the patient's dietary patterns. This assessment, in conjunction with information about laboratory testing, medications, and any other conditions that the RD would have received with the physician referral, allows the RD to evaluate such factors as the patient's macronutrient needs versus intake, the consistency of meals, the amount and consistency of carbohydrate intake, and the general quality of the patient's diet. The RD can then issue a coded, standardized nutrition diagnosis and tailor future individual counseling to help the patient meet goals related to the diagnosis and nutrition needs.

As an example, after reviewing a patient's physician referral information and dietary patterns, an RD might issue the following diagnostic statement: "Inconsistent carbohydrate intake related to inadequate meal planning as evidenced by meals and snacks containing a range from 0 to 150 g of carbohydrates on a daily basis." The nutrition intervention for this diagnostic statement might consist of the RD teaching the patient to plan meals and carbohydrate intake using the diabetes plate model or the carbohydrate exchange methods, helping the patient set goals related to consistent intake, and then using motivational interviewing or problem solving to address any barriers the patient has to meeting nutritional goals.[8,9]

As part of nutrition monitoring and evaluation, the RD monitors biochemical factors such as A1C and serum lipid levels, as well as lifestyle factors such as dietary intake. Depending on how many sessions the RD has with the patient, these factors are used to evaluate the effectiveness of interventions in meeting goals. Diagnoses and interventions might then be revised based on nutrition-related outcomes. Therefore, MNT for nutrition-related disorders such as

---

In Brief

Because of the direct correlation between diet and diabetes management, medical nutrition therapy (MNT) provided by a registered dietitian is a key complement to traditional medical interventions in diabetes treatment. This article describes MNT, summarizes evidence for the effectiveness of MNT in preventing and treating diabetes, and provides physicians with information about how to refer patients for MNT.

Table 1 MNT Provided by RDs

Application of Nutrition Care Process	MNT Provided by RD (for individual)
Nutrition screen/referral	The physician provider sends RD written referral for MNT for diabetes. The referral includes information regarding current laboratory test results, medications, and other medical diagnoses.
Nutrition assessment	The RD performs a comprehensive nutrition assessment utilizing the *Diabetes Type 1 and 2 Evidenced-Based Nutrition Practice Guideline for Adults* and Toolkit, as well as the best available current knowledge and evidence, client data, medical record data, and other resources.
Nutrition diagnosis	After analyzing assessment data, the RD makes initial nutrition diagnosis(es); for example, inconsistent carbohydrate intake (diagnosis code NI-5.8.4), inconsistent timing of carbohydrate intake throughout the day, day to day, or a pattern of carbohydrate intake that is not consistent with recommended pattern based on physiological or medication needs.
Nutrition intervention	The RD provides counseling and, with the client, determines interventions using the cognitive behavioral model, including problem solving, motivational interviewing, goal setting, and self-monitoring.
Nutrition monitoring and evaluation	The RD monitors A1C, microalbuminuria, BMI, serum lipid levels, goals for food plan/intake, activity, and other behavior changes.
	The RD implements changes to MNT (e.g., patient education goals, nutrition intervention, and counseling) in future visits based on outcomes and assessments at each visit.
Nutrition documentation (supports all steps of the Nutrition Care Process)	The RD documents MNT initial assessment, nutrition diagnosis(es), and intervention(s); shares with referring physician; and keeps a copy on file.
Outcome management systems	Based on RD analysis, critical thinking, and review of data from the patient's medical history and other health care professionals, the RD aggregates individual and population outcomes data; analyzes and shares with quality improvement department/group as indicated; and implements improvements to MNT services based on results.

Reprinted with permission from Ref.7.

diabetes is not necessarily a linear process. Counseling in MNT is individualized and tailored to a patient's clinical and lifestyle needs.

MNT is not synonymous with diabetes self-management training (DSMT). DSMT is an education and training program that helps patients manage their diabetes, whereas MNT consists of more individualized diagnosis, therapy, and counseling related to nutrition.[7]

According to the American Dietetic Association's 2009 White Paper on Nutrition Services that distinguishes between the delivery of DSMT and MNT services, the DSMT curriculum that is accredited by Medicare consists of 1 hour of individualized assessment and 9 hours of group classes.[7] One of the content areas outlined in the DSMT curriculum is "incorporating nutritional management into lifestyle," but others are less nutrition related, such as "using medications safely and for maximum therapeutic effectiveness."[7]

MNT for patients with diabetes is administered in several sessions with an RD, who provides "more intensive nutrition counseling and a therapy regimen that relies heavily on follow-up and feedback to assist patients with changing their behavior(s) over time."[7] Furthermore, although MNT for diabetes patients can be administered only by a licensed/certified RD or nutrition professional, DSMT can be administered by a registered nurse, dietitian, or pharmacist.[7]

Sometimes it may be most helpful for patients to receive DSMT first, followed by MNT. Other patients may receive DSMT and MNT simultaneously. Professional organizations such as the ADA, the American Dietetic Association, and the Centers for Medicare and Medicaid Services recognize both the distinctions between DSMT and MNT and the possibility that the two programs can complement each other.

Effectiveness of MNT in Diabetes Treatment and Prevention: A Summary of Current Literature

Research has shown MNT to be effective for the treatment of both type 1 and type 2 diabetes. The American Dietetic Association reviewed 18 studies that involved the provision of MNT by an RD as part of treatment for either type 1 or type 2 diabetes. Of these, all 8 of the randomized controlled trials reviewed demonstrated a positive effect of MNT in diabetes management, measured by improvement in A1C levels.[10] Based on this review, the American Dietetic Association reports strong support for the effectiveness of MNT provided by RDs in the management of type 1 and type 2 diabetes and suggests that MNT has the greatest potential for impact when diabetes is first diagnosed.[10]

Several studies in which MNT was included in diabetes treatment interventions that successfully affected metabolic and behavioral outcomes are summarized in Table 2, adapted from Pastors et al.[11–26] The studies varied in design and length (from 3 months to 9 years), and some were conducted before MNT for diabetes was as well defined as it is today. However, all studies listed implemented at least one study group that received individual counseling from an RD, with number of sessions ranging from two total sessions to monthly sessions for 6 months or more. These studies reported improvements in A1C ranging from 0.9 to 1.9% for groups receiving MNT and generally reported a greater improvement in A1C when compared to study groups that received less intensive or basic nutrition education.[11] In general,

Table 2 Summary of Evidence for Nutrition Therapy in Diabetes[11–26]

References	Study Length	Number of Subjects	Outcome
Randomized, Controlled Trials			
MNT only			
UKPDS Study Group, 1990[12]	3 months	3,042 patients with newly diagnosed type 2 diabetes	In 2,595 patients who received intensive nutrition therapy (447 were primary diet failures), A1C decreased 1.9% (from 8.9 to 7%) during the 3 months before study randomization
Franz et al., 1995[13]	6 months	179 people with type 2 diabetes; 62 in comparison group; duration of diabetes: 4 years	A1C at 6 months decreased 0.9% (from 8.3 to 7.4%) with nutrition practice guidelines care; A1C decreased 0.7% (from 8.3 to 7.6%) with basic nutrition care; A1C was unchanged in the comparison group with no nutrition intervention (from 8.2 to 8.4%)
Kulkarni et al., 1998[14]	6 months	54 patients with newly diagnosed type 1 diabetes	A1C at 3 months decreased 1.0% (from 9.2 to 8.2%) with nutrition practice guideline care and 0.3% (from 9.5 to 9.2%) in usual nutrition care group
MNT in combination with DSMT			
Glasgow et al., 1992[15]	6 months	162 type 2 diabetic patients > 60 years of age	A1C decreased from 7.4 to 6.4% in the control-intervention crossover group, whereas the intervention-control crossover group had a rebound effect; the intervention group had a multidisciplinary team with an RD who provided MNT
Sadur et al., 1999[16]	6 months	185 adult patients with diabetes	97 patients received multidisciplinary care, and 88 patients received usual care by primary care MD; A1C decreased 1.3% in the multidisciplinary care group compared with 0.2% in the usual care group; intervention group had a multidisciplinary team with an RD who provided MNT
Observational Studies			
Cross-sectional survey			
Delahanty and Halford, 1993[17]	9 years	623 patients with type 1 diabetes	Patients who reported following their meal plan > 90% of the time had an average A1C level 0.9% lower than subjects who followed their meal plan < 45% of the time
Expert opinion			
DCCT Research Group, 1993[18]			The Diabetes Control and Complications Trial (DCCT) research group recognized the importance of the role of the RD in educating patients on nutrition and adherence to achieve A1C goals; the RD was a key member of the team
Franz et al., 1994[19]			The DCCT made apparent that RDs and RNs were extremely important members of the team in co-managing and educating patients
Chart audit			
Johnson and Valera, 1995[20]	6 months	19 patients with type 2 diabetes	At 6 months, blood glucose levels decreased 50% in 76% of patients receiving nutrition therapy by an RD. Mean total weight reduction was ~ 5 lb
Johnson and Thomas, 2001[21]	1 year	162 adult patients	MNT intervention decreased A1C levels 20%, bringing mean levels to < 8% compared with subjects without MNT intervention, who had a 2% decrease in A1C levels

(continued)

Table 2   Summary of Evidence for Nutrition Therapy in Diabetes[11–26] *(continued)*

References	Study Length	Number of Subjects	Outcome
Retrospective chart review			
Christensen et al., 2000[22]	3 months	102 patients (15 type 1 and 85 type 2 diabetic patients with duration of diabetes > 6 months)	A1C levels decreased 1.6% (from 9.3 to 7.7%) after referral to an RD
Meta-Analyses of Trials			
Brown et al., 1996, 1990[23, 24]		89 studies	Educational intervention and weight loss outcomes; MNT had a statistically significant positive impact on weight loss and metabolic control
Padgett et al., 1988[25]		7,451 patients	Educational and psychosocial interventions in management of diabetes (including MNT, self-monitoring of blood glucose, exercise, and relaxation); nutrition education showed strongest effect
Norris et al., 2001[26]		72 studies	Positive effects of self-management training on knowledge, frequency, and accuracy of self-monitoring of blood glucose, self-reported dietary habits, and glycemic control were demonstrated in studies with short follow-up (< 6 months)

Adapted from Ref.11.

Pastors et al. summarize that "randomized controlled nutrition therapy outcome studies have documented decreases in [A1C] of ~ 1% in newly diagnosed type 1 diabetes, 2% in newly diagnosed type 2 diabetes, and 1% in type 2 diabetes with an average duration of 4 years."[11]

MNT has also been shown to be effective in the prevention of type 2 diabetes. Although the onset of type 1 diabetes is not considered to be preventable, the onset of type 2 diabetes can be delayed or prevented with lifestyle modifications, including changes to diet. Based on its review of seven studies that implemented nutrition therapy and 16 studies that implemented intensive lifestyle interventions including nutrition therapy, the American Dietetic Association concluded that MNT is "effective at reducing the incidence of type 2 diabetes."[10]

The same review concluded that pharmacotherapy is also effective at reducing incidence of type 2 diabetes, but stated that in all but one of seven studies reviewed, lifestyle interventions were more effective than pharmacotherapy at reducing incidence.[10] Regarding pharmacotherapy, it should be noted that although the ADA clinical practice recommendations do endorse the use of medications to treat certain complications of diabetes, they also endorse MNT and lifestyle changes as alternatives and complements to pharmacotherapy.[2]

A notable prevention study is the Diabetes Prevention Program, reported in 2002.[27] In this randomized, controlled trial of 1,079 participants aged 25–84 years, a 58% reduction in incidence of diabetes over 3 years was reported in subjects treated with an intensive lifestyle intervention that included MNT.

Based on the results of this prevention intervention, the Look AHEAD (Action for Health in Diabetes) study is currently underway to test the effects of similar lifestyle interventions in the treatment of 5,145 men and women who have already been diagnosed with type 2 diabetes.[28] Although this trial is planned to extend for 11.5 years, the researchers' recently published 1-year results reported that the intensive lifestyle intervention group lost 8.6% of initial weight compared to 0.7% in the control group.[28] The intensive lifestyle intervention included comprehensive diet counseling from an RD, whereas the control group received basic, or "usual," care for diabetes. The experimental group also had significantly greater decreases in A1C, systolic and diastolic blood pressure, and triglyceride levels in the first year, compared to the control group.[28]

The body of evidence pointing to the effectiveness of MNT in treating, and now preventing, adult type 2 diabetes continues to grow. However, research on the effectiveness of MNT for diabetes in children and adolescents is still limited. The Diabetes Control and Complications Trial, reported in 1993, did include subjects aged 13–39 years.[29] This randomized, controlled trial showed 34–76% reductions in complications of type 1 diabetes in the study group, which received both intensive insulin therapy and monthly clinic visits with an RD.

It is only within the past decade that the prevalence of type 2 diabetes among children and adolescents has been recognized, but studies currently underway are demonstrating that the incidence and prevalence of type 2 diabetes are increasing.[30,31] Therefore, there is a need for research to identify the most effective interventions to address the prevention and treatment of type 2 diabetes in children and adolescents. Even so, based on the existing evidence pertaining to adults with type 2 diabetes, the contribution of MNT to the prevention and treatment of this disease in children is promising.

MNT and Third-Party Coverage

Although the ADA and the American Dietetic Association are in agreement that patients with pre-diabetes and diagnosed diabetes should be referred for MNT, many primary care physicians are unsure of how to go about making these referrals.

Government Medicare insurance benefits cover Americans > 65 years of age, some disabled people > 65 years of age, and people of any age who have end-stage renal disease.[32] Since 2000, Medicare benefits have covered MNT for people with type 1 diabetes, type 2 diabetes, gestational diabetes, non-dialysis kidney disease, and post-kidney transplants who are otherwise eligible for Medicare insurance.[4]

Medicare generally serves as the standard for other third-party payers, so many private and state insurance plans also cover MNT for diabetes management. In fact, as of 2009, 46 states and the District of Columbia have legislation in place that requires insurance coverage for diabetes management, although the breadth of coverage offered varies greatly from state to state.[33]

However, although Medicare and other third-party payers cover diabetes screening for patients who have been diagnosed with pre-diabetes, Medicare does not currently cover MNT for patients with pre-diabetes.[34] The Diabetes Prevention Program study has done much to underscore the effectiveness and cost-effectiveness of lifestyle interventions, including MNT, in preventing or delaying the onset of type 2 diabetes.[27,35] In this time of insurance reform, it is possible that coverage for MNT for pre-diabetes, considered to be a preventive service, will become more widespread.

It is important to note that Medicare coverage of MNT for people with diabetes requires a physician referral. According to Medicare MNT legislation, patients with a physician referral can receive 3 hours of individual counseling with an RD during the first year of treatment and 2 hours of counseling each year after that.[4,5,7] Medicare also covers DSMT in addition to MNT without decreasing the benefit of either.[7]

Patients who are not covered under Medicare insurance should still be referred for MNT because many private insurers will cover MNT for diabetes and other conditions.[36] To ensure that patients will be eligible to receive MNT through third-party coverage, the American Dietetic Association suggests the following guidelines for physicians:[36]

- Take care to include the diagnosis and diagnosis code for diabetes on the referral form.
- Submit pertinent recent lab results and medication lists with referrals.
- Document patients' need for MNT in their medical charts.
- Encourage patients to make an appointment with an RD at a local hospital, clinic, or private practice office, and assist them in locating an RD if they are unsure how to do so.
- Provide another referral and make appropriate medical record documentation if and when patients need additional hours of MNT.

To assist patients in locating an RD, physicians can access the "Find a Nutrition Professional" section of the American Dietetic Association's website, www.eatright.org.

Some nutrition professionals hold the certified diabetes educator (CDE) credential in addition to the RD credential. CDEs are defined as "health care professionals who have defined roles as diabetes educators, not . . . those who may perform some diabetes-related functions as part of or in the course of other usual and customary occupational duties."[37] CDEs may be nurses, doctors, dietitians, or other health care providers. Whatever their discipline, they must have 2 full years of professional practice experience, perform 1,000 hours of diabetes self-management education (DSME), and take an accreditation examination to earn the CDE credential.

DSME is synonymous with DSMT. Therefore, the scope of diabetes self-management education provided by CDEs includes not only a nutrition education component, but also education and training in the effects of diabetes-related medicines, self-monitoring of blood glucose, and administering insulin.[6,37]

The difference between DSMT/DSME and MNT has been explained; the treatments are complementary but not identical in scope. Therefore, although a health care professional who is a CDE may be able to advise patients regarding some aspects of diabetes nutrition, only an RD can provide intensive MNT.

It is arguably ideal to refer patients who need diabetes MNT to RDs who are also CDEs, because these individuals would be experts in the nutrition aspect of diabetes and trained to educate patients in some non-nutrition aspects of diabetes management. However, the professional to whom patients are referred for MNT must, at the minimum, hold the RD credential.

Conclusion

MNT is an effective and increasingly affordable method to prevent type 2 diabetes and to treat both type 1 and type 2 diabetes. It is endorsed for the treatment of diabetes by the Institute of Medicine, the American Dietetic Association, and the ADA and is covered by Medicare. The provision of MNT by RDs, who are experts in offering individualized nutrition counseling, will improve the quality of counseling offered to patients and alleviate the burden on physicians to provide nutrition education. Primary care physicians should refer patients with symptoms of pre-diabetes and diabetes for MNT services, to be provided by an RD, to ensure the best care for their patients.

References

1. Gans KM, Ross E, Barner CW, Wylie-Rosett J, McMurray J, Eaton C: REAP and WAVE: new tools to assess/discuss nutrition with patients. *J Nutr* 133:556S–562S, 2003.
2. American Diabetes Association: Executive summary: standards of medical care in diabetes—2009. *Diabetes Care* 32:S6–S12, 2009.
3. Bantle JP, Wylie-Rosett J, Albright A, Apovian CM, Clark NG, Franz MJ, Hoogwerf BJ, Lichtenstein AH, Mayer-Davis E, Mooradian AD, Wheeler ML: Nutrition recommendations and interventions for diabetes: a position statement of the American Diabetes Association. *Diabetes Care* 31 (Suppl. 1):S61–S78, 2008.
4. U.S. Department of Health and Human Services: Final MNT regulations. CMS-1169-FC. *Federal Register,* 1 November 2001. 42 CFR Parts 405, 410, 411, 414, and 415.
5. American Dietetic Association: Comparison of the American Dietetic Association (ADA) Nutrition Care Process for nutrition education services and the ADA Nutrition Care Process for medical nutrition therapy (MNT) services [article online]. Available from www.eatright.org/advocacy/mnt. Accessed 6 November 2009.
6. Lacey K, Pritchett E: Nutrition care process and model: ADA adopts road map to quality care and outcomes management. *J Am Diet Assoc* 103:1061–1072, 2003.
7. Daly A, Michael P, Johnson EQ, Harrington CC, Patrick S, Bender T: Diabetes White Paper: Defining the delivery of nutrition services in Medicare medical nutrition therapy vs Medicare diabetes self-management training programs. *J Am Diet Assoc* 109:528–539, 2009.
8. Camelon KM, Hadell K, Jamsen PT, Ketonen KJ, Kohtamaki HM, Makimatilla S, Törmölö ML, Valve RH: The Plate Model: a visual method of teaching meal planning. *J Am Diet Assoc* 98:1155–1158, 1998.
9. Diabetes Care and Education Dietetic Practice Group: *American Dietetic Association Guide to Diabetes Medical Nutrition Therapy and Education.* Ross TA, Boucher JL, O'Connell BS, Eds. Chicago, American Dietetic Association, 2005.
10. Diabetes 1 and 2 Evidence Analysis Project: American Dietetic Association Evidence Analysis Library website. Available from www.adaevidencelibrary.com/topic.cfm?cat=1615. Accessed 6 November 2009.

11. Pastors JG, Warshaw H, Daly A, Franz M, Kulkarni K: The evidence of medical nutrition therapy in diabetes management. *Diabetes Care* 25:608–613, 2002.

12. UKPDS Study Group: Response of fasting plasma glucose to diet therapy in newly presenting type II diabetic patients. *Metabolism* 39:905–912, 1990.

13. Franz MJ, Monk A, Barry B, McClain K, Weaver T, Cooper N, Upham P, Bergenstal R, Mazze RS: Effectiveness of medical nutrition therapy provided by dietitians in the management of non-insulin-dependent diabetes mellitus: a randomized, controlled clinical trial. *J Am Diet Assoc* 95:1009–1017, 1995.

14. Kulkarni K, Castle G, Gregory R, Holmes A, Leontos C, Powers M, Snetselaar L, Splett P, Wylie-Rosett J: Nutrition practice guidelines for type 1 diabetes mellitus positively affect dietitian practices and patient outcomes. *J Am Diet Assoc* 98:62–70, 1998.

15. Glasgow RE, Toobert DJ, Hampson SE, Brown JE, Lewinsohn PM, Donnelly J: Improving self-care among older patients with type II diabetes: the "sixty-something" study. *Patient Educ Couns* 19:61–74, 1992.

16. Sadur CN, Moline N, Costa M, Michalik D, Mendlowitz D, Roller S, Watson R, Swain BE, Selby JV, Javorski WC: Diabetes management in a Health Maintenance Organization. *Diabetes Care* 22:2011–2017, 1999.

17. Delahanty LM, Halford BH: The role of diet behaviors in achieving improved glycemic control in intensively treated patients in the Diabetes Control and Complications Trial. *Diabetes Care* 16:1453–1458, 1993.

18. DCCT Research Group: Expanded role of the dietitian in the Diabetes Control and Complications Trial: implications for practice. *J Am Diet Assoc* 93:758–767, 1993.

19. Franz M, Callahan T, Castle G: Changing roles: educators and clinicians. *Clinical Diabetes* 12:53–54, 1994.

20. Johnson EQ, Valera S: Medical nutrition therapy in non-insulin-dependent diabetes mellitus improves clinical outcomes. *J Am Diet Assoc* 95:700–701, 1995.

21. Johnson EQ, Thomas M: Medical nutrition therapy by registered dietitians improves HbA1c levels (Abstract). *Diabetes* 50 (Suppl. 2):A21, 2001.

22. Christensen NK, Steiner J, Whalen J, Pfister R: Contribution of medical nutrition therapy and diabetes self-management education to diabetes control as assessed by hemoglobin A1c. *Diabetes Spectrum* 13:72–75, 2000.

23. Brown SA: Studies of educational interventions and outcomes in diabetic adults: a meta-analysis revisited. *Patient Educ Couns* 16:189–215, 1990.

24. Brown SA, Upchurch S, Anding R, Winter M, Ramirez G: Promoting weight loss in type II diabetes. *Diabetes Care* 19:613–624, 1996.

25. Padgett D, Mumford E, Hynes M, Carter R: Meta-analysis of the effects of educational and psychosocial interventions on management of diabetes mellitus. *J Clin Epidemiol* 41:1007–1030, 1988.

26. Norris SL, Engelgau MM, Venkat Narayan KM: Effectiveness of self-management training in type 2 diabetes. *Diabetes Care* 24:561–587, 2001.

27. Knowler WC, Barrett-Connor E, Fowler SE, Hamman RF, Lachin JM, Walker EA, Nathan DM, the Diabetes Prevention Program Research Group: Reduction in the incidence of type 2 diabetes with lifestyle intervention or metformin. *N Engl J Med* 346:393–403, 2002.

28. Wadden TA, West DS, Neiberg RH, Wing RR, Ryan DH, Johnson KC, Foreyt JP, Hill JO, Trence DL, Vitolins MZ, the Look AHEAD Research Group: One-year weight losses in the Look AHEAD study: factors associated with success. *Obesity* 17:713–722, 2009.

29. DCCT Research Group: The effect of intensive treatment of diabetes on the development and progression of long-term complications in insulin-dependent diabetes mellitus. *N Engl J Med* 329:977–986, 1993.

30. Mayer-Davis EJ, Bell RA, Dabelea D, D'Agostino R, Imperatore G, Lawrence JM, Liu L, Marcovina S, The SEARCH for Diabetes in Youth Study Group: The many faces of diabetes in American youth: type 1 and type 2 diabetes in five race and ethnic populations. *Diabetes Care* 32:S99–S101, 2009.

31. Fagot-Campagna A, Venkat Narayan KM, Imperatore G: Type 2 diabetes in children [Editorial]. *BMJ* 322:377–378, 2001.

32. U.S. Department of Health and Human Services Centers for Medicare & Medicaid Services: Medicare eligibility tool [online]. Available from www.medicare.gov/MedicareEligibility/. Accessed 6 November 2009.

33. National Conference of State Legislatures: State laws mandating diabetes health coverage [article online]. Available from www.ncsl.org/programs/health/diabetes.htm. Accessed 20 September 2009.

34. Centers for Medicare and Medicaid Services Medicare Learning Network: MLN matters, No. SE0660 [publication online]. Available from www.cms.hhs.gov/MLNMattersArticles/downloads/SE0660.pdf. Accessed 6 November 2009.

35. DPP Research Group: Costs associated with the primary prevention of type 2 diabetes mellitus in the Diabetes Prevention Program. *Diabetes Care* 26:36–47, 2003.

36. American Dietetic Association: Referrals to MNT as easy as 1-2-3 [article online]. Available from www.eatright.org/ada/files/0109_Referrals_1-2-3-_MNT_Flyer2.pdf. Accessed 6 November 2009.

37. National Certification Board for Diabetes Educators: *2009 Certification Handbook for Diabetes Educators* [publication online]. Arlington Heights, Ill., National Certification Board for Diabetes Educators, 2009. Available from www.ncbde.org/documents/HB2009Final.pdf. Accessed 6 November 2009.

Critical Thinking

1. What is the role of the Registered Dietician in the nutritional management of a diabetic patient?

SARA F. MORRIS, MAT, MPH, RD, is a research assistant in the Department of Nutrition at the University of North Carolina in Chapel Hill. JUDITH WYLIE-ROSETT, EdD, RD, is a professor and head of the Division of Behavioral and Nutritional Research in the Department of Epidemiology and Population Health at the Albert Einstein College of Medicine in Bronx, N.Y.

Understanding Hypovolaemic, Cardiogenic and Septic Shock

Sharon Garretson and Shelly Malberti

Aims and Intended Learning Outcomes

The aim of this article is to provide nurses with a comprehensive overview of three of the most common types of shock, namely, hypovolaemic, cardiogenic and septic shock. Reading this article will assist nursing staff to become more confident in the identification and care of this group of patients. After reading this article you should be able to:

- Define shock.
- Describe the stages of shock, the major causes and the clinical manifestations of hypovolaemic, cardiogenic and septic shock.
- Discuss the treatment options available for the various types of shock described.
- Outline the nursing measures required to manage these patients.

Introduction

Shock describes a life-threatening condition resulting from an imbalance between oxygen supply and demand, and is characterised by hypoxia and inadequate cellular function that lead to organ failure and potentially death (Kleinpell 2007).

Caused as a result of direct injury, or an underlying medical condition, shock can be life-threatening (Hand 2001, Chavez and Brewer 2002, Bench 2004). Shock occurs when the circulatory system is no longer able to complete one of its essential functions, such as providing oxygen and nutrients to the cells of the body or removing subsequent waste (Chavez and Brewer 2002). This results in inadequate tissue perfusion (Cottingham 2006, Justice and Baldisseri 2006, Medline Plus 2007), which triggers a cascade of events.

Shock has many causes, including sepsis, cardiac pump failure, hypovolaemia and anaphylaxis (Hand 2001, Justice and Baldisseri 2006), and is classified into various types. The three basic types of

shock are hypovolaemic, cardiogenic and distributive (Hand 2001). Distributive shock can be further classified as neurogenic, anaphylactic and septic shock. The purpose of this article is to identify and discuss three of the most common types of shock, in addition to reviewing the causes, treatment modalities and nursing considerations.

Stages of Shock

Research into shock has resulted in the classification of four distinct stages of shock (Chavez and Brewer 2002, Kleinpell 2007). These are: initial, compensatory, progressive and refractory (Kleinpell 2007).

Initial

In the initial stage of shock, the body experiences a reduced cardiac output (Box 1). During this stage, nurses should be aware that the cells switch from aerobic to anaerobic metabolism, which can lead to lactic acidosis, that is, excess acid resulting from a build up of lactic acid in the blood and lowering of the pH. Although clinical signs and symptoms may be subtle at this time, cellular damage can occur. A serum lactate level will provide an accurate assessment of acidosis because septic patients typically convert to anaerobic metabolism as a result of hypoperfusion. If the underlying cause of shock is not treated at this time, the patient will progress to stage two.

Compensatory

This stage is characterised by the body's attempt to regain homeostasis and improve tissue perfusion. Here, the sympathetic nervous system is stimulated resulting in catecholamine release (Box 1) (Chavez and Brewer 2002). This neurohormonal response causes increased cardiac contractility, vasoconstriction and a shunting of blood to the vital organs. The adrenal/renal system releases aldosterone, which promotes water conservation in an effort to maintain intravascular volume.

Progressive

In this third stage of shock, the body has lost its compensatory mechanisms, which have sustained tissue perfusion to this point. This decrease in perfusion results in metabolic acidosis, electrolyte imbalance and respiratory acidosis. The clinical symptoms will be such that there should be no doubt as to the severity of the patient's condition. The nurse will be able to observe severe hypotension, pallor, tachycardia and irregular rhythm, peripheral oedema, cool and clammy extremities and an altered level of consciousness. During this stage the blood pH decreases as the lactic acid production increases (Kleinpell 2007).

Time Out 1

The clinical condition of shock is one of complex physiology. To understand the chain of events surrounding the shocked state, review the basics of cardiac and vascular anatomy and physiology before reading on.

Box 1
Glossary

Term	Definition
Afterload	A resistance that the left ventricle must work against to pump blood through the aorta.
Arterial blood gases	A blood sample taken from an artery, that when analysed enables evaluation of gaseous exchange in the lungs by measuring the partial pressure of gases dissolved in arterial blood.
Cardiac output	The amount of blood ejected from the left ventricle per minute. Usual cardiac output is 4–8 L/min.
Catecholamine	Naturally occurring chemicals that stimulate the nervous system, constrict peripheral blood vessels, increase heart rate and dilate the bronchi.
Central venous pressure (CVP)	CVP reflects the amount of blood returning to the heart and the ability of the heart to pump the blood into the arterial system. It is a good approximation of right atrial pressure.
Colloids	A large molecule, such as albumin, that does not cross the capillary membrane in solution.
Crystalloids	A solute, such as sodium or glucose, that crosses the capillary membrane in solution, for example, sodium chloride 0.9% solution
Inotrope	A drug that affects the contraction of cardiac muscle.
Intra-aortic balloon pump	A balloon-type device inserted into the aorta, with the goal of being able to reduce the workload of the left ventricle and improving coronary perfusion.
Mean arterial pressure	The average arterial pressure during a single cardiac cycle.
Peripheral oedema	The accumulation of fluids in the interstitial tissues of those areas affected by gravity, such as the legs, feet and hands. Any oedema is an abnormal condition.
Pulmonary oedema	Fluid accumulation in the lungs due to failure of the heart to remove fluid from the lung circulation or following direct injury to the lungs. It leads to impaired gas exchange and may cause respiratory failure.
Preload	A stretching force exerted on the ventricle muscle by the blood it contains at the end of diastole.
$ScvO_2$	The oxygen saturation of venous blood as it returns to the heart, measured at the superior vena cava.
SpO_2	The oxygen saturation of peripheral blood, which can reflect respiratory status.
Thrombolysis	Dissolution or destruction of a thrombus.
Vasopressors	Drugs that stimulate cardiac contraction of the muscular tissues of the capillaries and arteries.

Time Out 2

Bill, a 52-year-old patient, returns from vascular surgery at 2 pm. He is in no pain and breathing normally with an oxygen saturation (SpO_2) of 98% on 40% oxygen (Box 1). His vital signs are stable, with a slightly elevated heart rate of 99 beats per minute (bpm). He is receiving sodium chloride 0.9% at 100 ml/hr and urine output has been greater than 50 ml/hr for the past three hours. By 4 pm, Bill's vital signs are as follows: heart rate 137 bpm, blood pressure 89/50 mmHg, temperature 36.9°C, SpO_2 92% on 55% oxygen. His urine output was 30 ml for the past hour. His abdomen Is firm and his skin looks pale. He is showing no electrocardiogram (ECG) changes, other than tachycardia. Given this clinical condition, which type of shock is Bill most likely to be experiencing? Provide a possible suggestion for the cause of shock. A suggested answer is provided on page 55.

Refractory

At this late stage, irreversible cellular and organ damage occur. Shock becomes unresponsive to treatment and death is likely (Hand 2001) (Table 1).

Hypovolaemic Shock

Hypovolaemic shock is different from cardiogenic and septic shock in that it has many varied and diverse origins, as opposed to a few defined and specific causes. It is characterised by an inadequate intravascular volume caused by significant blood and/or fluid loss (Hand 2001, Chavez and Brewer 2002, Bench 2004). This intravascular depletion can be caused by sustained vomiting, diarrhoea or severe dehydration (Hand 2001, Diehl-Oplingcr and Kaminski 2004), as well as burns, traumatic injury and surgery (Diehl-Oplinger and Kaminski 2004). Simple blood loss is often the most common source of hypovolaemic shock (Medline Plus 2007), which can occur from bleeding either inside or outside the body. Internal fluid collection, such as ascites and peritonitis, may also cause hypovolaemic shock (Hand 2001).

Table 1   The Stages of Shock

Initial stage	Compensatory stage	Progressive stage	Refractory stage

• Body switches from aerobic to anaerobic metabolism
 	
• Elevated lactic acid level
 	
• Subtle changes in clinical signs
	• Sympathetic nervous system stimulated
	  ↑ catecholamine release
	  ↑ cardiac contractility
	  ↓
	
	• Neurohormonal response: vasoconstriction and blood shunted to vital organs
	  ↓
	
	• Aldosterone released
	  ↓ urine output (<30 ml/hr)
	  ↓
	
	• ↑ Heart rate
	  ↓
	
	• ↑ Glucose levels
		• Electrolyte imbalance
		• Metabolic acidosis
		• Respiratory acidosis
		• Peripheral oedema
		• Irregular tachyarrhythmias
		• Hypotension
		• Pallor
		• Cool and clammy skin
		• Altered level of consciousness
			• Irreversible cellular and organ damage
			• Impending death

(Springhouse 2004)

Clinical Manifestations of Hypovolaemic Shock

Hypovolaemic shock has many clinical manifestations (Box 2), which coincide with the stages of shock as defined earlier. With a fluid loss of less than 750 ml, the body may enter a compensated state (Bench 2004), and changes to vital signs may be subtle and difficult to detect. At this point, the body essentially maintains homeostasis and the patient may be asymptomatic. For this reason, nurses should pay close attention to even the subtlest change in vital signs, while using their nursing judgement to summon help for the patient as appropriate. As fluid loss increases to more than 750 ml, cardiac output begins to fall (Hand 2001, Chavez and Brewer 2002), and changes in vital signs occur. As a result of the falling cardiac output, the sympathetic chain of the autonomic nervous system is initiated with the 'fight or flight' response (Hand 2001), and catecholamines are released. Vasoconstriction occurs as the nervous system endeavours to manoeuvre the blood away from the non-vital organs of the gut and extremities, and towards the central core. As a result, the patient's extremities may be cool and clammy and the patient may exhibit signs of anxiety (Hand 2001). Other compensatory mechanisms are initiated by the renal system (Bench 2004) and its release of renin. A cascade of events ensues with the production and release of the angiotensin-angiotensin II-aldosterone flow. This sequence promotes vasoconstriction and the reabsorption of sodium and water in an attempt to increase blood volume. Both of these mechanisms may increase blood pressure. Other clinical indicators that are noticeable in patients with hypovolaemic shock include increased respirations and decreased urine output (Chavez and Brewer 2002).

If the hypovolaemia remains undetected or untreated, the patient's clinical condition will worsen, the patient will become unstable and his or her blood pressure will drop significantly. As a result of tachycardia, cardiac arrhythmias and/or chest pain may occur due to the inability of the coronary arteries to fill adequately during diastole (American College of Surgeons Committee on Trauma: Shock 1997). Respirations will increase as the body tries to rid itself of accumulating lactic acid.

> ### Box 2
> ## Clinical Manifestations of Hypovolaemic Shock
>
> • Altered or decreased level of consciousness
> • Anxiety and restlessness
> • Decreased urine output
> • Delayed capillary refill
> • Increased heart rate
> • Increased respiratory rate
> • Pale, cool and clammy skin
> • Systolic blood pressure <90 mmHg or 40 mmHg below baseline

Eventually an altered level of consciousness ensues because of tack of tissue perfusion (Diehl-Oplinger and Kaminski 2004).

After approximately a 40% fluid loss, the situation can become life threatening. Multi-organ damage and cellular necrosis can occur with impending death likely (American College of Surgeons Committee on Trauma: Shock 1997).

Treatment

An essential aspect of treatment for patients with hypovolaemic shock is to restore fluid volume and blood pressure. This is usually achieved with intravenous fluids and vasopressors (Box 1).

Oxygen should be administered to counteract the respiratory effects of shock (Hand 2001). In some cases, it may be enough to administer oxygen by facial mask, but if shock progresses there may be a need for endotracheal intubation and mechanical ventilation. Whenever possible the head of the patient's bed should be elevated to promote comfort and adequate respirations. This is also a preventive measure against pneumonia, especially if the patient is ventilated (Craven 2006). Nevertheless, if the patient's condition is such that hypotension is pronounced, this may not always be possible. Nurses should use critical thinking skills and draw on their clinical knowledge to determine the appropriate position for the patient.

Hypovolaemic shock requires immediate fluid resuscitation (Diehl-Oplinger and Kaminski 2004), in conjunction with treating the underlying cause if the patient is to survive. There has been much debate in recent years as to whether colloids or crystalloids (Box 1) are the best choice for fluid resuscitation (Bench 2004). There seems to be growing consensus that crystalloids are the superior choice (Diehl-Oplinger and Kaminski 2004) because they mimic intracellular fluid more closely. Bench (2004) and Kirschman (2004) claim that colloids have been associated with multiple issues such as infections and reactions to the properties of the fluid(s). Common practice dictates that crystalloids are used for fluid volume loss of less than 1,500 ml (Hand 2001) while whole blood is used if the volume loss is greater, or if the sole cause of the hypovolaemia is blood loss (Chavez and Brewer 2002, Bench 2004, Diehl-Oplinger and Kaminski 2004). Crystalloids such as lactated Ringer's solution or sodium chloride 0.9% can address both hypovolaemia and electrolyte imbalances (Chavez and Brewer 2002). Treatment goals include raising mean arterial pressure (MAP) (Box 1) to 70 mmHg by infusing 1,000–2,000 ml of warmed crystalloid (Cottingham 2006). Additional fluids are administered based on the patient's clinical condition.

If colloids are used, they are used in smaller quantities than crystalloids (Diehl-Oplinger and Kaminski 2004) because they are a hypertonic solution. Blood transfusions may also be required clinically. However, regardless of the type of fluid infused, caution is necessary in patients with pre-existing heart failure, so as not to promote pulmonary oedema (Box 1). As with all large fluid infusions, blood samples should be drawn and tested at regular intervals to check haemoglobin levels, haematocrit, electrolytes and other tests as prescribed by the consulting physician.

It may also be necessary to induce vasoconstriction with inotropic and vasopressor medication (Box 1). The most common pharmacologic agents used in the critically ill shock patient are adrenaline (epinephrine), noradrenaline (norepinephrine) (Chavez and Brewer 2002) and dobutamine (Cottingham 2006). The goal of treatment is to increase cardiac output and MAP. Nurses should be aware that inotropic drugs or vasopressors should not be started until the patient has an adequate fluid volume or fluid volume has been replaced.

Cardiogenic Shock

Cardiogenic shock results in a decline in cardiac output and tissue hypoxia, despite adequate fluid volume. Tt remains the most serious complication of acute myocardial infarction (MI) (Sanborn and Feldman 2004), often resulting in death (Ducas and Grech 2003, Mann and Nolan 2006). Cardiogenic shock occurs in 5–10% of MI patients (Ducas and Grech 2003, Sanborn and Feldman 2004, Mann and Nolan 2006), and has a mortality rate of more than 50% (Sanborn and Feldman 2004). If revascularisation of the myocardium does not occur promptly, the outcome is usually fatal (Mann and Nolan 2006).

The causes of cardiogenic shock are limited, unlike that of hypovolaemic shock. It typically occurs following an MI and more commonly in ST-segment elevation MIs (STEMIs) (Holmes 2003) as a result of left ventricular failure (Ducas and Grech 2003, Sanborn and Feldman 2004, Bouki et al 2005). However, other causes include acute, severe mitral regurgitation and ventricular septal rupture. Mann and Nolan (2006) add cardiac tamponade as a further source. In some rare

cases medications have been known to cause this condition, specifically Metoprolol and clopidogrel (Mann and Nolan 2006). McLuckie (2003) also asserts that the older female patient is more at risk of developing cardiogenic shock, as are patients with diabetes and those with a history of previous MI.

Clinical Manifestations of Cardiogenic Shock

The clinical manifestations of cardiogenic shock can be similar to that of hypovolaemic shock, although in cardiogenic shock the patient's condition can deteriorate more quickly (Bench 2004). Essentially, the damaged left ventricle is unable to pump effectively (Chavez and Brewer 2002), and thus cardiac output is reduced. Characteristically, the cardiac output will be reduced to less than 2.2 L/min (usual cardiac output is 4–8 L/min). In an effort to compensate, the remaining non-ischaemic myocardium becomes hypercontractile (Ducas and Grech 2003). This action raises the oxygen demands of the heart, which further increases the workload. As blood pressure falls, catecholamines are released which cause vasoconstriction. In contrast to hypovolaemic shock, vasoconstriction can be detrimental in patients with cardiogenic shock, in that it forces the already damaged myocardium to work even harder (McLuckie 2003, Bench 2004). The failing myocardium cannot work as efficiently as usual, and so may not be able to clear blood quickly enough, which results in a build up of blood in the atrium. In turn, this can cause congestion in the lungs leading to pulmonary oedema and a compromised respiratory system, necessitating supplementary oxygen and potential mechanical ventilation.

Further symptoms include increased central venous pressure (CVP) (Box 1), chest pain as a result of the decreased coronary artery perfusion and low urine output (Bench 2004). The patient may also experience anxiety, as a result of pain, and feelings of doom and general demise (Hand 2001). The role of the nurse in this situation is to reduce the patient's pain and anxiety, thereby decreasing myocardial workload.

Treatment

Treatment options for cardiogenic shock are varied. An essential focus is on revascularisation of the damaged myocardium, and improving cardiac contractility and blood pressure (Bench 2004). Nevertheless, Ducas and Grech (2003) and Mann and Nolan (2006) maintain that respiratory function and oxygen delivery should be the first consideration. Oxygen is necessary to combat the effects of cardiac ischaemia and associated chest pain. Additional attention should be given to the management of pulmonary oedema, which can be addressed with diuretics. Evaluation of arterial blood gases (ABGs) (Box 1) and cardiac monitoring are important nursing responsibilities.

Reperfusion of the myocardium can occur as a result of thrombolysis (Box 1), or mechanical revascularisation by means of invasive procedures such as percutaneous coronary intervention (PCI) or coronary artery bypass grafting (CABG) (Mann and Nolan 2006). Thrombolysis involves the injection of pharmacologic agents such as streptokinase, urokinase or tissue plasminogen activator (TPA), which act to dissolve a clot in the affected coronary artery (American Heart Association 2007). However, one of the keys to their success is that they are used within a few hours of the initial insult and before cardiogenic shock sets in. Once cardiogenic shock develops, thombolysis may have little effect on the patient's clinical condition.

Mechanical reperfusion can be accomplished through PCI or CABG, and in patients under 75 years of age, the American College of Cardiology/American Heart Association have listed these procedures as Class I recommendations (Mann and Nolan 2006). A report from the National Registry of Myocardial Infarction reported improved

Table 2 Medications Used to Treat Patients in Cardiogenic Shock

Drug	Class	Dose	Effect
Dobutamine	Inotrope	2–40 mcg/kg/min	Increase cardiac contractility and cardiac output.
Dopamine	Inotrope	5–20 mcg/kg/min	Increase contractility and vasoconstriction.
Milrinone	Phosphodiesterase inhibitor	0.375–0.75 mcg/kg/min (reduce dose in renal failure)	Increase cardiac contractility and dilate vascular smooth muscle.
Noradrenaline	Catecholamine	2–30 mcg/min	Vasoconstriction. Increases peripheral vascular resistance.
Nitroglycerin	Vasodilator	Start at 5 mcg/min Maximum dose 200 mcg/min	Decreases preload and myocardial oxygen demand. Improves coronary artery blood flow.
Sodium nitroprusside	Vasodilator	0.5–6 mcg/kg/min. Maximum dose is 10 mcg/kg/min for <10 minutes	Reduces afterload in decreased cardiac output states.

(Adapted from Sasada and Smith 2003, Lynn McHale-Wiegand and Carlson 2005).

Time Out 3

List the most frequently used drugs to treat patients with cardiogenic shock. Discuss the dose range and the effects that treatment with these medications will have.

---

patient survival when PCI, such as coronary angioplasty, was used for cardiogenic shock in this age group (Babaev *et al* 2005). Survival rates from PCI and CABG are similar at 55.6% and 57.4% respectively (Sleeper *et al* 2005).

The use of inotropic agents and vasopressors in cardiogenic shock is widespread (Table 2), although they are usually viewed as a supportive measure rather than a curative intervention (Justice and Baldisseri 2006). When administered in the intensive care unit (ICU) usual medications include dopamine, dobutamine and noradrenaline (norepinephrine). Dupamine, however, has been associated with increased mortality (Bench 2004), and so should be used with caution. Vasodilators such as sodium nitroprusside and glycerin trinitrate (GTN) may also be used to reduce left ventricular afterload (Box 1). As with any medications that can potenrially affect blood pressure, frequent monitoring of the patient's vital signs is imperative (Bench 2004). Fluids may also be necessary in cardiogenic shock, but should be administered with extreme caution, especially in the presence of pulmonary oedema (Chavez and Brewer 2002, Ducas and Grech 2003).

In the most severely ill patients an intra-aortic balloon pump (IABP) may be used (Box 1). The IABP is another Class I recommendation (Chavez and Brewer 2002, Ducas and Crech 2003, Sanborn and Feldman 2004, Mann and Nolan 2006). This invasive balloon-attached catheter is inserted via the femoral artery, and sits in the descending thoracic aorta (Bouki *et al* 2005). The IABP is attached to an external machine which aids balloon inflation and deflation at exact moments in the cardiac cycle. Inflation occurs during diastole, with deflation following during systole. The goal ofthe IABP is to increase coronary artery perfusion during diastole and reduce systemic afterload during systole. The IABP is often used in conjunction with pharmacological agents and other interventions.

The ventricular assist device (VAD) is a final treatment option that may be used in the cardiothoracic ICU as a 'bridge to transplantation' (Cleveland Clinic Foundation 2004). This device is used in a last-stage effort to save the patient's life, when the damage from cardiogenic shock is so severe that only a cardiac transplant will prevent death.

The VAD is a mechanical pump that is attached to the patient's heart and is situated outside the body (Figure 1). It is used to circulate blood and assist the failing heart.

Septic Shock

In North America and Europe more than 750,000 individuals develop sepsis each year (Institute for Healthcare Improvement (IHI) 2005a). If septic shock develops, the mortality rate is estimated to be approximately 40–50% (Jindal *et al* 2000, Oppert *et al* 2005). Septic shock, the result of an overwhelming infection (Box 3), leads to hypotension, altered coagulation, inflammation, impaired circulation at a cellular level, anaerobic metabolism, changes in mental status and multi-organ failure (Kleinpell 2003a, 2003b, Rivers *et at* 2005). In septic shock, 'there is a complex interaction between pathologic vasodilation, relative and absolute hypovolaemia . . . direct myocardial depression' (Beale *et al* 2004). Although recognising the early signs of septic shock may be difficult, the nurse's role is pivotal in identifying these changes and facilitating immediate medical treatment (Bridges and Dukes 2005).

Guidelines, which closely mirror those traditionally used for other diseases such as acute MI, stroke and trauma, have now been developed for the early diagnosis and treatment of sepsis (Rivers *et al* 2005). The Surviving Sepsis Campaign, an international initiative, recommends a 24-hour sepsis pathway, with the therapeutic goal of improving survival of these patients and decreasing mortality (IHI 2005b).

Time Out 4

John is a 72-year-old patient admitted to the intensive care unit with a change in level of consciousness, blood pressure of 86/46 mmHg, respiratory rate of 36, urine output of 15 ml/hr for the past three hours, a rectal temperature of 38.6°C, and a heart rate of 140 beats per minute. The urinalysis completed in the accident and emergency department was positive for a urinary tract Infection. The patient has already received two litres of sodium chloride 0.9%, with no response in blood pressure. What two vasopressors are most likely to be prescribed by the consultant on the unit? A suggested answer is on page 55.

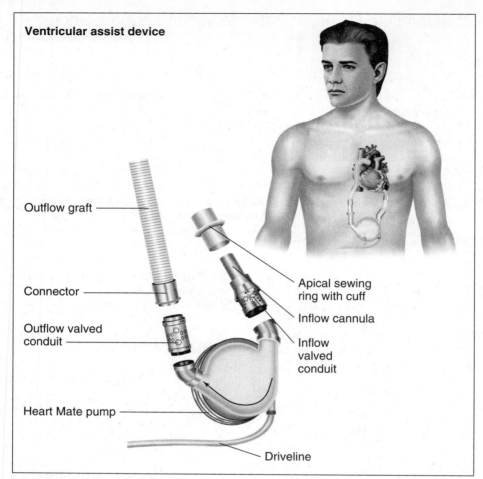

Ventricular assist device

Outflow graft

Connector

Outflow valved conduit

Heart Mate pump

Apical sewing ring with cuff

Inflow cannula

Inflow valved conduit

Driveline

Figure 1 Ventricular assist device.

Box 3
Sources of Infection in Septic Shock

Blood: bacteraemia.

Bone: osteomyelitis.

Cardiovascular: endocarditis and pericarditis.

Central nervous system: meningitis.

Intra-abdominal: diverticulitis, appendicitis and perforated or ischaemic bowel.

Invasive catheters: central venous or peripheral cannula

Pulmonary: community acquired or healthcare-associated pneumonia.

Soft tissue: cellulitis, skin and wound infections and necrotising fasciitis.

Surgical wounds: incision and deep infection.

Urinary tract: urinary tract and kidney infections.

Treatment

A classic sign of septic shock is the patient's development of both absolute and relative hypovolaemia (Hollenberg 2001, Beale *et al* 2004), and as such, fluid resuscitation is a primary element in the treatment plan. Absolute hypovolaemia can be the result of fluid loss due to vomiting, diarrhoea, sweating or oedema. Relative hypovolaemia occurs as a result of vasodilation and peripheral blood pooling (Vincent and Gerlach 2004). The type of fluid to be used during resuscitation is still under debate (Vincent and Gerlach 2004, Bridges and Dukes 2005); however, it is recommended that a minimum of 20 ml/kg of crystalloid (or colloid equivalent) is administered initially to patients who are hypotensive as a result of sepsis-related hypovolaemia (IHI 2005c). If fluid resuscitation efforts are unsuccessful at maintaining a MAP of 60–65 mmHg or if fluid resuscitation is in progress and hypotension remains life-threatening, then vasopressor therapy should be initiated (Bridges and Dukes 2005, IHI 2005c, Robson and Newell 2005). Indicators of adequate fluid resuscitation and tissue perfusion include a urine output greater than 0.5 ml/kg/hr, a decrease in serum lactate level, improved level of consciousness and a CVP ranging between 8–12 mmHg or 12–15 mmHg for patients receiving mechanical ventilation (Bridges and Dukes 2005, Shapiro *et al* 2006).

Commonly used vasopressor therapy includes dopamine, noradrenaline (norepinephrine), adrenaline (epinephrine) and phenylephrine (Bridges and Dukes 2005). The Surviving Sepsis Campaign guidelines recommend dopamine and noradrenaline (norepinephrine) as the first-line choice in septic shock (Bridges and Dukes 2005, IHI 2005d, Levy *et al* 2005). An arterial catheter is typically used for continuous and accurate blood pressure monitoring (Beale *et al* 2004). Careful consideration is required when implementing vasopressor treatment to ensure that adequate fluid volume resuscitation occurs; otherwise its use may be harmful and result in a further decrease in organ perfusion (Bridges and Dukes 2005).

Another resuscitation goal includes maintaining a central venous oxygen saturation (ScvO$_2$) greater than 70% (Box 1) (Robson and Newell 2005, Shapiro *et al* 2006). If the ScvO$_2$ is less than 70% and the haematocrit is less than 30%, then a blood transfusion can be considered. If the haematocrit is greater than 30%, dobutamine may be used (Shapiro *et al* 2006).

Once a diagnosis of sepsis has been determined, antibiotic therapy should be administered in a timely fashion—within minutes rather than hours. However, controversy exists regarding the timeframe in which antibiotic therapy should be initiated (Kumar *et al* 2006). The choice of antibiotic is dependent on the pathogen, drug tolerance and other underlying diseases. It is recommended that broad-spectrum antibiotics are administered within three hours for patients seen in the accident and emergency department and within one hour for ward and ICU patients (IHI 2005e).

One of the manifestations of septic shock can be an alteration in coagulation (Kleinpell 2003a, 2003b). This occurs as a result of an inflammatory response, stimulation of the coagulation cascade and a reduction in protein C and antithrombin III. These events produce an enhanced state of coagulation, sepsis-associated coagulopathy and even death (Kleinpell 2003a). Drotrecogin alfa (activated) is an adjunctive therapy used to treat patients with this type of enhanced state of coagulation (Kleinpell 2003a, Robson and Newell 2005). The Protein C Worldwide Evaluation in Severe Sepsis (PROWESS) trial indicted that the use of drotrecogin alfa (activated) or recombinant activated protein C improves survival in septic patients, and is recommended for septic patients at high risk of death (Kleinpell 2003a). Recombinant activated protein C can increase the risk of bleeding and is contraindicated in some patients (Robson and Newell 2005, IHI 2005f). As a consequence, specific nursing considerations are necessary when administering this drug (Box 4).

Corticosteroid therapy is an additional treatment option. The anti-inflammatory effect of glucocorticoids has meant that they have been used fordecades in the treatment of septic patients (Keh and Sprung 2004), yet high-dose corticosteroid therapy has not been shown to improve patient outcomes (Oppert *et al* 2005).

Glucocorticoids administered in high dosages, for example, 2–8 g methylprednisolone, may even be detrimental (Oppert *et al* 2005). More recently the use of low-dose glucocorticoids has been added to treatment plans (Keh and Sprung 2004). Adrenal function tests can be used to steer the decision regarding the use of corticosteroid therapy if adrenal insufficiency is suspected (Keh and Sprung 2004).

Nursing Considerations

Given the clinical complexity and potentially devastating consequences of shock, it is essential that the nurse remains diligent in the care of these patients. Understanding the clinical signs that the patient demonstrates during each stage of shock will assist nurses with patient assessments and in carrying out the treatment plan.

As with every clinical situation, the basics of nursing and medical attention should be paramount. Oxygenation and respiratory function are always a priority, whether patients are able tomaintain their own airway or mechanical ventilation is required. Proper positioning to promote respiratory function is essential and ABGs should be monitored as necessary. Circulatory function should be addressed with a combination of fluids and/or medications, which may be reliant on the type of shock involved.

As noted earlier in this article, clinical signs during the initial stage of shock may be cryptic, so the astute nurse should identify patients at risk, and monitor vital signs carefully, including body temperature, haemodynamic function, urine output, level of consciousness and laboratory values. Monitoring lactic acid levels is of primary importance, as in the initial stage the body is converting from aerobic to anaerobic metabolism, and this laboratory value may be one of the first signs of impending shock.

The established plan of care should focus on preventing the progression of shock. In the event that this goal cannot be accomplished and the patient advances to the compensatory stage or further, a transfer to the ICU is warranted.

The ICU nurse should focus on the maintenance of homeostasis through the use of fluid resuscitation, vasopressors and antibiotics, depending on the type of shock. In addition to continued diligence in vital sign monitoring and regular physical assessments, a central catheter should be inserted to facilitate rapid infusion of fluids and/or medications. Evaluation of a patient's fluid status should be monitored regularly by measuring the CVP. An indwelling urinary catheter should be inserted ro assist in maintaining accurate fluid balance. If the patient is in cardiogenic shock and IABP monitoring is necessary, close observation of the patient's limb is important, as ischaemia can be a complication of these devices. IABPs can reduce blood flow to the leg or thrombus formation can occur around the catheter. As more invasive devices are used to monitor and treat the patient in shock, strict adherence to aseptic technique is vital to prevent infection.

The nurse should also remain attentive to basic nursing measures, such as frequent and thorough mouth care, pressure area care and repositioning, pain control and emotional support. Emotional

---

Box 4
Nursing Considerations When Administering Drotrecogin Alfa (activated)

- Administer medication through a dedicated intravenous (IV) catheter.
- Administer continuously at a rate of 24 mcg/kg per hour for 96 hours or according to a specific hospital policy.
- Ensure a bedside risk assessment is completed to avoid administration to high-risk patients such as those with active or recent internal bleeding, recent haemorrhagic stroke, trauma with increased risk of bleeding or the presence of an epidural catheter.
- Administer with the use of an IV infusion pump.
- Discontinue infusion two hours before any procedure that may carry with it a risk of bleeding.
- Restart infusion one hour after an uncomplicated minor procedure or 12 hours after a major procedure or surgery.

(Kleinpell 2003a).

Time Out 5

Now that you have completed the article you might like to write a practice profile. Gtiiclelines to help you are on page 60.

support is vital for the patient, but it is also extremely important for nurses to recognise the turmoil that family members may be experiencing.

Conclusion

Shock is a complex clinical syndrome that, if not detected and treated promptly, can lead to death. Despite the many advances in medical and nursing care in recent years, the mortality rate remains high for patients who develop shock, and especially for those patients who develop cardiogenic or septic shock. For these reasons, it is essential for nursing staff to realise that patients may present in shock, yet provide little indication of this in terms of changes in vital signs or other outward deterioration, at least in the initial stages. It is essential that nurses caring for these patients understand the clinical manifestations of shock and how each type of shock differs from the other(s). Nurses should also be familiar with treatment options and best practices for this patient group. Dealing with the patient in shock is a challenge for critical care practitioners, yet basic assessment skills and a good knowledge of pathophysiology can assist the nurse to provide the best care possible for these patients. **NS**

Suggested Answers to Time Out Activities Time Out 2

Bill is apyrexial, which combined with very recent surgery and a firm abdomen would rule out septic shock at this point. He experiences no ECG changes and is not in any pain, cardiac or otherwise, which would mean that cardiogenic shock is unlikely. The most likely cause given these clinical circumstances is hypovolaemic shock as a result of the firm abdomen, increased heart rate and decreased blood pressure and urine output. These clinical indicators are suggestive of internal haemorrhage. A possible cause may be a bleeding blood vessel which occurred during surgery.

Time Out 4

Dopamine and noradrenaline (norepinephrine) are the two vasopressors most likely to be prescribed by the consultant. Dopamine is usually administered because of its ability to increase mean arterial pressure (MAP). Dopamine does cause an increase in heart rate, which may indicate a need to add a second type of vasopressor therapy. Noradrenaline (norepinephrine) is the second most likely choice of vasopressor. Noradrenaline also increases the MAP as a result of vasoconstriction but does not have an effect on the heart rate.

References

American College of Surgeons Committee on Trauma: Shock (1997) *ATLS: Advanced Trauma Life Support Program for Doctors.* Sixth edition. American College of Surgeons, Chicago II.

American Heart Association (2007) *Heart Attack Treatments.* www.americanheart.org/presenter.jhtml?identifier=4601 (Last accessed: July 27, 2007.)

Babaev A, Frederick PD, Pasta DJ et al (2005) Trends in management and outcomes of patients with acute myocardial infarction complicated by cardiogenic shock. *Journal of the American Medical Association.* 294, 4, 448–454.

Beale RJ, Hollenberg SM, Vincent JL, Parrillo JE (2004) Vasopressor and inotropic support in septic shock: an evidence-based review. *Critical Care Medicine.* 32, 11 Suppl, S455–465.

Bench S (2004) Assessing and treating shock: a nursing perspective. *British Journal of Nursing.* 13, 12, 715–721.

Bouki KP, Pavlakis G, Papasteriadis E (2005) Management of cardiogenic shock due to acute coronary syndromes. *Angiology.* 56, 2, 123–130.

Bridges EJ, Dukes S (2005) Cardiovascular aspects of septic shock: pathophysiology, monitoring, and treatment. *Critical Care Nurse.* 25, 2, 14–16, 18–20, 22–24 passim.

Chavez JA, Brewer C (2002) Stopping the shock slide. *RN.* 65, 9, 30–34.

Cleveland Clinic Foundation (2004) *Implantable Ventricular Assist Device (VAD).* www.clevelandclinic.org/heartcenter/pub/guide/disease/heartfailure/lvad.htm (Last accessed: July 27 2007.)

Cottingham CA (2006) Resuscitation of traumatic shock: a hemodynamic review. *AACN Advanced Critical Care.* 17, 3, 317–326.

Craven DE (2006) Preventing ventilator-associated pneumonia in adults: sowing seeds of change. *Chest.* 130, 1, 251–260.

Diehl-Oplinger L, Kaminski MF (2004) Choosing the right fluid to counter hypovolemic shock. *Nursing.* 34, 3, 52–54.

Ducas J, Grech ED (2003) ABC of interventional cardiology. Percutaneous coronary intervention: cardiogenic shock. *British Medical Journal* 326, 7404, 1450–1452.

Hand H (2001) Shock. *Nursing Standard.* 15, 48, 45–52.

Hollenberg SM (2001) Cardiogenic shock. *Critical Care Clinics.* 17, 2, 391–410.

Holmes DR Jr (2003) Cardiogenic shock, a lethal complication of acute myocardial infarction. *Reviews in Cardiovascular Medicine.* 4, 3, 131–135.

Institute for Healthcare Improvement (2005a) *IHI Contributing to Bold International Campaign to Dramatically Reduce Mortality from Sepsis.* www.ihi.org/IHI/Topics/CriticalCare/Sepsis/ImprovementStories/IHIContributingtoBoldInternationalCampaigntoDramaticallyReduceMortalityfromSepsis.htm (Last accessed: July 27, 2007.)

Institute for Healthcare Improvement (2005b) *Sepsis Care Enters New Era.* www.ihi.org/IHI/Topics/CriticalCare/Sepsis/ImprovementStories/SepsisCareEntersNewEra.htm (Last accessed: July 27 2007.)

Institute for Healthcare Improvement (2005c) *Implement the Sepsis, Resuscitation Bundle: Treat Hypotension and/or Elevated Lactate with Fluids,* www.ihi.org/IHI/Topics/CriticalCare/Sepsis/Changes/IndividualChanges/TreatHypotensionandorElevatedLactatewithFluids.htm (Last accessed: July 27 2007.)

Institute for Healthcare Improvement (2005d) *Implement the Sepsis Resuscitation Bundle: Apply Vasopressors for Ongoing Hypotension.* www.ihi.org/IHI/Topics/CriticalCare/Sepsis/Changes/IndividualChatiges/

ApplyVasopressorsforOngoingHypotension.htm (Last accessed: July 27 2007.)

Institute for Healthcare Improvement (2005e) *Implement the Sepsis Resuscitation Bundle: Improve Time to Broad-Spectrum Antibiotics.* www.ihi.org/IHI/Topics/CriticalCare/Sepsis/Changes/IndividualChanges/ImproveTimetoBroadSpectrumAntibiotics.htm (Last accessed: July 27 2007.)

Institute for Healthcare Improvement (2005f) *Implement the Sepsis Management Bundle: Administer Drotrecogin Alfa (Activated) by a Standard Policy.* www.ihi.org/IHI/Topics/CriticalCare/Sepsis/Changes/IndividualChanges/AdministerDrotrecoginAlfaActivatedbyaStandardPolicy.htm (Last accessed: July 27 2007.)

Jindal N, Hollenberg 5M, Dellinger RP (2000) Pharmacologic issues in the management of septic shock. *Critical Care Clinics.* 16, 2, 233–249.

Justice JR, Baldisseri MR (2006) Early recognition and treatment of non-tranumatic shock in a community hospital. *Critical Care.* 10, 2, 307.

Keh D, Sprun CL (2004) Use of corticosteroid therapy in patients with sepsis and septic shock: an evidence-based review. *Critical Care Medicine.* 32, 11 Suppl, S527–533.

Kirschman RA (2004) Finding alternatives to blood transfusion. *Holistic Nursing Practice.* 18, 6, 277–281.

Kleinpell RM (2003a) Advances in treating patients with severe sepsis. Role of drotrecogin alfa (activated). *Critical Care Nurse.* 23, 3, 16–29.

Kleinpell RM (2003b) The role of the critical care nurse in the assessment and management of the patient with severe sepsis. *Critical Care Nursing Clinics of North America.* 15, 1, 27–34.

Kleinpell RM (2007) *Recognizing and Treating Five Shock States.* www.nurse.com/ce/course.html?CCID=3723 (Last accessed: July 27 2007.)

Kumar A, Roberts D, Wood KE *et al* (2006) Duration of hypotension before initiation of effective antimicrobial therapy is the critical determinant of survival in human septic shock. *Critical Care Medicine.* 34, 6, 1589–1596.

Levy B, Dusang B, Annane D, Gibot S, Bollaert PE; College Interregional des Reanimateurs du Nord-Est (2005) Cardiovascular response to dopamine and early prediction of outcome in septic shock: a prospective multiple-center study. *Critical Care Medicine.* 33, 10, 2172–2177.

Lynn McHale-Wiegand DJ, Carlson KK (EDS) (2005) *AACN Procedure Manual for Critical Care.* Fifth edition. Elsevier Saunders, Missouri MO.

Mann HJ, Nolan PE Jr (2006) Update on the management of cardiogenic shock. *Current Opinion in Critical Care.* 12, 5, 431–436.

McLuckie A (2003) Shock: an overview. In Oh TE, Bernsten AD, Soni N (Eds) *Oh's Intensive Care Manual.* Fifth edition. Butterworth-Heinemann, London, 71–77.

Medline Plus (2007) *Shock.* www.nlm.nih.gov/medlineplus/ency/article/000039.htm#Definition (Last accessed; July 27 2007.)

Oppert M, Schindler R. Husung C *et al* (2005) Low-dose hydrocortisone improves shock reversal and reduces cytokine levels in early hyperdynamic septic shock. *Critical Care Medicine.* 33, 11, 2457–2464.

Rivers EP, McIntyre L, Morro DC, Rivers KK (2005) Early and innovative interventions for severe sepsis and septic shock: taking advantage of a window of opportunity. *Canadian Medical Association Journal.* 173, 9, 1054–1065.

Robson W, Newell J (2005) Assessing, treating and managing patients with sepsis. *Nursing Standard.* 19, 50, 56–64.

Sanborn TA, Feldman T (2004) Management strategies for cardiogenic shock. *Current Opinion in Cardiology.* 19, 6, 608–612.

Sasada M, Smith S (2003) *Drugs in Anaesthesia and Intensive Care.* Third edition. Oxford Medical Publications, New York NY.

Shapiro NI, Howell MD, Talmor D *et al* (2006) Implementation and outcomes of the Multiple Urgent Sepsis Therapies (MUST) protocol. *Critical Care Medicine.* 34, 4, 1025–1032.

Sleeper LA, Ramanathan K, Picard MH *et al* (2005) Functional status and quality of life after emergency revascularization for cardiogenic shock complicating acute myocardial infarction. *Journal of the American College of Cardiology.* 46, 2, 266–273.

Springhouse (2004) *Critical Care Nuning Made Incredibly Easy!* Lippincott, Williams and Wilkins, Philadelphia PA.

Vincent JL, Gerlach H (2004) Fluid resuscitation in severe sepsis and septic shock: an evidence-based review. *Critical Care Medicine.* 32, 11 Suppl, S451–454.

Critical Thinking

1. What is the difference among hypovolaemic, cardiogenic, and septic shock?

SHARON GARRETSON is nurse manager and SHELLY MALBERTI is clinical co-ordinator, Intensive Care Unit and Step-Down Unit, University Hospitals Richmond Medical Center, Richmond Heights, Ohio, United States.

Acknowledgements—The authors would like to acknowledge the invaluable assistance of Heather Kish, medical librarian, and Mary Beth Rauzi, learning services manager, University Hospitals Richmond Medical Center.

Tracheostomy: Facilitating Successful Discharge from Hospital to Home

BEN BOWERS AND CLAIRE SCASE

The preparation for a client to be discharged home with a tracheostomy will need to address the physical changes and psychological responses to this life-changing health condition (Scase, 2004). Having a tracheostomy will impact on an individual's general health, psychological wellbeing, lifestyle choices and relationships. These considerations define how well each person adjusts to having a long-term tracheostomy and indicates the future level of support they will need. An individual who has psychological issues associated with caring for their tracheostomy will need extra support when they go home (Mason et al, 1992).

Ideally, in order for a client to be discharged home with their tracheostomy they should be independent with their care needs. They will have to perform intricate and complex physical tasks to care for their tracheostomy safely and effectively (Schreiber, 2001). The individual's manual dexterity should be assessed to ascertain if this is achievable. Furthermore, the person must understand the management issues associated with their tracheostomy tube if they are to provide their own care (Lewarski, 2005). It is essential for the multidisciplinary care team to consider the individual's intellectual ability in light of this. Cognitive impairment or any anticipated decline in function may bring into question their ability to self-care. Similarly, the client's psychological wellbeing also needs to be considered. A successful lifestyle with a tracheostomy is dependant on the individual's aptitude, motivation and attitude towards managing their day-to-day routine and responding to complications.

An alternative to self-care, in the event of the person not being able to fulfil these criteria, is to identify a key carer who will be responsible for meeting the client's tracheostomy care needs. This individual will need to be capable of providing the physical care needs of the client with a tracheostomy, available to commit to meeting their complex needs and, most importantly, be willing to adopt the role (Krouse et al, 2004). The client or the healthcare team should not assume this is a long-term answer, as the commitment can be intense, demanding and can overwhelm any carer. Caring for someone with a tracheostomy causes a substantial amount of caregiver strain, and this must be recognized by everyone involved (Ferrario et al, 2001). The healthcare team may suggest respite options to offer the key carer, in order to avoid exhaustion and potential failure to cope. Depending on the level of input the individual requires, the carer may have a financial impact to consider. If the individual requires almost constant supervision, the carer concerned may have to give up work.

Having established who will be providing the individual's tracheostomy care, inpatient training will be required to ensure the individual is capable and competent to be discharged. The individual, or their carer, must be able to perform all aspects of tracheostomy care and know how to cope with complications and emergency situations. They should feel confident with when and how to change and clean the inner cannula, apply humidification and suction, undertaking stoma and skin care (Mason et al, 1992).

Preparing Clients for Going Home

It is important to define and clarify early in the discharge planning stage the role of the community nursing team in supporting each client with their tracheostomy care. Nurses on the ward should discuss what expectations the individual and their family have of the community nursing team and ensure these remain realistic. It is important to emphasize to the patient and carers the of the service the community team can provide. Community nurses are there to offer guidance and support individuals and their carers so that they can retain the best quality of life. As previously discussed, it will be the patient's and/or carer's responsibility to perform the care needs of the tracheostomy at home. Certain aspects of tracheostomy care, including suctioning and humidification, will need to be provided by the client or their carer without delay, as they may be required at any time during a 24-hour period. They will need to respond instantly to the changing needs of the tracheostomy at any time of day or night, a provision which the community team will be unable to fulfil.

Communication between Services

Good communication between the ward staff and the community nursing team is essential in building up the team's confidence in meeting the client's needs at home. Community nurses may be anxious and concerned about the thought of caring for an individual with a tracheostomy once they return home (Barnett, 2005). Clients with newly-formed tracheostomies are infrequently seen in the community setting and this can understandably lead to nurses feeling they lack the skills and confidence to care for this client group. However, through hospital and community-based services working closely together the client's care is more likely to remain coordinated and person-centred (Department of Health, 2005).

Barnett (2005) suggests that the client's care should be discussed with the district nursing team 1–2 weeks before their discharge home in order to allow staff to become appropriately prepared and organized. In our experience, community nurses feel more confident and prepared if they are made aware of the individual as soon as they are identified as requiring a long-term tracheostomy. This gives the district nursing team time to visit the ward, introduce themselves to the client, become familiar with the type of tracheostomy tube the individual has and identify what level of support they should provide. Likewise, clients benefit from building up a relationship with the staff who will be supporting them at home.

Community Team Preparation and Education

Supporting an individual with a tracheostomy requires an understanding of the impact the tracheostomy tube has on the client's airway and knowing how to manage potential complications (Serra, 2000). Therefore, the community team will require background knowledge on the tracheostomy. The indication for the tracheostomy will have a direct influence on the content and emphasis of support and education required following the discharge (Wilson and Malley, 1990). As each tracheostomy is different, community nurses should be informed why the tracheostomy is required and what the function of the tube will be (*Table 1*). This will confirm whether the tube serves as a primary airway for the individual or whether they still have a functioning upper airway. In the event of an airway emergency this knowledge will be vital to the community team.

The type of tracheostomy tube will vary between individuals (e.g. plastic, silicone, silver) and style (e.g. cuffed or uncuffed). This information will influence how the client's airway is managed once at home. From an early stage in the discharge process the community nursing team needs to know

Table 1   Reasons Why a Long-term or Permanent Tracheostomy May Be Required

- The individual is dependent on mechanical ventilation for respiratory support
- To bypass a long-term or permanent upper airway obstruction (for example, congenital abnormalities, tumour, vocal cord palsy)
- To provide access to chest secretions in the event of respiratory insufficiency
- To protect from aspiration in the event of impaired swallow reflex (for example, neuromuscular disorders)

Sources: Serra (2000): Barnett (2005).

what type of tracheostomy tube the individual will have. This knowledge will help the team determine the management plan and possible care requirements of the individual (Tamburri, 2000; Docherty and Bench, 2002). Whenever possible, a double-lumen tube with a removable inner-cannula should be used. The regular removing and cleaning of the inner cannula will prevent the build-up of secretions and may reduce the risk of tube blockage.

Hospital-based staff will need to ensure the community nurses have a knowledge and skills base to oversee routine care needs including stoma care, dressings and tape collar changes. The hospital visit by the community nurses provides an opportunity for ward staff to establish the level of support and training their community colleagues will require. Witnessing and discussing the process of providing tracheostomy care with nursing staff on the ward will reassure community nurses and enable them to feel more confident in providing care. Practicing tracheal suctioning on a specialised mannequin and examining the tracheostomy tubes can also be beneficial (Woodrow, 2002). Ideally, if possible, one member of the community team should return to the ward to participate in the client's next tube change. Not only does this allow the nurse to familiarize herself with the process, it also gives her an ideal opportunity to practice suctioning and experience best practice in maintaining a secure and open airway.

The community nursing team will also require knowledge and understanding of what is 'normal' for the client with a tracheostomy and how to identify complications or an unexpected deterioration in the person's clinical condition (*Table 2*). It is essential that the community nurses are aware of these potential complications and know what to do when they arise. They can then, in turn provide effective support and advice for the individual and their carer. Clear guidance on when and how to contact hospital based support services for further intervention is also paramount. Both the client and the community team will require access to the hospital team for information and advice.

Table 2 Potential Complications Associated with a Long-term Tracheostomy

- Wound infection
- Hypergranulation to stoma
- Chest infection
- Haemorrhage
- Aspiration
- Tracheoesphageal fistula
- Granuloma of stoma or tracheal lumen

Sources: Schrelber (2001): Price (2004).

Tube Changes

Arrangements for changing the client's tracheostomy tube following discharge should be agreed before they go home. The frequency of tube changes will depend on the individual's condition, clinical needs and the type of tracheostomy tube used. Whenever possible, a double lumen tube with a removable inner cannula should be used. A fundamental consideration will be to determine who will be performing the tube changes. The tube change may be straightforward and can be carried out by the district nurse, carer or even the individual concerned. However, if the procedure is complex, clinical practitioners will need to possess advanced clinical skills and experience to be able to perform the tracheostomy tube changes. Barnett (2005) observes that this is the area which causes the most concern for community nurses managing patients with a tracheostomy tube.

The history of the client's previous tube changes and current clinical needs should be considered to identify the suitability for community-based tube changes and the potential risk of the procedure (*Table 3*). In view of these considerations it may not be appropriate for community nurses to carry out an individual's tracheostomy tube change, particularly as they can not access immediate clinical support in the event of a difficult tube change. It may be more appropriate for the individual to return to the ear, nose and throat (ENT) unit for tube changes or have it done at home by a specialist nurse. Agreeing with the care plan for each client before discharge saves confusion and ensures continuity in tracheostomy tube changes.

Equipment and Supplies

Tracheostomy equipment needs to be ordered well in advance of the client coming home. An appropriate portable suction unit and portable nebuliser machine will need to be available in the client's home in time for their discharge (Tattersall, 2005). Following the discharge, the community team will be responsible for ordering ongoing supplies. Community nurses are often unfamiliar with the equipment and ward staff need to give clear guidance on what is needed

Table 3 Questions to Ask When Assessing the Suitability of Carrying Out a Client's Tracheostomy Tube Change in the Community Setting

- How often does the particular tube need to be changed? (This should be in accordance with the manufacturer guidelines)
- Has there been any difficulty or incident during previous tube changes, which may alert the team to the particular risks associated with the individual patient?
- Is there any anticipated deterioration in the individual's clinical condition which will cause increasing difficulty during tube changes? (for example increasing turnour size or tumour haemorrhage)
- Do the clinicians undertaking the procedure have the knowledge, experience and equipment to manage tube change complications in the home setting?

Source: Russell (2004).

and the rationale and best practice on when to replace supplies. We suggest that ward staff supply the community team with a list of the stock items required for the individual early on in the discharge planning process (*Table 4*). The provision of this list will save time, reduces the risk of ordering incorrect items and gives clinical practitioners the opportunity to identify the best uses of limited financial resources.

Table 4 Tracheostomy Equipment and Supplies List

- Tracheostomy tubes (of current size)
- Tracheostomy tube (smaller size*)
- Cuff pressure manometer (cuffed tubes)
- 10 ml syringes (cuffed tubes)
- 2 Inner cannula's
- Control-tip suction catheters
- Suction tubing
- Nebuliser dispenser
- Tracheostomy mask
- Heat moisture exchanger/Buchanan bibs
- Tracheostomy dressings
- Velcro collars/tracheostomy tapes
- Inner cannula cleaning brushes
- Speaking valve (if required)
- Lubricating jelly (single use tubes or sachets)

*For use in the event of being unable to insert current size tube.
Note: Tracheal dilators have not been included within this list. Unless the dilators are used correctly, there is a risk of causing damage to the trachea. An alternative, yet effective method of maintaining a tracheostomy opening is to hold an inner cannula to the a tube can be reinserted.

Some tracheostomy care items routinely used in the hospital setting remain unavailable on prescription and will need to be specially ordered from suppliers. This can take several weeks and often involves negotiation between primary care service budget holders as to who will pay for the items. Availability of supplies and funding can delay the provision of equipment. In our experience, this can be a very real obstacle in promptly getting the individual home. To ensure a smooth transition from hospital to home the discharging ward should provide an agreed quantity of supplies for the client to take home on the day of discharge. We recommend at least a 7-day supply.

Informing Other Agencies

Effective communication with other agencies is vital during the discharge planning process (Barnett, 2005). Emergency support services should be contacted in preparation for the client's safe discharge home. The Local Ambulance Control should be informed that the individual is a 'neck breather'. This will then be highlighted against the person's address and emergency crews will have advance knowledge of the individual's altered airway.

If the local electricity supplier is aware of the individual's clinical needs, they can keep the client informed in advance of planned power cuts to ensure their suction and nebuliser units are fully charged.

Supporting Self-Care at Home

Individuals who are quite competent with self-caring for their tracheostomies in hospital can become less motivated and reduce their standard of self-care once they are at home. Once discharged home, individuals may become more socially withdrawn and depressed leading to reduced levels of self-care (Mason et al, 1992; Hutton and Williams, 2001). As Serra (2000) identifies, having an artificial tube sticking out of a hole in the neck through which sputum is expectorated is unlikely to reduce a person's anxiety or enhance their self-esteem. As one individual reported, he avoided social activities outside of his home as it was 'too embarrassing both for other people and me if I had [to give myself] suction' (Scase, 2006). The community nurse visiting the client at home is in the ideal position to monitor just how effectively they are managing to self-care. Identifying potential problems early on can help individuals maintain an effective and trouble-free airway (Docherty and Bench, 2002).

Possible signs of ineffective standards of self-care can vary but practitioners should be aware of the more common airway issues. For example, if an individual with a double lumen tracheostomy tube is not cleaning the inner cannula frequently enough it is more likely to occlude (Docherty and Bench, 2002). Some clients have been known to permanently remove their inner-tube rather than cleaning them. This situation can potentially lead to a critical incident if the outer cannula becomes blocked with secretions and occludes. Clients may simply need re-educating on why they should maintain their inner tube. However, such behaviour could be an indication that the person is not coping with own care.

Clients may also present with insecure or soiled tracheostomy tapes/tubeholders and foam dressings. This can also indicate that they have an inadequate level of self-care. Loose tapes/tubeholders can result in the tracheostomy tube becoming displaced (Serra, 2000). If the tapes/tubeholders or dressing are soiled they can cause function and quickly lead to skin breakdown (Harkin, 1998: Tamburri, 2000). Individuals tend to look in a mirror to ensure they have the correct tension in tapes or to change their tracheostomy dressings. Therefore, if these items are poorly maintained it could be an indication that the individual is avoiding looking at their reflection and is not coping with their altered body image (Dropkin, 1989).

If a client has any self-care deficits then it is important to encourage them to vocalize their concerns and seek psychological counselling if it is needed (Minsley and Wrenn, 1996). Re-education and encouragement on tracheostomy self-care can help maintain a client's independence but this may have to be combined with more frequent support visits. Equally, the local head and neck services may be able to offer clients further assistance and specialist support (National Institute for Clinical Excellence, 2004).

Conclusion

Excellent communication and coordination of care between hospital and community services from the earliest opportunity in the discharge process is crucial in enabling a good quality of care to continue once a client returns home. The experience of being discharged home with a tracheostomy often remains daunting process for the client and their carers. By providing ongoing, individualized education and support, hospital and community nursing staff can ensure the client and their carer are able to manage safely at home. However, if community nurses are to effectively provide clients support at home, they need appropriate training and ongoing access to advice from head and neck specialist services.

References

Barnett M (2005) Tracheostomy management and care. *Journal of community Nursing* **19**(1): 4–8

Docherty B. Bench S (2002) Tracheostomy management for patients in general ward settings. *Prof Nurse* **18**(2): 100–4

Dropkin MJ (1989) Coping with disfigurement and dysfunction after head and neck cancer surgery: a conceptual framework. *Semin Oncol Nurse,* **5**(3): 213–9

Ferrario S, Zotti A, Zaccaria S, Donner C (2001) Caregiver strain associated with tracheostomy in chronic respiratory failure. *Chest* **119**(5): 1498–1502

Harkin H (1998) Tracheostomy management Nurs *Times* **94**(21): 56–8

Hutton JM, Williams M (2001) An investigation of psychological distress in patients who have been treated for head and neck cancer *Br J Oral Maxillofac Surg* **39**: 333–9

Krouse H, Rudy S. Vallerand A et al (2004) Impact of tracheostomy or laryngertony on spousal and caregiver relationship. *Head* **22**(1): 10–25

Lewarski J (2005) Long-term care of the patient with a tracheostomy *Respir Care* **50**(4): 534–7

Mapp C. (1988) Trach care: Are you aware of the dangers? *Nursing* **18**(7): 34–42

Mason J, Murry G, Foster H, Bradley P (1992) Tracheostomy self care the Nottingham system. **106**: 723–4

Minsley M. Wrenn S (1996) Long-term care of the tracheostomy patient from an outpatient nursing perspective. ORL **14**(4): 18–22

National Institute for Clinical Excellence (2004) *On Cancer Services Improving Outcomes In Head And Neck The Manual NICE,* London.

Department of Health (2005) *National Service Framework for Long-term conditions* DH, London

Price T (2004) Surgical tracheostomy, In: Russell C. Matta, B. eds. *Tracheostomy A Multiprofessional Handbook.* Greenwich Medical Media Ltd. London 56–7

Sease C (2004) Long-term tracheostomy and continuing care. In Russell C., Matta B, eds. *Tracheostomy A Multiprofessional Handbook.* Greenwich Medical Media Ltd, London: 288

Scase C (2006) Tracheostomy questionnaire (Unpublished Audit) Cambridge University Hospitals National Health Service Foundation Trust, Cambridge.

Serta A (2000) Tracheostomy care. *Stand* **14**(42): 45–55.

Schreiber D (2001) Trach care at home. **64**(7): 43–4

Tatterall S (2005) Choosing a suction machine for use patients in the community setting *Prof* **20**(5): 50–1

Tamburri I (2000) Care of the patient with a tracheostomy **19**(2): 49–59

Wilson E, Malley N (1990) Discharge planning for the patient with a new tracheostomy, *Crit Care Nurse* **10**(7): 73–

Woodrow P (2002) Managing patients with a tracheostomy in acute care. *Nurs Stand* **16**(44): 39–48

Critical Thinking

1. What is the nurse's role in facilitating a successful transition from the hospital to homecare?

From *British Journal of Nursing—BJN,* 16:8, April 26–May 9, 2007, pp. 476–479. Copyright © 2007 by MA Healthcare Limited. All rights reserved. Reprinted by permission.

Are You Ready to Care for a Patient with an Insulin Pump?

Meet the challenge of caring for a hospitalized patient who uses an insulin pump by following these guidelines.

David K. Miller, RN-BC, CDE, CPT, MSED

More than 375,000 Americans with diabetes are using insulin pumps, according to manufacturers' estimates.[1] In the past, insulin pumps were used primarily by patients with type 1 diabetes. Now patients with type 2 and gestational diabetes are also choosing continuous subcutaneous insulin infusion (CSII) or insulin pump therapy because of its advantages compared with multiple daily insulin injection therapy.[2] Used properly, insulin pump therapy can decrease the frequency and severity of hypoglycemia. (See *Looking at the pros and cons of insulin pumps.*)

This article will explain how to care for a hospitalized patient with an insulin pump, who's likely to be strongly motivated and involved in self-care. First, consider some background on these devices.

Boost Your Pump I.Q.

An insulin pump is a small battery-powered device, about the size of a small cell phone, that continuously delivers a subcutaneous infusion of short- or rapid-acting insulin 24 hours a day. Most insulin pumps are attached to an infusion set, which includes a soft plastic cannula inserted under the skin. This set must be changed every 2 or 3 days or according to the manufacturer's recommendation.

An insulin pump has a reservoir that's filled with insulin and a microcomputer that allows adjustment of the amount of insulin delivered to the patient. This basal insulin infusion keeps your patient's blood glucose levels in target range between meals and overnight. Additional insulin dosages can be programmed to be delivered at different times of the day and night (bolus insulin); for example, to cover carbohydrates in meals or to treat high blood glucose levels.[3] The patient programs the pump according to the healthcare provider's orders. The patient then administers the insulin according to the amount of carbohydrates consumed, blood glucose level, or both.

When the basal rate is set correctly, the patient's blood glucose won't rise or fall between meals. Some insulin pumps can calculate bolus insulin doses after the user provides blood glucose readings or carbohydrates eaten.

Looking at the Pros and Cons of Insulin Pumps

Here are some advantages of insulin pump therapy:

- tighter glycemic control, which decreases the risks of long-term complications
- variable basal rates can be used to accommodate fluctuations in insulin requirements during the night caused by hormonal releases
- reducing basal rates during low physiologic requirements may lessen the frequency and severity of hypoglycemia
- improved convenience, flexibility, satisfaction, and lifestyle
- meals and snacks can be customized
- insulin needs can be tailored to changes in schedules

- more precise dosing because basal and bolus doses can be delivered in 0.5 units.

Here are some disadvantages of pump therapy:

- the pump is a constant reminder of the disease
- frequent self-blood glucose testing is required
- skin irritations and infections can occur if the patient fails to use proper insertion and skin care techniques
- possible technical and mechanical failures
- diabetes ketoacidosis can occur very quickly in patients with type 1 diabetes
- the insulin pump and supplies are costly.

Basal rates are set as units/hour. The average rate is between 0.4 and 1.6 units/hour.[1] Variable basal rates can be set to accommodate fluctuations in insulin requirements during the night.

The correction factor is the amount of blood glucose (mg/dL or mmol/L) that will be reduced by one unit of insulin. To calculate the correction bolus, use the 2,000 rule if the basal rate is 50% of the patient's daily dose. Divide 2,000 by the total units of insulin per day. (If the patient's basal rate is 40% of the total daily dose, use the 1,800 rule. Divide 1,800 by the total units of insulin per day.)[1] The healthcare provider makes these calculations when determining initial orders.

The second type of bolus covers the carbohydrates eaten at a meal or snack. The pump uses an insulin-to-carbohydrate ratio to determine the bolus amount. Carbohydrate bolus doses are established to match the carbohydrate content of foods. Usually, one unit of insulin will cover 8 to 20 grams of carbohydrates for most adults and adolescents, but this will vary based on the patient's insulin sensitivity and weight.

The insulin-to-carbohydrate ratio is based on the 500 (or 450) rule. Divide 500 (or 450) by the total units of insulin per day to get the number of grams of carbohydrate covered by one unit of insulin.[1] The ratio may differ at different times of the day.

All insulin pumps provide the same basic functions, but features vary among models available. The pump keeps a history of insulin used.

While using these pumps, the patient needs to check blood glucose levels according to the healthcare providers' instructions; for example:

- before meals and at bedtime
- 2 hours after meals
- between 2 and 4 A.M. weekly
- before driving or operating machinery
- when the patient has signs or symptoms of hypoglycemia or feels nauseated or sick.[4]

Some pumps take blood glucose measurements every minute and then display a 5-minute average. Fingerstick measurements with a standard blood glucose meter are still needed before adjusting therapy and calibrating the system.[4]

Some meters can work with real-time continuous glucose monitoring. Data from a tiny glucose sensor is sent to a transmitter, a small attached device. Using advanced radio frequency wireless technology, the transmitter sends the glucose data to the insulin pump. The glucose sensor, which can be worn for up to 3 days, is inserted with an automatic insertion device.[5]

How to Assess the Patient

Many hospitals have consents in place stating the nursing responsibilities as well as the patient's responsibilities during the inpatient stay. If you're caring for a hospitalized patient with an insulin pump, you'll need to assess the patient's ability to manage self-care to determine the appropriateness of continuing to use the device in the hospital. Relay the results of your assessment to the healthcare provider.

A Tall Order for Insulin Pumps

The healthcare provider's orders should include the following information for a patient in the hospital with an insulin pump:

- type of insulin
- a schedule for monitoring blood glucose (the recommended testing for intensive insulin therapy is fasting, premeal, 2-hour postprandial, bedtime, and occasionally at 3 A.M.)
- when to test for ketones
- diet
- insulin injection schedule if pump failure occurs
- basal rate (specific time frames and the number of units for each basal rate)
- blood glucose targets
- correction factor, which is also called the insulin sensitivity factor
- insulin-to-carbohydrate ratio.

Patients using pump therapy must possess good diabetes self-management skills. They must also have a willingness to monitor their blood glucose frequently (at least four times per day) and record blood glucose readings, carbohydrate intake, insulin boluses, and exercise. Standards of practice include letting the patient self-monitor blood glucose in the hospital, but you should monitor it as well.

Standards of practice include letting the patient self-monitor blood glucose in the hospital, but you should monitor it as well.

Besides assessing the patient's physical and mental status, review and record pump-specific information, such as the pump's make and model. Also assess the type of insulin being delivered and the date when the infusion site was changed last.

Assess the patient's level of consciousness and cognitive status. If the patient doesn't seem competent to operate the pump, notify the healthcare provider and document your findings. Consider any possible impairment related to the patient's medical condition or prescribed medications that prevent competent use of the pump.

An insulin pump doesn't guarantee good insulin control, so routinely evaluate self-care behaviors to determine the patient's willingness to manage the pump correctly. Some pump users get tired of managing the pump, which can be evident if the patient has been hospitalized for episodes of unstable blood glucose, such as diabetic ketoacidosis or severe hypoglycemia. Also assess the condition of the infusion site. If it's poor, this can also indicate lack of self-care.

Assess whether the patient needs more education about how to use the pump properly, especially if there have been problems in the past. Alert the healthcare provider if the patient doesn't seem competent to use it properly.

Counting Carbohydrates

To regulate the insulin dosage delivered by a pump, your patient must be able to count carbohydrates. The first level is *basic carbohydrate counting,* which usually consists of carbohydrate servings, not grams of carbohydrates, as well as how to manage patterns of blood glucose levels. That is, the patient is able to match food eaten with glucose levels. For instance, the patient may say that after eating pizza, glucose levels are always higher.

The second level is *advanced carbohydrate counting.* In this level, the patient matches the prescribed insulin dose to personal carbohydrate consumption. The patient may also develop an understanding of the basal-bolus concept and provide food records to help demonstrate the ability to count carbohydrates. To use a pump, the patient must be able to count grams of carbohydrates and understand the basal-bolus concept.

Your Role

If using the insulin pump in the hospital is appropriate for the patient, you'll need to obtain orders from the healthcare provider. See *A tall order for insulin pumps.*

Once you receive the orders, ask the patient to show you the set basal rates, how to suspend delivery, time and amount of last bolus, and review of history. The patient needs to demonstrate how to administer a bolus dose, change the infusion set, and rotate sites. Observe the patient administering all bolus insulin doses. Verify the amount of the bolus and that the patient is following orders.

You'll need to make arrangements with the patient or the patient's support person to bring in supplies needed. With all the equipment available, it's not practical for hospitals to stock pump supplies other than insulin and batteries (AA and AAA). Make sure the patient has enough supplies to change his infusion sites every 2 or 3 days throughout his hospitalization.

The insulin pump should be discontinued for certain tests, including magnetic resonance imaging and computed tomography scans. The patient can safely disconnect the pump and most infusion sets for up to 1 hour but he should check his blood glucose before disconnecting and after reconnecting.[4] Never place the insulin pump in the direct line of X-rays. Take special care to avoid dislodging the catheter when transferring the patient for these procedures.

Finally, document all of your interventions. See *Documenting the details.*

Solving Problems

Potential acute problems for patients with insulin pumps include hypoglycemia, hyperglycemia, diabetic ketoacidosis, and skin infections. Patients using insulin pumps don't seem to have a higher incidence of hypoglycemia than those managing their diabetes with multiple daily injections.[2] Blood glucose levels should always be checked prior to giving any insulin (bolus or correction dose).

If the patient develops hypoglycemia, treat it according to hospital protocol. If the patient can swallow, the recommended treatment is 15 grams of carbohydrate (4 oz. [120 mL] of

---

Documenting the Details

Your documentation should include the following information:

- type and amount of insulin
- basal rate or rates
- bolus insulin doses (number of doses and units given)
- any supplemental insulin given by injection
- blood glucose levels, including any the patient checked with a personal monitor
- condition of the infusion site
- change of infusion site
- the make and model of the insulin pump
- when the insulin pump is suspended or removed, such as for showers or procedures, the time it was removed, and where the pump was placed while it was discontinued.

Looking for Trouble

If the pump isn't delivering insulin as expected, check for the following:

- erythema, edema, or tenderness of the site
- leakage, breakage, or kinking of the tubing
- battery failure
- empty reservoir or cartridge
- improper basal rate programming
- air in the tubing
- omitted bolus or improper amount given
- ineffective insulin (past expiration date, exposure to heat or cold)
- crimped catheter or needle not penetrating the skin
- insulin leaks at site.

Source: Walsh J, Roberts R. *Pumping Insulin.* San Diego, CA: Torrey Pines Press; 2006.

---

orange juice or 6 oz.[180 mL] of regular (not diet) soda or two glucose tablets or two doses of glucose gel). Repeat the blood glucose testing in 15 minutes. Repeat the same steps if necessary. If the patient is N.P.O. or can't swallow, administer 50 mL D50 I.V. (1 amp) and start I.V. D_5W.

If insulin delivery is interrupted for any reason, hyperglycemia and diabetic ketoacidosis can occur within 2 hours. Treat the patient according to hospital protocol and assess the catheter site for problems. (See *Looking for trouble.*) Provide supplemental insulin by subcutaneous injection. If the patient has ketoacidosis, he'll need I.V. insulin.

If you encounter problems with pump operation, obtain a consult with a certified diabetes educator who's also a certified insulin pump trainer. The manufacturers of most insulin pumps provide a toll-free telephone number on the back of the device for round-the-clock technical support.

Being Ready

When patients who use an insulin pump are admitted to your unit, you'll now be better prepared to help them continue using the pump throughout hospitalization, if appropriate. This treatment option lets them continue to manage their own condition and may increase patient satisfaction.

References

1. Walsh J, Roberts R. *Pumping Insulin.* San Diego, CA: Torrey Pines Press; 2006.

2. American Diabetes Association. Continuous subcutaneous insulin infusion. *Diabetes Care.* 2004; 27 (suppl 1):S110.

3. Sanofi Aventis. Frequently asked questions about Apidra and insulin pump therapy. http://www.apidra.com/pump_friendly/pump_faq.aspx.

4. Animas Corporation. Insulin pumping education. http://www.animascorp.com/ViewArticles.aspx?Subcategory=1.

5. Medtronic. REAL-time continuous glucose monitoring. http://www.medtronicdiabetes.com/products/insulinpumps/components.cgm.html.

Resources

American Diabetes Association. Insulin pumps. http://www.diabetes.org/type-1-diabetes/insulin-pumps.jsp.

Mensing C. ed. *The Art and Science of Diabetes Self-Management Education: A Desk Reference for Healthcare Professionals.* Chicago, IL: American Association of Diabetes Educators; 2006.

Critical Thinking

1. What is the nurse's role in assessing the patient who is receiving insulin via an insulin pump?

DAVID K. MILLER is the president of Health Education and Life Promotion in Hope, Ind.

From *Nursing,* October 2009, pp. 57–60. Copyright © 2009 by Lippincott, Williams & Wilkins/Wolters Kluwer Health. Reprinted by permission via Rightslink.

Head Attack

You're late, the traffic is a nightmare and you're yelling at the kids to stop fighting in the back. Is your mental stress putting you at greater risk for a heart attack?

MICHAEL FELD AND JOHANN CASPAR RÜEGG

Gerry suddenly clutched at his chest. His heart was racing, and he could barely breathe. Ten minutes after the call to 911, he was on his way to the nearest emergency room in an ambulance. There an electrocardiogram and blood tests provided the big shock: Gerry hadn't suffered a heart attack at all. The hospital doctor reassured him: "Physically, you are fine. Your problems are psychological in origin."

Gerry's experience is not unusual. For at least a quarter of all patients who enter hospitals with suspected heart attacks, physicians can find no physical cause for their symptoms. But it is a mistake to dismiss such occurrences as "just psychosomatic," because that minimizes the importance of the mind's effects on the body's well-being. Studies in psychosomatics, the area of medicine that deals with diseases and complaints that are at least partly psychologically based, find that one everyday aspect of modern life stands out in a startling variety of physical ailments: stress. [*For a list of related ills, see box.*] Worse, extreme emotional distress—caused by the death of a spouse, a furious quarrel, a natural disaster such as an earthquake, even looming heavy deadlines at work—can trigger a real heart attack in a person who is already at risk.

In the United States alone, 1.5 million people suffer heart attacks every year, and more than 200,000 die. It is difficult to determine how many of those incidents might be attributed to stress, but it is clear that duress plays a role. Andrew Steptoe and Philip C. Strike of University College London recently reviewed a number of medical studies conducted between 1974 and 2004 that examined what people were doing and feeling in the hours before they had a heart attack. Emotional stress was one of the most common triggers, they reported in the March/April issue of *Psychosomatic Medicine*. For example, in one study of 224 patients, more than half said they had been very upset or under stress in the 24 hours before their heart attack.

Mind over Matter

How can your head hurt your heart? To answer that question, it helps to take a look at what happens in the body when you are experiencing stress. Imagine you are ambling across a street when a car unexpectedly rounds the corner without stopping, barreling toward you. Heart pounding, legs pumping, you dash out of harm's way. What just happened?

As your brain recognizes imminent danger, your body undergoes several changes. Stress hormones—epinephrine, norepinephrine, glucocorticoids—pour into your bloodstream, preparing you for a "fight or flight" response. To conserve energy for your leg muscles,

nonessentials such as your digestive tract shut down. Your heart rate increases, to deliver oxygen and energy to your thighs and calves. Veins throughout the circulatory system constrict, as when you squeeze a water hose, propelling blood back to the heart more vigorously. That returning blood slams into heart walls, which in turn snap back with greater force, like a stretched rubber band. Arteries relax, increasing blood flow from the heart to those needy muscles.

Such physical reactions are helpful when you are bolting from a careless driver—or when early humans had to flee a hungry predator. And small stresses actually have an upside, because they sharpen our attention, making us feel focused and alert. (Think of playing a challenging quiz game or watching an exciting whodunit.)

But stress also arises frequently from the everyday hassles of modern life, as we run late to that meeting, fret about getting the kids to a play date across town or worry about getting all the details just right in time for tonight's dinner party. We are especially susceptible when we feel that conditions are out of our control despite our struggles. The result is that our bodies keep working in overdrive far more than our evolutionary history has shaped us to do. Chronic stress can lead to high blood pressure. This hypertension, in turn, adds to a vicious cycle of physical changes that can tip the balance for people at risk, contributing to the onset of arrhythmia (irregular heartbeat, in which distended muscle chambers cannot efficiently pump out blood) or heart attack. In a heart attack, a clump of plaque lodges in a small vessel in the heart. The resulting blockage deprives nearby cells of nutrients and oxygen, starving them.

Stress experiments have revealed the mental mechanisms involved. In the 1990s James E. Skinner, now at the Vicor Technologies laboratory in Bangor, Pa., investigated which brain regions play a role. He worked with pigs, beginning by tying off one coronary artery to imitate the condition of a patient with coronary artery disease. Then he implanted cooling elements at specific spots to block nerve impulses running from the frontal lobe, the location of higher-reasoning centers in the brain, to areas involved in emotional reactions and in mediating excitatory hormones: the amygdala, hypothalamus, brain stem and sympathetic nervous system. When the pigs without nerve blocks were exposed to severe psychosocial stress—such as being put in entirely new, alarming surroundings—they often experienced fatal fibrillation, a condition in which the heart contracts erratically and does not pump blood. Similarly, electrical stimulation of certain parts of the frontal lobe in the pigs elicited a rapid heart rate and arrhythmias, in some cases leading to cardiac arrest. The pigs whose nerves had been blocked by cold, however, were spared.

The Hostile Heart

To evaluate your overall tendency toward stressful hostility, use the Minnesota Multiphasic Personality Inventory Anger Content Scale. Answer true or false to each question:

T F

○ ○ **1.** At times I feel like **swearing**.

○ ○ **2.** At times I feel like **smashing** things.

○ ○ **3.** Often I can't understand why I've been so **irritable and grouchy.**

○ ○ **4.** At times I feel like picking a **fistfight** with someone.

○ ○ **5.** I easily become **impatient** with people.

○ ○ **6.** I am often said to be **hotheaded.**

○ ○ **7.** I am often so **annoyed** when someone tries to get ahead of me in a line of people that I speak to that person about it.

○ ○ **8.** I have at times had to be **rough** with people who were rude or annoying.

○ ○ **9.** I am often sorry because I am so **irritable and grouchy.**

○ ○ **10.** It makes me **angry** to have people hurry me.

○ ○ **11.** I am very **stubborn**.

○ ○ **12.** Sometimes I get so **angry and upset** I don't know what comes over me.

○ ○ **13.** I have gotten **angry** and broken furniture or dishes when I was drinking.

○ ○ **14.** I have become so angry with someone that I have felt as if I would **explode.**

○ ○ **15.** I've been so angry at times that I've **hurt someone** in a physical fight.

○ ○ **16.** I almost never lose **self-control.**

0-5 Anger is not a problem.

6-10 Anger level is moderate: work on ways to relax.

11-16 Anger level is a concern; your health may suffer the consequences if corrective measures are not taken.

Heart at Risk

Chronic stress leads to high blood pressure, which in turn causes a cycle of physical changes in the body that contribute to the risk of heart attack. Among the changes is a narrowing of coronary arteries in part from the buildup of plaque. A rupture in diseased blood vessels can result in a blood clot, which can lodge in a narrowed artery, causing a heart attack. Heart cells near the blocked vessel are deprived of nutrients and oxygen, and they may die.

Head to Heart

So what are the important emotional factors? In the early 1900s, Hungarian-American psychoanalyst and psychiatrist Franz Gabriel Alexander, now often called the father of psychosomatics, played a leading role in identifying emotional tension as a significant cause of physical illness. Alexander and other pioneers in the field believed that disorders such as ulcers, high blood pressure, neurodermatitis and asthma were the body's reaction to chronic tension and psychological stress. Following in the footsteps of psychoanalysts, they held that certain individuals—who suppressed conflicts and emotions—were predisposed to develop ailments as a result. This point of view has fallen out of favor today, as purported links between certain personality types and diseases have been refuted. For example, many studies have shown that the melancholy "cancer personality" is just a myth. On the other hand, a given person's style of dealing with problems does matter.

That is what heart specialists-Meyer Friedman and Ray Rosenman concluded in 1974, after conducting a multiyear study of people with so-called Type A personalities. They claimed these individuals—whose behavior is characterized by ambition, competitiveness and impatience—have a considerably higher risk of heart attacks. In several additional studies, researchers sought a comprehensive evaluation of Friedman and Rosenman's belief; they were not able to provide confirmation. Yet the aggression and hostility exhibited by Type As contribute to higher levels of stress and its deleterious effects. And although Type As do not necessarily have an increased lifetime risk of having a heart attack, their short-tempered, impatient behavior makes it more likely that they will have a heart attack sooner, according to a study in the May/June 2003 issue of *Psychosomatic Medicine* by John E. J. Gallacher of the University of Wales College of Medicine.

Mental Illnesses

According to current scientific thinking, the following disorders are among those believed to be at least partially caused psychosomatically:

Gastrointestinal system: Eating disorders (anorexia, bulimia, psychogenic obesity), constipation, irritable bowel syndrome, gastric ulcers

Cardiovascular system: High blood pressure (hypertension), syncope (fainting spells), cardiovascular heart disease, arrhythmia, heart attack

Airways: Asthma, nervous cough

Psychosomatic pains: Headaches, abdominal pains, soft-tissue rheumatism, certain muscular pains (myalgia)

Ear, nose and throat: Dizziness, hearing problems, tinnitus, swallowing problems

Endocrine system: Diabetes, psychosomatic dwarfism

Reproductive system: Male dysfunction, menstrual cycle disturbances, false pregnancy

Skin: Neurodermatitis, psychogenic pruritus, possibly psoriasis

Many top executives may be Type As, but simply being a Master of the Universe does not raise the risk of heart attack, perhaps because those at the pinnacle of the corporate hierarchy have greater control over their day-to-day working lives than their minions. Middle-ranking employees are more likely to suffer a special kind of stress, called the negative affect. People with this sensitivity disorder exhibit above-average levels of anxiety and depression. After a multiyear study of the negative affect in men and women, Bruce C. Jonas and James F. Lando of the Centers for Disease Control and Prevention reported in the April 2000 *Psychosomatic Medicine* that such chronically stressed people are twice as likely to have hypertension as normal individuals.

Men who explode with anger or expect the worst from people may punish their own bodies as well. Such men are more likely to develop a type of arrhythmia, says an article in the March 2004 issue of *Circulation*. Feelings of hostility, for example, made men 30 percent more likely to develop the condition. Other studies have shown that a strong adverse emotion such as anger doubles the risk of heart attack during the next couple of hours. [*See* [The Hostile Heart] *to find out if hostility might be a problem for you.*]

Men who explode with anger or expect the worst from people may punish their own bodies as well.

Irritation and fury are not the only threats to diseased coronary arteries. Nancy Frasure-Smith of McGill University believes that depression also seriously prejudices the chances of heart patients for recovery. Depression, in turn, can result from chronic uncontrollable stress, as well as from a previous heart attack. Victims often suffer from inner hopelessness, such as fears of being unable to meet challenges in their work or personal lives. And loss of a beloved and trusted partner can literally break someone's heart: as long ago as 1969 Colin Murray Parkes, a British doctor, showed that widows and widowers suffered greatly increased mortality.

Looking on the Bright Side

As the work with the pigs showed, the frontal brain seems important in fibrillation and apparently is connected to the nerve cell bodies of the sympathetic nervous system in the spinal cord. Through this connection, the human mind ought to be able to influence heart function in a positive manner. Relaxation techniques such as autogenic training may possibly utilize this mechanism. Along with targeted stress management, such methods may improve the survival chances of heart patients more than daily exercise, as suggested by James A. Blumenthal of the Duke University Medical Center in 1997.

Psychotherapy's positive influence on bodily processes is especially evident in studies of pain patients. Neuropsychologist Pierre Rainville of the University of Montreal set up a therapeutic study based on suggestion, called guided imaging. Using positron-emission tomography (PET) imaging, he discovered that a brain region responsible for the conscious awareness of pain, the anterior cingulate gyrus, would become less active—merely because of spoken words.

Another means to break free of the self-reinforcing cycle of heart disease, stress and depression is cognitive behavioral therapy. Patients learn to give more weight to positive events in their lives than to negative ones [see "Treating Depression: Pills or Talk?" by Steven D. Hollon, Michael E. Thase and John C. Markowitz; SCIENTIFIC AMERICAN MIND, Premier Issue, 2004]. A strong social network, as well as contact with trusted individuals, helps people to overcome stress, too. Heart disease patients who are married or in stable relationships have longer average life expectancies.

Two other important ingredients to reversing cardiovascular disease are developing more healthful eating habits and exercising regularly. Dean Ornish and his colleagues at the University of California at San Francisco tracked the progress of patients with coronary artery disease who ate low-fat vegetarian diets and got regular exercise. The subjects stopped smoking, and they sought to bring calm to their lives through stress management training and group therapy. After a year, the condition of their coronary arteries had improved noticeably.

Is the power of the brain supreme when it comes to affecting physical well-being, or does the body's health sway our mental states? Both usually go hand in hand: body and mind are bound up, inseparably, in a continual feedback loop. The scientific knowledge gained in recent years teaches us that just as corporeal phenomena can change our minds and spirits, it works in the other direction as well: thoughts and emotions can cause real changes to our bodies.

Critical Thinking

1. What is the connection between stress and physical illness?

MICHAEL FELD is a physician and freelance science writer in Cologne, Germany. **JOHANN CASPAR RIIEGG** is professor emeritus of physiology at the University of Heidelberg in Germany.

UNIT 5

Nursing Practice Areas/ Specialties

Unit Selections

Learning Outcomes

After reading this unit, you should be able to:

- List the differences between pediatric hospice and adult hospice.

- Discuss the role the Public Health nurse plays in the community.

- Describe the recent disasters that brought the Mercy Ship out of a 14-year retirement.

- Discuss how the role of public health nurses has evolved in the post 9/11 world.

- Explain the advantages of the Assisted Living setting for the geriatric population.

Student Website

www.mhhe.com/cls

Internet References

Air & Surface Transport Nurses Association
www.astna.org
APHA Public Health Nursing
www.csuchico.edu
The Hospice Care Network
www.hospice-care-network.org/HospiceWeb/pediatricprogram.cfm
The International Association for Forensic Nurses
www.iafn.org
Mercy Ships
www.mercyships.org
National School Nurse Association
www.nasn.org
New Careers in Nursing
www.newcareersinnursing.org
Pfizer
www.pfizercareerguides.com

The purpose of this unit is to reveal the many options available to individuals who have made a conscious decision to follow their dream of becoming a nurse and make a positive social change within their community.

A number of nursing specialty areas are introduced in this unit. For those of you entering nursing, consider combining your career with your interests, likes, and personality. If you begin to succumb to "burnout," consider a new practice area for your nursing skills. There are opportunities to work with patients of all ages, from the tiny neonate, to the child in need of end-of-life care, to the healthy adult on the vacation of a lifetime, to an older adult in need of assistance. Adventurous types can consider a career in forensics, or emergency preparedness, or in a far away jungle. Wherever you choose to use your skills, you will always be a nurse ministering to the minds and bodies of those patients entrusted to your care.

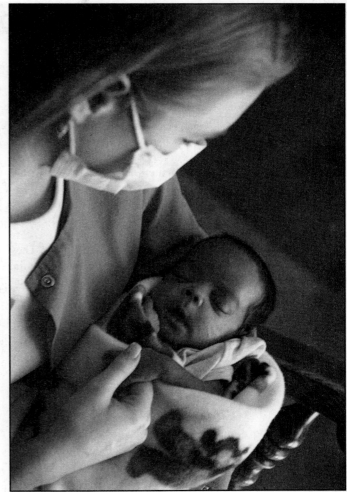

Pediatric Hospice: Butterflies

Christine Contillo, RN, BS, BSN

Mary Ann West has an office filled with framed photos of children, both her own teenagers as well as the kids she's taken care of as a pediatric nurse. Her desktop reflects the passionate way she feels about her life-long career, but not many nurses would be able to handle the emotional roller coaster she deals with daily. Mary Ann is the coordinator of The Valley Hospital's Butterflies, a program dedicated to pediatric hospice and palliative care for children with life limiting or life threatening illness. They handle hospital visits to smooth the transition from hospital to home or back to re-admission, and from curative to palliative care. They provide home care services with an interdisciplinary team specially trained in pediatric pain and symptom management.

Hospice for children is unusual for a number of reasons. As adult hospice begins to become more widely accepted, far fewer children contract fatal illnesses. Parents are often unwilling to enroll children, wanting to take every step possible to improve their chances of success before acknowledging that they've reached the limit of what's possible to cure. Through genetic counseling, some serious genetic diseases can be detected before birth. Since modern medicine has eliminated many of the childhood diseases that in previous generations had sometimes wiped out siblings in the same family, many parents are unprepared for the death of a child and have no role models to look to. Children with illnesses that will eventually prove fatal, such as brain tumors or AIDS, may still have young and relatively healthy organs, and with good care can continue to rally, passing their physician's expectations, making the decision of when to stop care almost impossible.

But there are will always be a few children who deserve the kind of end-stage care that hospice entails, and adult hospice nurses are not always willing or prepared for such an assignment, which is why the Butterflies program is so valuable to the community. It allows these children and their families to become familiar over a time period which sometimes lasts years and to develop a relationship with the nurses involved. They can then help them make and support decisions which may be overwhelming, such as whether or not to transfer to the hospital, how to get appropriate pain relief, and when or if to bring in a chaplain. An art therapist is available to the team, and will see the patient if appropriate, or can even work with

siblings. Speech, physical, and occupational are all options, as is home health aid, and dietician, and, of course, the child's physician.

But there are will always be a few children who deserve the kind of end-stage care that hospice entails, and adult hospice nurses are not always willing or prepared for such an assignment, which is why the Butterflies program is so valuable to the community.

Her core staff consists of four nurses, two social workers and a chaplain, but each also has responsibilities to the larger maternal-child health visiting nurse service which sees children that are acutely ill, or with ongoing chronic problems such as asthma or diabetes. Currently there are nine children in the Butterflies program, and the referrals come from a wide-ranging area, since most other visiting nurse services in the area don't offer comparable care. That's fine with Ms. West, because it allows her nurses to stay with hospice, keep up their skills, but not become burned out. Their aim is to support the entire family. Often as the illness increases, they find that it's the parents who need more emotional care. The nurses step in to help them recognize when the child is dying so that they can refocus their attention and use the time left in a way that will be beneficial to all.

None of this is easy. While it's always wrenching to see sick kids, she admits that it's often the parents that are more difficult to deal with. Ms. West relates that parents as full time caregivers are stressed and their emotions are being stretched to the limit. Often, though they are grateful for the care, they still resent the hospice nurse for the inevitable event she represents. Parents will make appointments, cancel, fail to open the door and in almost every way make the nurse's job more difficult than it already is. Mothers may make the decision to have the child die at home, only to decide at the last minute to transfer to the hospital. Even then, though, the results can be rewarding. One particularly difficult mom found that after her child had died that she could not leave the hospital room knowing the child was alone. The stricken mother asked the hospice nurse to hold the child in her arms until she was gone. "What more could you want?" asked Ms. West. "You help them so that they never regret their decisions. It's not your agenda, it's theirs."

Finances inevitably enter the picture, and Butterflies tries to work with all insurance plans for the benefit of the patient. They can accept Medicaid and managed care, but children are accepted regardless of their ability to pay. Often the program has to absorb costs that they find right for the child but can't convince the insurance manager of the benefit. They are so committed to the program that they find a way. While fund-raising isn't a priority, they do receive well-deserved donations from the community.

Finally, Butterflies has trained volunteers and offers anticipatory grief counseling and a bereavement follow up program. Their goal is open and honest communication that acknowledges the child's life and its impact on the family.

Critical Thinking

1. What are the differences between pediatric hospice and adult hospice?

CHRISTINE CONTILLO holds BS and BSN degrees. She has worked as a nurse since 1979. Christine has written extensively for various nursing publications as well as the *New York Times*.

Doing More with Less
Public Health Nurses Serve Their Communities

Cathryn Domrose

Sometimes Lisa Leighton's patients surprise her.

One, a 27-year-old Native-American woman, had almost died from an abdominal abscess. Like many who Leighton, RN, BSN, PHN, visits in their homes as a public health nurse on the San Francisco Department of Public Health's chronic care team, this patient had multiple health and social problems. She was overweight, had diabetes, and drank too much.

But she is a very spiritual person who saw her near-death experience as a sign to turn her life around. With Leighton helping her find the right resources, she has stayed sober for several months, been eating a nutritional diet, and is staying healthy.

She is the rare success story for the chronic care team, Leighton says, and an obvious answer to the question: Why pay for public health nurses?

But most of Leighton's patients are like a less hopeful case, an elderly, long-term alcoholic who is severely psychotic and takes 23 pills a day. He has alcoholic dementia and serious hypertension and is heading into renal failure.

Leighton fills his prescriptions, arranges his medical appointments, schedules a van to take him to appointments, calls and reminds him on the morning of an appointment or he won't go. This man lives alone, has no money and no one to look after him. He says he has a son, but can't remember his phone number.

Leighton's efforts are keeping him alive and functioning, out of the emergency room and in his own home, using fewer resources than he would in a hospital or nursing home or jail. But he's not going to get better. He's not going to get sober. He is a less obvious answer to the public health nurse question.

"The real crux of what we do is harm reduction," says Leighton, who has worked in public health for 25 years. "A lot of what we're doing is silent. It's hard to explain to people. There's no equivalent to public health nursing in the private sector. We're sort of the only game in town."

Budget Constraints

Over time, budget cuts and the nursing shortage have forced many public health departments to eliminate nursing positions, especially those of field nurses such as Leighton. A large segment of experienced public health nurses will soon be eligible to retire, with few new candidates to replace them.

Public health nurses are looking for ways to make the best use of the dwindling resources they have. They are partnering with other community groups, documenting how their work benefits entire communities, and exploring ways to bring in new nurses and develop and encourage new leaders. But ultimately, they say, the future of public health nursing depends on making voters and politicians understand that the lives at stake may be their own.

Thirty of 37 states responding to a survey last year by the Association of State and Territorial Health Officials reported shortages of public health nurses, at least twice as many as for any other public health profession. The public health nursing shortage has been exacerbated by severe state budget cuts in 2002 and 2003, the deepest in 60 years, the survey reported.

Health departments also say they have trouble attracting qualified candidates because they can't compete with hospitals offering sign-on bonuses and higher salaries.

"Even rural hospitals pay better than most public health departments," says Debbie Lee, RN, PhD, local public health nurse liaison for the Washington State Department of Health. "The positions either fade away or they convert them to some other position such as a health educator or support staff. When what we really need is a nurse."

Washington County, Ore., has been lucky enough to avoid a reduction in its field nursing staff in recent years, but demand for services has rapidly increased, says Sue Omel, RN, BSN, MPH, MS, field team supervisor for the county Department of Health and Human Services. This means nurses must put patients on waiting lists or decide to limit visits to first-time teen moms or high-risk pregnancies.

Faced with their own shortages, other social agencies including child welfare and Head Start are asking public health nurses to take on such duties as monitoring children in their homes or visiting a mother too depressed to look after her child and connect her with mental health services, Omel says.

At-risk Patients on Waiting List

The San Francisco Department of Public Health had more than 80 field nurses in the 1980s. It now has around 30. The department's maternal-child health nurses have a waiting list of referrals, says Martha Hawthorne, RN, a public health nurse on the maternal-child team.

Hawthorne worries constantly about patients falling through the cracks. The public health nurses used to see new mothers within a few days after delivery. Now it can take a week or more. One of her recent patients, a 15-year-old mother on probation, had moved out of the city by the time her referral came to Hawthorne.

Another, a Spanish-speaking mother with tuberculosis, gave her infant six times the amount of prescribed medication for two weeks because she didn't understand the English instructions on the label.

"Fortunately this baby came through with no liver damage," Hawthorne says, but a public health nurse visiting a day or two after delivery could have caught the mistake much earlier.

Public health nurses expect the shortage to worsen within the next 10 years, when many will be eligible to retire. According to data from the 2000 National Sample Survey of Registered Nurses, the age of the average public health nurse was 49, nearly four years older than the average age of all nurses.

"In the next five years, I can't even think about how many of these people are going to be gone," Lee says. "I don't know that we'll ever build back up to what we were before."

Despite the frustration of budget cuts and impending shortages, public health nursing has had high retention rates in most areas. Public health nurses say over and over that they see themselves as a last resort for their patients and that they believe in the mission of public health.

Public health nurse leaders say their profession is resilient, and they believe it will survive in some form. But to do so, it must change with the times, they add.

Changing Roles

In some cases, they say, this means letting other community agencies do some of the clinical work done by public health nurses, with the nurses taking on a coordination role, identifying what the community's health needs are and figuring out ways to meet them.

"Health departments can't do it all, and we've really become aware of that," says Betty Bekemeier, MPH, MSN, chair of the Public Health Nursing Section of the American Public Health Association and deputy director of The Turning Point National Program Office at the University of Washington School of Public Health.

"I don't want to see public health nurses not doing home visits, but there might be other providers doing them. Our main function is seeing that the needs of the entire population are being met."

Many public health nurses already work with other agencies. Leighton, the San Francisco chronic care nurse, is part of a collaboration between private and public health care organizations

looking for new ways of delivering care to diabetes patients. The collaboration could serve as a role model for all chronic illness care.

One of the most important jobs for public health nursing in the future, say public health nurses, will be to collect data that will show how it is not just individual patients who benefit from the work of the public health nurse, but the entire community.

Mentors Pass the Baton

Some health departments are working with nursing schools, offering sites for clinical practice and instructors to introduce students to public health nursing. Bekemeier suggests mentoring programs that pair experienced retired nurses with new public health nurses. Nurse leaders say health departments also need to actively recruit new leaders who can be mentored by current leaders before they retire.

Recent national and international events, such as the fear of bioterrorism, the SARS outbreak, and last year's flu vaccine fiasco have made more people aware of the importance of a good public health system, says Philip A. Greiner, RN, DNSc, associate professor of the School of Nursing and director of the Health Promotion Center at Fairfield University in Connecticut, and president of the Connecticut Association of Public Health Nurses.

This awareness has brought some new money to public health, especially for bioterrorism protection programs, he says. But it isn't enough. People need to see public health nursing as they would a police or fire department, he says, with a base of public health nurses prepared to handle emergencies such as a pandemic or a bioterrorist attack, but in the meantime doing preventative education, advocating for healthier environments, and managing care for those who cannot care for themselves. Most health departments now don't have the nurses to do either properly, he says.

The future of public health nursing is tied to the public's commitment to health care in general and to how much people value the health of their communities, Greiner says. "If we value it, we have to be able to put the money toward it."

SARS came close to showing us what we didn't have, he says. Next time we might not be so lucky.

"It might take an event where tens of thousands of people die because we don't have the health care infrastructure."

"I hope to God it doesn't."

Critical Thinking

1. What role does the public health nurse play in the community?

CATHRYN DOMROSE is a staff writer for NurseWeek. To comment on this story, send e-mail to editorca@nurseweek.com.

A New Way to Treat the World

As the U.S. Navy ship *Mercy* carried out its mission in Indonesia, PARADE's Health Editor found devastation, heroism—and new hope for the people of both nations.

DR. ISADORE ROSENFELD

My first view of the *USNS Mercy* was from a Seahawk helicopter off the coast of Alor, Indonesia, as I was about to land on its deck. I had been invited by the Navy to observe this ship's mission in the troubled region.

The *Mercy* is nearly 900 feet long—the length of three football fields—and gleaming white. It can accommodate 1000 patients, has 12 well-equipped operating rooms, an intensive-care unit and the latest medical equipment necessary to perform everything from preventive health measures to the most complicated surgical operations.

Since the end of the Gulf War in 1991, the *Mercy* had been stationed in San Diego's harbor and used only for the occasional training exercise. But after a horrific tsunami rocked Southeast Asia on Dec. 26, 2004, she was ordered to sail to the Indian Ocean. Given the magnitude of the disaster, the ship needed many more doctors, nurses and medical technicians than its normal Navy staff.

At the Navy's request, Project HOPE—a Virginia-based health education and humanitarian aid organization involved in medical programs in 35 countries on five continents—asked for volunteers. More than 3000 doctors and nurses responded to its nationwide call. Of these, 210 were selected to join their Naval colleagues aboard the *Mercy* and commit themselves to a 30-day tour away from their jobs, homes and families.

On Feb. 3, 2005, the *Mercy,* with her complement of 518 medical and support personnel, arrived at her destination in the Indian Ocean near Banda Aceh, Indonesia, the area hardest hit by the tsunami.

Voyage of Mercy

When a tsunami devasted the Indonesian region of Banda Aceh (map omitted) in December, the U.S. Navy and a dedicated group of doctors and nurses volunteered to sail to Southeast Asia on the hospital ship *Mercy*. After tending to the tsunami victims, they provided much-needed medical assistance to other Indonesians and residents of East Timor on stops during the voyage home.

During the next 40 days, the *Mercy's* staff treated more than 9500 patients, ashore and afloat, and performed nearly 20,000 medical procedures, including 285 surgical and operating room cases. The ship's teams provided water and sanitation, rewired hospital equipment, repaired oxygen tanks, immunized hundreds of men, women and children, and established other public-health measures.

They performed many miracles. In one particularly dramatic case, Iqbal, an 11-year-old boy, had been found clinging to a piece of driftwood about a mile out in the ocean. His immediate family had perished. He suffered respiratory arrest and required a respirator because of "tsunami lung"—a severe bacterial infection caused by swallowing muddy water. A helicopter transported him to the *Mercy*. A week later, after intensive therapy, Iqbal was off the respirator and watching *Spider-Man* movies. In less than four weeks, he was able to leave the hospital ship.

After most of the acute medical problems in the Banda Aceh area had been dealt with, the *Mercy* set sail. I boarded her in mid-March, which is when my personal account of her next mission began.

She roamed the Flores Sea off the coast of Indonesia, her goal to help any man, woman or child who needed medical care. That need, in areas that had been spared nature's violence, turned out to be far greater than anything we had imagined.

I have never in my life seen so much rampant, unrecognized, undiagnosed and untreated disease—entire communities riddled by malaria, tuberculosis, pneumonia and other chronic lung diseases, malnutrition, blindness, cancers, abscesses that were spreading bacteria throughout their victims' bodies. I also witnessed unbelievable rates of maternal mortality and skyrocketing infant mortality. Village after village had no access to any medical care whatsoever: Their kids had never been vaccinated and never used (or even seen) a toothbrush; infections were rampant among thousands of people, young and old.

The personnel aboard the *Mercy* responded with energy, compassion and professionalism. They set fractures, extracted hundreds of abscessed teeth, repaired hernias, removed cataracts, provided reading glasses and even created new prescription lenses for those who needed them. They treated life-threatening infections and instituted public-health measures to reduce the incidence of malaria, dengue fever and cholera. They repaired broken water lines and sprayed mosquito repellents.

The Mercy staff set fractures, extracted teeth, repaired hernias, removed cataracts and instituted public-health measures to prevent outbreaks of malaria and cholera.

The photojournalist and war correspondent who accompanied me on the trip, Karen Ballard, captured dozens of scenes of hope and heartbreak.

There were many moments, not recorded on film, that seared my memory: a doctor who persisted in examining scores of children in 100-degree heat until he was on the verge of collapse from dehydration; Dr. Dana Braner, a pediatric specialist for intensive care from Oregon, carrying one child after another onto the ship, many of whom would have died within days had they not undergone emergency surgery or been treated for their high fevers and serious infections; Dr. Jim Pressly, an ophthalmologist from Charlotte, N.C., removing cataracts from the eyes of "blind" people, who shouted as they left the ship, "I can see, I can see!"

One 7-year-old boy who had suffered severe burns on his knees and had not walked in years was able to prance from the ship after scar tissue had been removed.

Suffering is the "norm" in these very poor regions of the world. They had not qualified for disaster relief because no disaster had struck. But they were lucky that a group of American volunteers had stopped by to help on their way home from the tsunami disaster area.

There are important lessons for us to learn from this experience. We can reverse the hostility that much of the world feels against America simply by performing such humanitarian acts. According to a BBC poll, almost 70% of the people in Indonesia, the most populous Muslim nation, viewed our country with hostility before the tsunami. Today, according to a poll released by the Heritage Foundation, almost 70% think more favorably of us.

Listen to what Tamalia Alisjahban, an Indonesian interpreter, said in her "thank you" speech to the ship's staff as the *Mercy* departed Banda Aceh:

"You were first greeted with suspicion, then puzzlement and then great fondness. And nearly all the patients are saying how grateful they are and that we really can't thank you enough. There's nothing we could give to you to repay your kindness and care, and it will have to be God who repays you.

"I don't know how we can ever thank you. In Indonesia we say *terima kasih,* which means 'accept love.' Because to thank someone is to give a bit of love. Please do accept our love."

I disembarked from the *Mercy* the evening before my flight home from Dili, East Timor. From my hotel room on the beach I could see the beautiful ship with the large red crosses emblazoned on her brilliant white hull. She was scheduled to leave early the following morning to continue her humanitarian mission in other "nondisaster" areas as she sailed home.

At 6 P.M., I received a call from the mission commander, Capt. Timothy McCully, who had planned to come ashore for a farewell dinner that evening. "We've just been ordered back to Sumatra," he said. "They've had an 8.7 magnitude earthquake that has already taken hundreds of lives. They need us. I expect a large-scale onboard medical-treatment operation, with multiple operating rooms working simultaneously."

Fifteen minutes later, from my hotel window, I saw the *Mercy* set sail. I later learned that most of Project HOPE's doctors and nurses, who had been scheduled to leave in the next few days, had decided to stay aboard. New volunteers would join the *Mercy* near Sumatra.

Officers and deckhands; nurses, doctors and pharmacists; helicopter pilots and boiler-room engineers—I had never met such a dedicated and hardworking group of men and women. And now they were going back to the center of the storm. Their only concern was to care for the sick. I have never felt more proud to be a physician and an American.

I have never felt so proud to be a physician and an American.

Before the tsunami, the *Mercy* remained in San Diego Harbor for 14 years. Would it not have been more humane for her and other hospital ships to sail during that time on missions to improve the lives of the world's poor and downtrodden?

I cannot imagine any of the children or parents whom I met on the Mercy embracing Osama bin Laden's anti-American vitriol or wishing harm on a single American or supporting the cause of our terrorist enemies.

Can we earn the respect—and accept the love—of millions of other poor people around the world who are so often turned against us?

I think we can. I believe we must.

I hope we will.

Critical Thinking

1. What recent disasters brought the Mercy Ship out of a 14-year retirement?

DR. ISADORE ROSENFELD, PARADE's Health Editor, is Rossi Distinguished Professor of Clinical Medicine at New York Presbyterian Hospital/Weill Cornell Medical Center. He's medical consultant for Fox News Channel's "Sunday Housecall" and the author of 11 books, including "Breakthrough Health 2005."

Emergency Preparedness

Public health nurses take center stage in a post-9/11 world.

CHRISTINE CONTILLO, RN, BS, BSN

By the year 2000, Public Health in America had become a victim of its own success. Many serious childhood diseases had been eliminated through vaccinations. Clean water and sewage treatment had become mandated, and the rapid population growth that had once resulted in crowded and unhealthy conditions had been slowed. Chronic diseases like tuberculosis still existed, but were being held in check with testing and medication. It followed that funding was steadily decreased until staffing was at a minimum.

Enter terrorism, and with it the recognition that emergency preparedness is crucial to our safety. Fears of bioterrorism and the anthrax scare in 2001 led to Public Health emerging as the system likely to be able to take on a leadership role. Public Health nurses already administered millions of flu shots each year, and were capable of the kind of mass inoculation clinics that could be required in an emergency.

Nurses in annual flu clinics already had written protocols for steps to be taken, and supplies to be ordered. Public Health nurses dealt daily with communicable disease epidemiology and chronic disease screening, so again the investigative skills and written protocols necessary were already in place.

Other Public Health skills like disease surveillance, monitoring, and timely reporting are crucial to the ability to evaluate clusters of symptoms in a geographic area. A few cases of respiratory illness in one emergency room might not be significant; several cases in a number of institutions at the same time might point to a deliberate event. Data from a wide area must be evaluated, and again it was Public Health that was called on for help. Disaster readiness became part of our vocabulary.

The Center for Disease Control has contributed multiple pages on its website to emergency preparedness, relying on the Public Health system to help get its message out. Every state or community now has an emergency operations coordinator, and nurses everywhere should be familiar with the levels of organization that go into disaster readiness.

An emergency is any problem, natural or man-made, that has the potential for causing harm to people or to property. The events of 9/11, serious earthquakes on the west coast, or the multiple Florida hurricanes of fall 2004 are all examples of disasters. These are emergencies that overwhelm the local workers and supplies available.

The purpose of any emergency plan is to keep damage at a minimum while ensuring the safety of the public health workers involved in the recovery. Nurses assess the health or safety risks in advance, prepare for unexpected emergencies, implement the plan, and, when recovery is complete, evaluate its effectiveness. These steps can be followed whether developing a plan for an entire community, a hospital or health facility, physician's office, or even a private home.

Chain of Command

Being familiar with the terminology is a first step. The Incident Command System by which many government agencies practice enables everyone involved to speak the same language. Each incident has a commander, and the channels of responsibility to him become straightforward. Using the ICS enables personnel involved in an event to know the chain of command; when a problem is detected the system can be mobilized and coordinated in a way that minimizes panic. In order for any emergency plan to be successful it must include a way to obtain supplies and a way to get them to where they are best needed and the ICS allows for this by having specific systems managers. ICS can also help reduce response time, facilitate recovery, and hopefully contain the damage.

Tabletop exercises allow all concerned to air their concerns. Logistical problems will emerge, and time spent on preparation will provide a framework for future procedures. Turf battles can be held to a minimum with everyone taking on the role for which they are best suited.

Chemical Spill: Who Responds First?

As an example, picture a chemical spill on a major highway in which the driver of the vehicle is injured. Who can provide the best response? Medical personnel who can help the driver but may become contaminated themselves in the process?

Police who may be able to reroute the traffic but will be unable to get to the injured person without the proper equipment? Or firefighters who have the necessary chemical knowledge and equipment to handle the spill but may be unprepared for the injuries of the driver?

In this scenario only one person should be coordinating services and giving orders, and emergency personnel must be prepared to come onto the scene knowing who that will be. Using the Incident Command System and thinking through the scenario in advance allows for a speedier, safer response. Many state police organizations offer training in its use.

Just like the communities themselves, organizations falling within their boundaries need to have their own written emergency plans and keep them updated. Cell phone, fax, e-mail and beeper or blackberry numbers need to be kept up to date. Lists of individuals needing special help should be constantly evaluated, evacuation routes considered, and emergency equipment available and functioning.

Nurses need to be comfortable with their own level of skill and decide where their services will be a good fit. If they have not worked for many years but want to volunteer they should consider updating their health assessment skills with a professional development class offered at local community colleges or through nursing schools.

Departments need to plan for "surge capacity" or the ability to pull in extra, competent help when needed. The Red Cross Disaster unit accepts nurses for two-week assignments, but only after they have passed a training program. Disaster work often relies on nursing staff to evaluate not just the physical but also emotional distress of victims. The Medical Reserve Corps was established after 9/11 and is a way for medical volunteers to respond to local emergencies. There are currently 212 units across the country; all are locally organized and managed. In many places it is the state or county departments of health that offer the recruitment opportunities, and their chain of command leads them to the Office of the Secretary of Health and Human Services.

Your Personal Emergency Plan

Nurses themselves should have personal emergency plans or detailed written instructions for those who will care for their children, parents and pets if they are called to additional overtime shifts or simply can't leave their unit or are prevented from traveling home. Always be prepared to reach out to a neighbor and have a number where they can reach you. Agree in advance with your family on a meeting place and time if necessary. Anyone working in a potentially dangerous situation needs to carry identification, and have information readily available about any medical conditions, allergies and medications taken. Bottled water and a three-day supply of canned food, along with candles, and a battery-powered radio stored at home will be welcome.

We may have come a long way since our days spent in Girl Scouts, but their motto is still relevant: Be Prepared.

Critical Thinking

1. How has the role of public health nurses evolved in the post 9/11 world?

CHRISTINE CONTILLO holds BS and BSN degrees. She has worked as a nurse since 1979. Christine has written extensively for various nursing publications as well as the *New York Times*.

A Nursing Career to Consider Assisted Living

CHRISTINE CONTILLO, RN, BS, BSN

Nancy Ullanday, RN, is the Director of Nursing at an Assisted Living facility in Paramus, New Jersey. She's a geriatric nurse practitioner by training and a bedside nurse at heart. As today's senior citizens live longer with more medical conditions, she couldn't be happier to take her place by their side as their 24/7 advocate when they look for a home at her facility.

Assisted Living facilities were originally conceived as a long-term retirement option for those who might need more help with their daily activities than is available in a standard retirement community. Each resident has their own unit, much like a studio apartment, and it is treated entirely as their own home. They can come and go as they wish, or have friends and family visit, but still get help with personal care, eat their meals in a family-style atmosphere and contract for services such as beauty care and laundry.

The residents, no longer isolated, gain additional socialization opportunities with the availability of companionship and planned recreation activities, and benefit from no longer having a home of their own to manage.

Good medical care continues to extend the life expectancy of Americans, which presents a unique set of problems for Ms. Ullanday at Care One at the Cupola. Residents are older, spend a longer period of time in her care, and can become what she terms "medically disabled." They may suffer from a number of chronic conditions requiring multiple medications, but still not necessarily be ready for the level of care found in nursing homes. In her opinion it's the perfect setting for nurses to practice assessment skills and utilize the educational training they've received.

When asked what kind of duties nurses perform in an Assisted Living facility, medication issues immediately come to mind. Each resident is independent and sees the physician of their choice. Therefore, unlike nurses in a hospital found wheeling a med cart down the center of a long hall, nurses in Assisted Living may instead spend their time monitoring medications—making sure that the residents can read the label and take it correctly, and they're not mixing it with over-the-counter or alternative treatments.

Nurses may wait in a day room and have the patients come to them if that's a service contracted for. They may even need to count the doses to make sure that the right amount of medication is being taken at the correct time. It's a tricky business; patients are free to "self-medicate" and may be found to abuse alcohol or sleeping aids. If patients become non-compliant, Ms. Ullanday says that a dialogue with the physician and family can be initiated, but only with the patient's consent.

The nursing staff needs to be alert to physical and behavioral changes that aging brings on. Many of her residents still drive, and it may be the nursing staff that begins to notice the deterioration in driving skills. As Director Of Nursing she becomes "the bad guy" by initiating conversations with the family so that her nursing staff can continue their good relationship with the resident. The staff may pick up clues of impending dementia, and the facility has recently initiated a locked dementia unit.

Just like in the hospital, the advances in medical technology constantly bring new challenges to the table. Nurses must keep up with the market in terms of hardware, like pumps, shunts, and catheters. Patients may have reasons for acute hospital stays but then return again to their own unit post-hospital discharge still needing ongoing nursing care. And while home care nurses may come into the facility when ordered by a physician, they're not likely to make more than one visit a day. Any treatments ordered more often than once daily, such as catheter care or dressing changes, need to be picked up by the facility nursing staff.

One recent change at The Cupola brought on by a string of unusually bad winters was the increase in physicians and nurse practitioners who make home visits to Care One. Again, patients are encouraged to voice their preferences—some would rather be seen in the privacy of their own room while others consider their rooms private and would prefer to meet the practitioner in a day room. Nurses then stay on hand to assist the practitioner with the examinations. Even ancillary professions, such as ophthalmology and podiatry, are able to schedule visits to residents in the facility.

When asked what's best about her job, Ms Ullanday said "Really, all of it." As a geriatric Nurse Practitioner, she loves the hands-on nursing care she's called on to give, but also enjoys the education piece given to both staff and patients and has lectured at a nearby college. She spends some of her time in infection control dealing with the rashes common to the elderly and staving off viral infections brought in by visitors. Even nursing management issues, like staffing, make the job special for her.

When there's an infrequent opening to fill on her staff, she looks for nurses with good general experience. Almost any background will fit the bill because medical, surgical, acute care, or even same day surgery skills could be put to use. Working in Assisted Living might seem to some to be a step down from the excitement and pace of a hospital position, but it's far from that, with the additional benefit of being a way to practice nursing with the autonomy we're all looking for.

Critical Thinking

1. What are the advantages of the Assisted Living setting for the geriatric population?

CHRISTINE CONTILLO holds BS and BSN degrees. She has worked as a nurse since 1979. Christine has written extensively for various nursing publications as well as the *New York Times*.

UNIT 6

Nutrition and Weight Management

Unit Selections

Learning Outcomes

After reading this unit, you should be able to:

- List the complications that can occur due to poor nutrition during pregnancy.

- Explain the symptoms prevalent in advanced cancer and how nutrition is impacted at the end of life.

- State the impact nutrition can have on a patient and his/her family during palliative care.

- List three strategies to improve nutrition in the elderly population.

- List some of the post-op deficiencies a patient may experience after a gastric bypass.

Student Website

www.mhhe.com/cls

Internet References

National Institute of Children and Human Development
www.nichd.nih.gov/health/topics/Diet_and_Nutrition.cfm
Nutrition.gov
www.nutrition.gov/nal_display/index.php?info_center=11&tax_level=1
Obesity Help
www.obesityhelp.com
United States Department of Agriculture
www.mypyramid.gov
WebMD Medical Reference
http://my.webmd.com

This units examines the impact poor nutrition can have on individuals throughout their life cycle. A person's nutritional status can often be directly linked to their health and well-being. Moreover, the impact of how nutrition or malnutrition will affect a person can also be linked back to preconception. The essence of good nutrition begins at preconception and can affect an individual throughout the life cycle. The National Institute of Child Health and Human Development (NIH) indicates that good nutrition means the body is getting all the nutrients, vitamins, and minerals it needs to work at its best level. Eating a healthy diet is the main way to maintain good nutrition. Most people know that a balance of good nutrition and physical activity can help them reach and maintain a healthy weight, but the benefit of good nutrition goes beyond weight. Good nutrition can also improve cardiovascular health and other body system functions, mental well-being, school/cognitive performance, wound healing or recovery from illness or injury, and reduce the risk for diseases, including obesity, diabetes, stroke, some cancers, and osteoporosis. It can increase energy and the body's ability to fight off illness. Article 37 indicates that poor nutrition prior to conception can lead to many complications. Numerous studies have identified a direct correlation between poor nutrition and poor fetal development. Poor nutrition can impact an individual's ability to conceive. However, once conception occurs, it is essential that the mother begin a healthy diet. The maintenance of good nutrition and proper weight gain during the prenatal period is necessary for the health and well-being of both the fetus and the mother. Poor nutrition during pregnancy can lead to many complications such as gestational diabetes, hypertension, anemia, toxemia, and still birth. Poor nutrition can increase risk of a cesarean section and poor fetal weight. After birth, good nutrition is necessary during the infancy stage to help promote growth and development.

Furthermore, nutritional care at the end of life is just as essential as nutritional care at the beginning of life. Although each person's nutritional needs may differ, it is important to assess the physiological needs and one's ability to tolerate oral nutrition. Nutritional care can assist in optimizing quality of life and a sense of wellbeing, as well as helping to alleviate unpleasant symptoms which occur at the end of life (Acreman, 2000). A person's nutritional status at the end of life may not only impact the patient, but it may leave family members with ethical questions about their loved ones' nutrition, hydration, and quality of life. Educating the patient and family members is imperative during this time period. The nurse plays a vital role in supporting both the patient and family during palliative care. One of the nurse's primary challenges will be to provide patient support relating to nutritional needs and ensuring the patient's and family's needs are met. It is often difficult for families to fathom the fact that a person's nutritional needs will change during the final stage of life. Addressing the family's primary concerns often facilitates a smoother understanding for both the family and patient during terminal conditions.

© Jack Hollingsworth/Getty Images

The prevalence of malnutrition continues as the aging process occurs. Malnutrition in the elderly has become a global concern. Poor nutritional status in the elderly population can relate back to many contributing factors such as illnesses, lack of financial resources, and the normal aging process associated with pathological and physiological changes. Dining alone has also been identified as a prevalent factor of malnutrition in the elderly population. Nutritional assessment by the appropriate professional is necessary to decrease the incidence of malnutrition in this growing population. Appropriate nutritional supplements can be incorporated into the daily diet to ensure that the nutritional needs are met. Finally, nutritional deficiencies can also occur as a result of surgical procedures such as the patient who has just experienced post-gastric bypass. Adjustment in a patient's diet and the implementation of the appropriate supplements can help to decrease the occurrence of post surgical nutritional deficiencies.

The articles in this unit will discuss the aforementioned topics in greater detail and will describe the role of the nurse as he/she cares for the patient who is experiencing nutritional deficiencies.

Fetal Nutrition and Adult Hypertension, Diabetes, Obesity, and Coronary Artery Disease

Joan Nalani Thompson, RNC, MSN, NNP

The association between low birth weight and adult disease is not well understood. The exact cause for increased adult morbidity seen with poor nutrition *in utero* and in early infancy is still being investigated. Many studies support the fetal-origins-of-adult-disease or Barker hypothesis (the fetal-origins hypothesis), which suggests that nutrition *in utero* determines the development of risk factors associated with heart disease later in life.[1-3] The purpose of this review of the literature is to synthesize epidemiologic and experimental studies that focus on low birth weight and its association with adult diseases such as hypertension, diabetes, obesity, and coronary artery disease (Figure 1).

Most studies reviewed for this article were published between 1995 and 2005. The databases used included PubMed, Ebscohost, OVID, and the American Academy of Pediatrics. A total of 105 papers was reviewed, of which 43 investigated the relationship between fetal and infant nutrition and adult disease and thus were selected for this article. The main keywords included the author Barker, IUGR (intrauterine growth restriction) and diabetes, obesity, hypertension, and fetal nutrition. Low birth weight and malnutrition were initially used as keywords, but the results included numerous studies not specific to the relationship between birth weight and adult diseases.

The Fetal-Origins Hypothesis

The fetal-origins hypothesis, proposed by Professor David Barker, describes an adaptive phenomenon whereby the physiology and metabolism of a human fetus may change as it adapts to decreased or limited nutrients and oxygen by slowing its rate of cell division. Lasting memories of undernutrition may permanently reduce the number of cells in particular organs and change the distribution of cell types, patterns of hormonal secretion, metabolic activity, and organ structure. This change in the body's memory may become translated into pathology, which is unsurprising because undernutrition *in utero* has led to persistent changes in blood pressure, cholesterol metabolism,

and insulin response to glucose in many animal studies. For example, Barker states, "[C]oronary heart disease is associated with specific patterns of disproportionate fetal growth that results from fetal undernutrition in middle to late gestation" (p. 171).[4] Other diseases, including stroke, diabetes, and hypertension, may also have their origins in these "programmed" changes.[5] Undernutrition *in utero* affects fetal growth.[3] Low birth weight and disproportionate head circumference, length, and weight may indicate a lack of nutrients during gestation.[6] Uteroplacental insufficiency results in a decreased transfer of nutrients and oxygen through the uterus and to the fetus. The developing fetus subjected to decreased nutrients necessary for growth *in utero* adapts to the limited nutrition, causing permanent changes in physiology and metabolism that predispose the infant to increased morbidity as an adult.[5] Many studies support the fetal-origins hypothesis, identifying a positive association

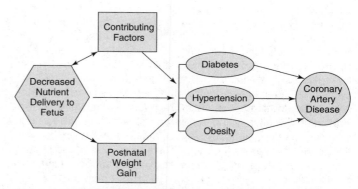

Figure 1 Fetal/newborn nutrition and adult morbidity. Depicted is the relationship between birth weight and hypertension, diabetes, obesity, and coronary artery disease. The ovals and hexagon represent the negative association found between fetal undernutrition and adult diseases supported by numerous studies. The squares represent potential areas of research investigating the influence of postnatal weight gain among infants exposed to undernutrition in utero and other contributing factors that may impact fetal growth or may affect later adult health.

between low birth weight and adult diseases like hypertension, diabetes, obesity, and coronary artery disease.[3, 7, 8]

According to Barker, "the fetal origins hypothesis states that fetal undernutrition in middle to late gestation, which leads to disproportionate fetal growth, programmes later coronary heart disease" (p. 171).[4] In a review, Henriksen suggests that first epidemiologic evidence for support of the Barker hypothesis was provided by Rose in 1964, Forsdahl in 1977, and then Barker and Osmond in 1986.[9] Rose investigated the existence of familial aggregation for ischemic heart disease and found increased fetal and infant mortality among siblings with coronary artery disease. Families of individuals with heart disease had stillbirth and infant mortality rates twice those of the control group (patients of the same age and sex without arterial disease or hypertension). One possible explanation for the difference observed between the two groups may be that heart disease "tends to occur in individuals who come from constitutionally weaker stock more liable to succumb to a variety of diseases" (p. 80).[10] Forsdahl found that in "countries where infant mortality was high, the same generation also had a higher mortality in middle age" (p. 91), and poverty during early life followed by affluence in later life was found to be associated with higher mortality rates from arteriosclerotic heart disease. Forsdahl speculated that the relationship between higher rates of ischemic heart disease and highest infant mortality rates in Norway may be caused by "some form of permanent damage caused by a nutritional deficit" (p. 95).[11] Barker and Osmond reported that lower birth weights in England and Wales in the first decades of this century were correlated with an increased risk of coronary vascular disease in adults, suggesting that the association may be related to prenatal and early postnatal nutrition.[12] Barker and colleagues have published many papers in support of the hypothesis.[2,5,6,13] It has also been supported by animal studies and studies on various populations throughout the world.

Animal Studies

Animal studies support the hypothesis that poor fetal growth effects negative changes in metabolism and blood pressure. It appears to cause diabetes and obesity in adults.

Vickers and coworkers studied the effects of maternal undernutrition on fetal development in rats. Rats were fed either a standard *ad libitum* or undernutritional (30 percent of *ad libitum*) diet throughout gestation. The study found that rats fed the undernutritional diet had no change in litter number, but pups were growth restricted with higher blood pressures at 100 days of age. All pups were then breastfed by mothers that had been fed the standard diet, divided into eight pups per litter. After weaning, two diets, control versus hypercaloric (equal amounts of minerals and vitamins), were introduced, and pups were observed for associated hyperphagia, obesity, and hypertension. Fetal undernutrition was found to induce adult hyperphagia, and fetal undernutrition followed by postnatal hypercaloric feeding augmented hyperinsulinism, hyperlipidemia, hypertension, and obesity.[14]

Franco and associates also found that undernutrition during gestation had no effect on litter number, but off-spring had decreased birth weights, higher blood pressure levels (males at 6 weeks, females at 9 weeks), higher blood glucose concentrations, and a decreased number of nephrons at 14 weeks postnatal age compared to offspring not exposed to undernutrition *in utero*.[15] Brawley and coworkers found that adult offspring of rats on protein-restricted diets during gestation were smaller, had elevated systolic blood pressures, and had some vascular dysfunction.[16]

In contrast, offspring of sheep undernourished for 10–20 days during late gestation (restricted diet started on gestational day 105; term gestation is 146 days) weighed less, but there was little effect on glucose tolerance or blood pressure.[17] It is possible that the different outcomes among the sheep offspring may be related to the timing of undernutrition, which occurred very late in gestation, or that the effects of undernutrition vary between species.

Human Populations

Research involving various human populations has been extensive, mostly in the form of retrospective epidemiologic studies. The studies of human populations in different countries have investigated the association between low birth weight and adult diseases. Some studies have already been cited, but another was among the papers considered in this review: A study of a large birth cohort consisting of 12,150 individuals born in Scotland between 1950 and 1956 found that birth weight was inversely associated with stroke and coronary heart disease, not dependent on adjustment for size in later childhood, and independent of social class or maternal and pregnancy characteristics.[18]

Six studies in our review identified an association between low birth weight and factors increasing the risk for heart disease and diabetes. A U.S. study comparing 195 very low birth weight (VLBW) and 208 normal birth weight babies found that adults born VLBW had a higher systolic blood pressure at 20 years of age. The increase in systolic blood pressure was significant in women after adjusting for maternal race and education with and without adjusting for later size, whereas in men there was significance only after adjusting for later size (height and weight at 20 years of age).[19] A review of 22,846 questionnaires providing information on U.S. men between 40 and 75 years of age found adult hypertension and diabetes associated with low birth weight and obesity associated with high birth weight.[20] A study in Japan examining birth and health records of 135 men and 148 women demonstrated an inverse relationship between birth weight and adult blood pressure and cholesterol.[21] Elevated blood pressure and plasma cortisol levels were found among low birth weight adults in Adelaide, Australia, and Hertfordshire, UK.[22] Blood pressure, glucose, and adrenocorticotrophic hormone-stimulated cortisol levels were elevated in 73 South African low birth weight adults, compared to 64 adults who had been born at an appropriate weight for gestational age.[23] A study in East Hertfordshire, UK, found that low birth weight was related to increased systolic blood pressure, higher plasma glucose levels, higher triglycerides, and higher hip-to-waist ratios in 297 women aged 60–71, allowing for body mass, and independent of social class.[24]

A study that reviewed detailed birth and school health records from 1924 to 1933 in Helsinki, Finland, and hospital admission and death records between 1971 and 1995 identified men born with low birth weight and low ponderal index (weight divided by length). Members of this population with two patterns of growth, thinness at birth followed by rapid weight gain in childhood and failure of infant growth followed by persistent thinness in childhood, had a higher incidence of coronary heart disease in adulthood. The detailed birth records (birth weight, crown-to-heel length, head circumference, and placental weight) were useful for identifying possible indicators of fetal undernutrition.[25] Among women born in Helsinki during the same time period, short body length was a stronger predictor for coronary heart disease.[26] A later study that contacted 500 people from this cohort (mean age of 70) still living in Helsinki found that smaller body size at birth was associated with higher systolic pressures among elderly men and women (mean age of 70) and that body mass index (BMI), waist circumference, and percentage body fat were unrelated to systolic pressure among people diagnosed as having hypertension.[27]

Not all studies are consistent with the fetal-origins hypothesis. One of these investigated the effects of poor maternal nutrition on a population exposed to the World War II siege of Leningrad (now St. Petersburg) between September 8, 1941, and January 27, 1944. Most deaths (750,000 out of a population of 2.4 million) occurred from starvation between November 1941 and February 1942, when the average daily rations were around 300 calories with almost no protein. Anthropometric measurements and blood samples were taken on 361 people exposed to the siege (169 exposed to undernutrition *in utero* and born between November 1, 1941, and June 30, 1942, and 192 infants born in the first half of 1941 who were exposed to the siege after birth) and 188 people born outside Leningrad who were not exposed to malnutrition. Exposure to the siege was used as an indicator of maternal nutrition in the absence of reliable birth weight data. This study did not find intrauterine malnutrition associated with glucose intolerance, dyslipidemia, hypertension, or cardiovascular disease in adulthood, but did identify a stronger effect of obesity on hypertension for those exposed to the siege *in utero*. When the relationship between siege exposure combined with obesity and its effects on blood pressure was examined, blood pressure was positively related to obesity, suggesting that exposure to conditions of malnutrition with subsequent adult obesity may act synergistically to increase susceptibiliy to hypertension.[28]

A prospective study on 19-year-olds living in the Netherlands sought to determine the effects of IUGR on blood pressure among 418 people who were born before 32 weeks gestation (group 1) and 170 people born at or after 32 weeks gestation but weighing less than 1,500 gm at birth (group 2). The mean systolic blood pressure was high for both groups, and current weight and BMI were identified as the best predictors for blood pressure. According to the authors, these findings do not support the hypothesis that poor nutrition *in utero* negatively impacts adult health because blood pressure was unrelated to the extent of IUGR, birth weight, or gestational age. Possible causes identified by the authors for the negative

association may be the young age (19 years), selection bias (64 percent response), or a nonexistent relation between birth weight and blood pressure. The authors of this study did not specifically state that members of the cohort born at less than 32 weeks gestation were not IUGR.[29]

Contributing Factors

Many researchers have attempted to identify other factors that may contribute to adult diseases. Maternal characteristics and diet have been implicated. A Finnish study that looked at 3,302 men born at Helsinki University Hospital during 1924–1933 found that a high maternal BMI was strongly associated with coronary heart disease in their offspring.[30] A study in the UK found a relationship between higher birth weight and head circumference of mothers and lower systolic and diastolic blood pressures in their offspring, indicating that poor birth weight may affect more than one generation.[31] Another UK study identified higher systolic and diastolic blood pressures in 11-year-old children whose mothers had below-median triceps skin fold thickness and poor pregnancy weight gain in early gestation independent of birth weight, ponderal index at birth, or gestational age at birth.[32] Two studies that took blood pressures and reviewed birth records of adults exposed to the Dutch famine during 1944–1945 identified higher blood pressures associated with low birth weight and a low protein-to-carbohydrate ratio diet in the third trimester. Whereas other studies have looked at low birth weight, these studies have identified an association between maternal characteristics and high blood pressure.[33,34]

Postnatal Weight Gain

Various studies have focused on associations between chronic adult illnesses and postnatal weight gain. One of these evaluated preterm infants and the effects of rapid postnatal weight gain on brachial flow-mediated endothelium-dependent dilation (FMD). Low brachial FMD is a measurement of endothelial dysfunction and "is closely related to dysfunction in the coronary vessels" and "linked to later development of adverse clinical cardiovascular events" (p. 1112). The researchers found lower brachial FMD among adolescents with low birth weights and among those with rapid early postnatal (two weeks) growth independent of birth weight.[35]

Another study on preterm infants found that, at 38.6 weeks, preterm infants differed from full-term infants in adipose tissue distribution. Preterm infants with accelerated postnatal growth had increased intra-abdominal adiposity, which is associated with insulin resistance and dyslipidemia, and illness severity was a strong determinant for this type of adipose distribution. Although this study looked at preterm and therefore small infants, the authors stated that IUGR infants may have their own distinct risks; however, the risks may overlap.[36] The studies by Singhal and colleagues and Uthaya and coworkers suggest that rapid catch-up growth in the early postnatal period among preterm infants is associated with potential adverse effects on later cardiovascular health.[35,36]

Other research has looked at weight gain during infancy and early childhood. One study examined the health records of 4,630 men in Finland and found that low birth weight was associated with later coronary heart disease, but increased weight at one year was associated with decreased risk of coronary artery disease, independent of birth size. Two growth patterns were associated with coronary heart disease: thinness at birth followed by an increased weight gain from age three years onward and poor infant growth followed by persistent thinness during childhood.[25] A study of 290 men living in East Hertfordshire, UK, found that men with signs and symptoms of coronary heart disease had lower mean weight at one year of age (one pound lower) than men without coronary heart disease, but the prevalence of coronary heart disease was not related to birth weight.[37] A study of 297 women from that area found no association between weight at one year and increased risk for coronary heart disease; however, low birth weight was associated with risk factors for coronary heart disease: higher glucose plasma levels, higher systolic blood pressures, and higher serum triglyceride levels.[24]

Two studies documented that individuals born with a low birth weight followed by rapid growth in early childhood had increased incidence of coronary heart disease and hypertension. In 3,641 men living in Helsinki, the highest death rate from coronary heart disease was among men who were thin at birth and had accelerated weight gain (above-average BMI) by age 7. The authors suggested that coronary artery disease may be a consequence of the combination of prenatal undernutrition followed by improved postnatal nutrition and weight gain.[38] A study of 2,026 Filipino adolescents (ages 14–16) born thin but heavy as adolescents found that a high BMI and rapid weight gain after 8 years of age increased the risks for high blood pressure only among boys in the lower two-thirds of the BMI index distribution at birth. The authors stated that small birth size with rapid childhood growth is evidence of fetal programming.[39]

Studies on Children

Four investigations of the effects of fetal growth on later adult health have been done on children. These do not consistently support the fetal-origins hypothesis. A study of 1,570 children aged 3–6 who were born at term found blood pressure inversely related to current weight in Chile, China, Guatemala, and Sweden, but found no association in Nigeria.[40] No association between birth weight and blood pressure was found in 214 7- to 8-year-olds living in India.[41] A study of children aged 13–16 in the UK found an association between higher birth weight and lower body fat, suggesting that low birth weight programs a smaller proportion of lean mass later in life, which may influence both obesity and cardiovascular disease.[42] Because blood pressure and blood total cholesterol are strong determinants of coronary heart disease risk, an investigation studied these factors and birth weight among 13- to 16-year-olds living in ten British towns. A very weak correlation was found between low birth weight and total cholesterol, and a stronger relationship was found between current BMI and cholesterol, suggesting that childhood obesity was a stronger determinant for coronary heart disease risk than birth weight.[43]

Summary of the Literature

Research indicates that poor fetal nutrition may be but is not always associated with increased risk of chronic adult disease. Some indicators of decreased nutrient delivery to the fetus that affect fetal growth are low birth weight, low ratio of birth weight to head circumference, and low ponderal index.

Animal studies examining the effects of undernourishment during gestation demonstrate that offspring had increased blood pressures, glucose levels, obesity, and other factors associated with coronary artery disease.

Many, but not all, studies involving human populations are consistent with the fetal-origins hypothesis. Although many studies have focused on fetal growth, others have identified maternal characteristics (low maternal birth weight, poor diet, low triceps skin fold test, and poor weight gain) that increase the risk for adult diseases in offspring. Low birth weight followed by rapid early postnatal weight gain among preterm infants, low weight at one year, and high BMI in childhood (ages seven to eight) have also been associated with adverse effects on later cardiovascular health. Studies of young populations do not consistently support the fetal-origins hypothesis, possibly because these diseases generally afflict older adults and are not frequently seen until later adulthood.

One weakness identified in the literature is that many studies are retrospective, cross-sectional, or epidemiologic studies associated with disease, not causation. Another is limited incorporation of environmental or socioeconomic influences or other confounders that may affect results. Another weakness is that many studies place babies in the "low birth weight" category without consideration of gestation; pathophysiology differs among preterm versus term, small for gestational age, and IUGR infants.

While acknowledging methodologic weaknesses, many studies done on various populations have identified an association between low birth weight and higher risk for adult diseases. Research in this area has progressed, beginning with epidemiologic studies, then expanding its scope of investigation in attempting to identify causes of associations and continuing with studies that have expanded on previous studies.

Summary

If we are to contribute to the current body of knowledge and promote positive adult health, more research is needed to investigate the relationship between fetal nutrition and adult health. Many studies have focused on birth weight, which is a poor marker of fetal growth. Studies should include infant body proportions, as well as the identification of IUGR versus small for gestational age infants and term versus preterm infants. Another area for research is an examination of the timing and quality of maternal undernutrition, as well as other factors that affect nutrient delivery to the fetus, to identify specific critical times of development that may affect adult health.

Various patterns of postnatal weight gain have also been implicated as synergistically contributing to an increased risk for hypertension, obesity, and coronary artery disease. Current practice supports weight gain that mimics *in utero* growth of preterm infants, as well as infant and childhood weight gain following growth charts. Research needs to be conducted to identify whether current recommendations for weight gain are in the best interest of public health.

References

1. Barker, D. J. P., & Lackland, D. T. (2003). Prenatal influences on stroke mortality in England and Wales. *Stroke, 34,* 1598–1603.

2. Godfrey, K. M., & Barker, D. J. P. (2000). Fetal nutrition and adult disease. *The American Journal of Clinical Nutrition, 71,* 1344–1352.

3. Szitanyi, P., Janda, J., & Poledne, R. (2003). Intrauterine undernutrition and programming as a new risk of cardiovascular disease in later life. *Physiological Research, 52,* 389–395.

4. Barker, D. J. P. (1995). Fetal origins of coronary heart disease. *British Medical Journal, 311,* 171–174.

5. Barker, D. J. P. (1997). The long-term outcome of retarded fetal growth. *Clinical Obstetrics and Gynecology, 40,* 853–863.

6. Barker, D. J. P., & Clark, P. M. (1997). Fetal undernutrition and disease in later life. *Journal of Reproduction and Fertility, 2,* 105–112.

7. Reynolds, R. M., & Phillips, D. I. W. (1998). Long-term consequences of intrauterine growth retardation. *Hormone Research, 49,* 28–31.

8. Waterland, R. A., & Garza, C. (1999). Potential mechanisms of metabolic imprinting that lead to chronic disease. *The American Journal of Clinical Nutrition, 69,* 179–197.

9. Henriksen, T. (1999). Foetal nutrition, foetal growth restriction and health later in life. *Acta Paediatrica, 429* (Suppl.), 4–8.

10. Rose, G. (1964). Familial patterns in ischaemic heart disease. *British Journal of Preventive & Social Medicine, 18,* 75–80.

11. Forsdahl, A. (1977). Are poor living conditions in childhood and adolescence an important risk factor for arteriosclerotic heart disease? *British Journal of Preventive & Social Medicine, 31,* 91–95.

12. Barker, D. J., & Osmond, C. (1986). Infant mortality, childhood nutrition, and ischaemic heart disease in England and Wales. *Lancet, 1,* 1077–1081.

13. Barker, D. J. P., & Lackland, D. T. (2003). Prenatal influences on stroke mortality in England and Wales. *Stroke, 34,* 1598–1602.

14. Vickers, M. H., Breier, B. H., Cutfield, W. S., Hofman, P. L., & Gluckman, P. D. (1999). Fetal origins of hyperphagia, obesity, and hypertension and postnatal amplification by hypercaloric nutrition. *American Journal of Physiology, Endocrinology and Metabolism, 279,* E83–E87.

15. Franco Mdo, C., Arruda, R. M., Fortes, Z. B. de, Oliveira, S. F., Carvalho, M. H., Tostes, R. C., et al. (2002). Severe nutritional restriction in pregnant rats aggravates hypertension, altered vascular reactivity, and renal development in spontaneously hypertensive rats offspring. *Journal of Cardiovascular Pharmacology, 39,* 369–377.

16. Brawley, L., Itoh, S., Torrens, C., Barker, A., Bertram, C., Poston, L., et al. (2003). Dietary protein restriction in pregnancy induces hypertension and vascular defects in rat male offspring. *Pediatric Research, 54,* 83–90.

17. Oliver, M. H., Breier, B. H., Gluckman, P. D., & Harding, J. E. (2002). Birth weight rather than maternal nutrition influences glucose tolerance, blood pressure, and IGF-I levels in sheep. *Pediatric Research, 52,* 516–524.

18. Lawlor, D. A., Ronalds, G., Clark, H., Smith, G. D., & Leon, D. A. (2005). Birth weight is inversely associated with incident coronary heart disease and stroke among individuals born in the 1950s: Findings from the Aberdeen children of the 1950s prospective cohort study. *Circulation, 112,* 1414–1418.

19. Hack, M., Schluchter, M., Cartar, L., & Rahman, M. (2005). Blood pressure among very low birth weight ($<$ 1.5 kg) young adults. *Pediatric Research, 58,* 677–684.

20. Curhan, G. C, Willett, W. C., Rimm, E. B., Spiegelman, D., Ascherio, A. L., & Stampfer, M. J. (1996). Birth weight and adult hypertension, diabetes mellitus, and obesity in US men. *Circulation, 94,* 3246–3250.

21. Miura, K., Nakagawa, H., Tabata, M., Morikawa, Y., Nishijo, M., & Kagamimori, S. (2001). Birth weight, childhood growth, and cardiovascular disease risk factors in Japanese aged 20 years. *American Journal of Epidemiology, 153,* 783–789.

22. Phillips, D. I. W., Walker, B. R., Reynolds, R. M., Flanagan, D. E. H., Wood, P. J., Osmond, C., et al. (2000). Low birth weight predicts elevated plasma cortisol concentrations in adults from 3 populations. *Hypertension, 35,* 1301–1306.

23. Levitt, N. S., Lambert, E. V., Woods, D., Hales, C. N., Andrew, R., & Seckl, J. R. (2000). Impaired glucose tolerance and elevated blood pressure in low birth weight, nonobese, young South African adults: Early programming of cortisol axis. *The Journal of Clinical Endocrinology and Metabolism, 85,* 4611–4618.

24. Fall, C. H. D., Osmond, C., Barker, D. J. P., Clark, P. M. S., Hales, C. N., Meade, T., et al. (1995). Fetal and infant growth and cardiovascular risk factors in women. *British Medical Journal, 310,* 428–432.

25. Eriksson, J. G., Forsen, T., Tuomilehto, J., Osmond, C., & Barker, D. J. P. (2001). Early growth and coronary heart disease in later life: Longitudinal study. *British Medical Journal, 322,* 949–953.

26. Forsen, T., Erikssen, J. G., Tuomilehto, J., Osmond, C., & Barker, D. J. P. (1999). Growth *in utero* and during childhood among women who develop coronary heart disease: Longitudinal study. *British Medical Journal, 319,* 1403–1407.

27. Yliharsila, H., Eriksson, J. G., Forsen, T., Kajantie, E., Osmond, C., & Barker, D. J. P. (2003). Self-perpetuating effects of birth size on blood pressure levels in elderly people. *Hypertension, 41,* 446–450.

28. Stanner, S. A., Bulmer, K., Andres, C., Lantseva, O. E., Borodina, V., Poteen, V. V., et al. (1997). Does malnutrition *in utero* determine diabetes and coronary heart disease in adulthood? Results from the Leningrad siege study, a cross sectional study. *British Medical Journal, 315,* 1342–1348.

29. Keijzer-Veen, M. G., Finken, M. J. J., Nauta, J., Dekker, F. W., Hille, E. T. M., Frolich, M., et al. (2005). Is blood pressure increased 19 years after intrauterine growth restriction and preterm birth? A prospective follow-up study in the Netherlands. *Pediatrics, 116,* 725–731.

30. Forsen, T., Erikssen, J. G., Tuomilehto, J., Teramo, K., Osmond, C., & Barker, D. J. P. (1997). Mother's weight in pregnancy and coronary heart disease in a cohort of

Finnish men; Follow up study. *British Medical Journal, 315,* 837–840.

31. Barker, D. J. P., Shiell, A. W., Barker, M. E., & Law, C. M. (2000). Growth *in utero* and blood pressure levels in the next generation. *Journal of Hypertension, 18,* 843–846.

32. Clark, P. M., Atton, C., Law, C. M., Shiell, A., Godfrey, K., & Barker, D. J. P. (1998). Weight gain in pregnancy, triceps skinfold thickness, and blood pressure in offspring. *Obstetrics and Gynecology, 91,* 103–107.

33. Roseboom, T. J. van der, Meulen, J. H. P., Ravelli, A. C. J. van Montfrans, G. A., Osmond, C., Barker, D. J. P., et al. (1999). Blood pressure in adults after prenatal exposure to famine. *Journal of Hypertension, 17,* 325–330.

34. Roseboom, T. J.van der Meulen, J. H. P. van Montfrans, G. A., Ravelli, A. C. J., Osmond, C., Barker, D. J. P., et al. (2001). Maternal nutrition during gestation and blood pressure in later life. *Journal of Hypertension, 19,* 29–34.

35. Singhal, A., Cole, T. J., Fewtrell, M., Deanfield, J., & Lucas, A. 2004). Is slower early growth beneficial for long-term cardiovascular health? *Circulation, 109,* 1108–1113.

36. Uthaya, S., Tomas, E. L., Hamilton, G., Dore, C. J., Bell, J., & Modi, N. (2005). Altered adiposity after extremely preterm birth. *Pediatric Research, 57,* 211–215.

37. Fall, C. H. D., Vijayakumar, M., Barker, D. J. P., Osmond, C., & Duggleby, S. (1995). Weight in infancy and prevalence of coronary heart disease in adult life. *British Medical Journal, 310,* 17–20.

38. Eriksson, J. G., Forsen, T., Tuomilehto, J., Winter, P. D., Osmond, C., & Barker, D. J. (1999). Catch-up growth in childhood and death from coronary heart disease: Longitudinal study. *British Medical Journal, 318,* 427–431.

39. Adair, L. S., & Cole, T. J. (2003). Rapid child growth raises blood pressure in adolescent boys who were thin at birth. *Hypertension, 41,* 451–456.

40. Law, C. M., Egger, P., Dada, O., Delgado, H., Kylberg, E., Lavin, P., et al. (2000). Body size at birth and blood pressure among children in developing countries. *International Journal of Epidemiology, 29,* 52–59.

41. Kumar, R., Bandyopadhyay, S., Aggarwal, A. K., & Khullar, M. (2004). Relation of birthweight and blood pressure among 7–8 year old rural children in India. *International Journal of Epidemiology, 33,* 87–91.

42. Singhal, A., Wells, J., Cole, T. J., Fewtrell, M., & Lucas, A. (2003). Programming of lean body mass: A link between birth weight, obesity, and cardiovascular disease? *The American Journal of Clinical Nutrition, 77,* 726–730.

43. Owen, C. G., Whincup, P. H., Odoki, K., Gilg, J. A., & Cook, D. G. (2003). Birth weight and blood cholesterol level: A study in adolescents and systematic review. *Pediatrics, 111,* 1081–1089.

Critical Thinking

1. What is the relationship between fetal nutrition and adult health?

JOAN NALANI THOMPSON is an assistant professor of nursing at the University of Hawaii at Hilo. She is currently enrolled in the PhD program at the University of Hawaii at Manoa.

Nutrition through the Life-Span. Part 1: Preconception, Pregnancy and Infancy

ALISON ANNE SHEPHERD

Concern over poor nutrition has become more evident in the past decade, with both obesity and malnutrition reaching unprecedented levels globally. Both of these conditions pose a tremendous economic burden on the UK healthcare system. It is estimated that malnutrition alone costs the NHS £7.3 billion per year (European Nutrition For Health Alliance, 2007) and National Statistics (2008) estimate that treating obesity currently costs £3 billion per year with a predicted rise in 2010 to £3.6 billion.

Obesity is a significant health problem in children; latest figures show that 24% of 2–3-year-olds are described as 'overweight' with 11% of these being classed as morbidly obese (Harrod-Wild, 2007). In both men and woman obesity is thought to increase with age (Arterburn et al, 2004), and the latest figures show that among adults aged between 25–74 years, 71.68% of men and 61.1% of woman are either overweight or obese (British Heart Foundation, 2007). Obesity is a major risk factor in the development of chronic diseases, including type 2 diabetes and cardiovascular disease, and is strongly associated with insulin resistance, dyslipidaemia and hypertension (McPherson et al, 2007).

It is estimated that one in seven people aged 65 years or over have a high or medium risk of malnutrition (Elia, 2007). The prevalence is higher in institutionalized people than those who are in their own homes. Malnutrition is a serious issue and predisposes individuals to disease, delays recovery from illness and adversely affects body function, wellbeing and clinical outcome (Elia, 2007).

The literature also shows that malnutrition is a growing concern in infants and children. In the UK, 16% of children in hospitals were found to have severely stunted growth, 14% demonstrated a degree of muscle wasting and 20% were deemed to be at risk of developing malnutrition secondary to metabolic stress (Borton, 2007). Current research also proposes that 70% of malnutrition in the UK goes unrecognized and untreated (Elia, 2007), which is a grave cause for concern (Schenker, 2006).

Nurses and nursing support workers are in a prime position to identify nutritional problems and take appropriate steps to rectify these (Royal College of Nursing [RCN], 2007). Therefore, this two-part article seeks to educate healthcare professionals by giving some guidelines on how individuals may achieve a healthy nutritional status right through from preconception, pregnancy infancy/childhood to adulthood and old age. The issues surrounding the evidence of nutrition supplementation of the older adult in hospital is also addressed.

Preconception Nutrition

Achieving a healthy nutritional status during the preconceptual period is critical for enhanced fertility, maintenance of a viable pregnancy (Ruder et al, 2008) and long-term adult health (McMillen et al, 2008).

Early life nutrition has the potential to change chromatin structure, leading to an alteration in gene expression and therefore individual phenotype which will modulate health throughout the course of life (Mathers, 2007). Recent evidence has proposed that human exposure to environmental contaminants in air, food and water may lead to altered gene expression. This can cause metabolic and hormonal disorders in both men and woman causing infertility and the risk of developing uterine, cervical, or prostate and testicular cancer later in life (Woodruff et al, 2008). Reproductive function in females has been shown to be impaired by lead exposure, pesticides and the consumption of contaminated fish (Mendola et al, 2008).

Evidence suggests that nutritional changes in the maternal preconceptual period may also result in altered development of the fetal hypothalamo–pituitary axis. This is potentially serious as it may alter the ability of the fetus to respond to acute stressors and can potentiate the development of cardiovascular and metabolic health problems in later life (MacLaughlin and McMillen, 2007).

Obesity is a significant global health issue present in both men and woman and can also lead to infertility problems (Metwally et al, 2007; Pauli etal, 2008). These include abnormal semen parameters in obese men, secondary to endocrine dysfunction (Hammoud et al, 2008). In females the literature suggests that endocrinological changes in obese females, for example changes in circulating adipokines (hormones from adipose tissue) sex steroids as well as insulin resistance, has been shown to contribute to their inability to conceive (Metwally et al, 2008).

Increasing adherence to a 'fertility diet' pattern (*Table 1*) may favourably influence fertility in healthy woman. However, it is also proposed that the majority of infertility cases secondary to ovulatory disorders may also be preventable through modification of diet and lifestyle (Chavarro et al, 2007b).

Regular prenatal exercise is also deemed to be important both prior to and during pregnancy (Weissgerber et al, 2006). The most suitable forms of sports are aerobic types, including swimming, cycling, or aerobic training. Exercise in a fitness studio is permissible provided there have not been any contraindications during pregnancy (Lochmuller and Friese, 2005).

Infertility in Males

Research has proposed that there is also now an increasing incidence of male reproductive abnormalities and falling sperm counts (West et al, 2005). The true aetiology is unknown but it is thought that reproductive dysfunction in males may be either biochemical or nutritional (Ebisch et al, 2007). See *Table 2* for nutritional factors which may affect male fertility.

Table 1   The Female Fertility Diet

Food Groups	Number of Servings Per Day
Low glycaemic carbohydrates, including porridge oats, wholemeal pasta, rice and bread, potatoes	Five to seven servings per day
Fruit and vegetables: these are high in antioxidants which may influence time and viability of a healthy pregnancy (Ruder et al, 2008)	At least five portions per day One portion (85g) can be obtained through: • One apple, orange, banana, peach, pear or fresh fruit in season • Two small plums, apricots, satsumas • Three heaped tablespoons of fresh, frozen or tinned vegetables • One heaped tablespoon of dried fruit • One glass (150ml) of fresh fruit juice only counts as a maximum of one portion per day
High-fat dairy products (high intake of low fat dairy foods may promote anovulatory infertility) (Chavarro et al, 2007a)	Two to three servings per day of yoghurt, milk and cheese (not unpasteurized products)
Meat, fish and vegetarian alternatives (increasing vegetable protein may reduce ovulatory infertility risk) (Chavarro et al, 2007b)	Two servings per day of lean meat, chicken, fish or textured vegetable protein, peas, beans and pulses
Dietary supplements	A folate-supplemented diet with folic acid 5 mg in combination with a multivitamin should begin at least 3 months prior to conception to prevent neural tube defects (Wilson et al, 2007). Consumption of folate-rich foods, including spinach, lentils, chick peas, asparagus, broccoll;, oranges and fortified grains, are also recommended

Nutrition in Pregnancy

The unborn child requires adequate nutrients from birth and through their life-span, to enable them to achieve optimal physical health and mental wellbeing (Katzen-Luchenta, 2007). A healthy diet is important at all times in life but even more so during pregnancy. The maternal diet must provide sufficient energy and nutrients to meet both the mother's usual requirements, as well as the growing needs of the fetus (Williamson, 2007).

It has been proposed that alterations in fetal development due to impaired, excessive or imbalanced growth *in utero* can lead to the development of obesity and associated metabolic diseases later in life (Cottrell and Ozanne, 2008). Maternal under-nutrition is also associated with the potential development of cardiovascular disease, obesity and the metabolic syndrome in the adult years (McMillen et al, 2008).

Obesity during pregnancy has been linked to several complications affecting both mother and fetus. Maternal complications include hypertension, gestational diabetes mellitus and delivery by caesarean section (Yogev and Visser, 2008). For the fetus, the potential risks are more serious, and maternal obesity has been associated with an increased risk of intrauterine death and still birth (Satpathy et al, 2008). The fetus is also at risk from developing macrosomia (secondary to mother being obese), a condition which describes a fetus that is larger than normal (Ricart et al, 2008). As a result of macrosomia, the child may require admission to a neonatal intensive care (Arendas et al, 2008). A recent study has also shown that high maternal body mass may increase the risk of neonatal development of early onset group B streptococcal diseases (Håkansson and Källen, 2008).

Low maternal weight and poor weight gain in pregnancy have also been shown to carry the risks of preterm birth, low birth weight and failure to initiate breast-feeding (Viswanathan et al, 2008). To this end, Derbyshire (2007) suggests that woman need to be provided with clear, consistent information about how much weight they should gain throughout pregnancy, and be aware of the adverse effects of both under- and over-nutrition.

Currently there are no official UK recommendations for weight gain during pregnancy. However, research has shown that woman with a healthy pre-pregnancy weight, and an average weight gain through pregnancy of 12–14 kg, have an associated reduced risk of complications during pregnancy, labour and regarding the long-term health of the infant (Williamson, 2007).

Essential Fatty Acids

Docosahexaenoic acid (DHA) and arachidonic acid (ARA) are two long-chain polyunsaturated fatty acids (LC-PUFA) which play a significant role in the development of the central nervous system, intellectual and visual development (Vidailhet, 2007). The supply of LC-PUFA during pregnancy is important, particularly with respect to DHA, which is a structural component of cell membranes within the central nervous system. Furthermore, it has also been proposed that adequate DHA supply in pregnancy reduces the risk of pre-term delivery (Hanebutt et al, 2008).

The accumulation of these fatty acids in the brain begins early in pregnancy and increases rapidly during the first trimester. It has been proposed that the likely rate of utilization of DHA during this stage in pregnancy cannot be met through dietary sources alone, leading to both shortfalls for mother and child (Haggarty, 2004).

Results of LC-PUFA supplementation studies in humans are equivocal. Seafood is the predominant source of omega 3 fatty acids. However, recently there is a growing concern that fish may be contaminated with methyl mercury which has the potential to cause behavioural and neurodevelopmental problems in children (Oken and Bellinger, 2008). The Food Standards Agency (FSA, 2007) suggest that the organic pollutants in oily fish can build up within the body and have therefore recommended that pregnant, breast-feeding woman, and those who may

Table 2 Nutritional Factors Affecting Male Fertility

Nutrients which may enhance fertility in males

- Selenium is shown to enhance spermatogenesis (Flohe, 2007), and although the evidence is weak, it proposes that selenium supplementation may enhance male fertility (Keskes-Ammar et al, 2003)
- Foods rich in selenium include mushrooms, nuts, fish, poultry and eggs
- Folate and zinc are key nutrients which are important for sperm release (Ebisch et al, 2007)
- Dietary sources of zinc include beef, lamb, pork, oatmeal, jacket potatoes and salmon
- Dietary lycopene (rich sources in tomato-based foods) or lycopene supplementation may reduce oxidative stress in males and promote fertility. This requires further research but is promising (Goyal et al, 2007)

Nutrients which may compromise fertility or cause infertility in males

- Soy foods and foods high in soy isoflavones are associated with low sperm production (Chavarro et al, 2008)
- Tobacco smoking can potentially lower vitamin C levels and cause oxidative stress, which is thought may lead to reduced semen production (Mostafa et al, 2006)
- Metabolic syndrome, which is linked with hypertension, obesity, insulin resistance and dyslipidaemia, has recently been shown to have a negative impact on male reproduction (Kasturi et al, 2008)
- Alcohol consumption should be discouraged as it has been shown to impair the production, release and or motility of sperm (Emanuele and Emanuele, 1998)

want children in the future, should have not more than two portions (140 g per portion) each week. These can include mackerel, salmon, trout or sardines.

General Guidelines

The key to a successful pregnancy is to find a balance between consumption of a wide variety of foods, appropriate weight gain and physical exercise (Kaiser and Allen, 2008). *Table 3* gives a guide as to portion sizes and the types of foods that it is recommended to consume during pregnancy.

Physical exercise is still deemed to be important through pregnancy. Recent research by Clapp (2008) shows that woman who maintain weight-bearing exercise during pregnancy maintain their long-term fitness and have a lower risk of developing cardiovascular disease during the menopause than those women who do not.

Foods to Avoid

Maternal diet during pregnancy has the potential to affect fetal airway development and to promote an exaggerated immune response which can lead to the development of asthma or allergy in childhood (Chatzi et al, 2008). It is suggested that nut consumption during pregnancy can lead to an increased incidence of asthma, so it may be beneficial to avoid nuts during this time (Willers et al, 2008). Further research is recommended before this may be an official recommendation.

Partly cooked or raw egg and foods which contain these, for example mousses and mayonnaise, should be avoided owing to the risks of salmonella infection. Any salad vegetables should be washed prior to use and any unpasteurized or blue-veined cheeses should not be consumed as these may potentiate the risk of developing listeria infection. The evidence surrounding the consumption of caffeine and the risk of miscarriage is equivocal (Savitz et al, 2008), but consuming more than four cups of caffeinated tea or coffee per day should be avoided (FSA, 2007).

According to the National Institute for Health and Clinical Excellence (NICE, 2008) pregnant women should avoid consumption of alcohol during the first trimester of pregnancy to avoid the potential risk of miscarriage. If a woman wishes to continue to consume alcohol during pregnancy, then NICE (2008) recommends no more than 1–2 units once or twice a week.

Nutrition throughout Infancy

The first years of life mark a time of rapid development and dietary change, as children move from an exclusive milk diet to a modified adult diet. It has been proposed that children's nutritional status not only has an impact on the health of the child at the present time, but also in their future years. However, recent evidence proposes that many toddlers within the UK do not consume a diet which promotes an adequate nutritional status. Many toddlers are seen to be developing nutritional problems, including iron deficiency anaemia, constipation, and dental caries with a subsequent rise in the levels of obesity (Turnbull et al, 2007; More, 2008).

Nutrition for Babies

Breast-feeding has been described as the superior infant feeding method from birth, with research consistently demonstrating its numerous short- and long-term health benefits for both mother and infant (Ip et al, 2007; Lowdon, 2008). Therefore, the World Health Organization (Kramer and Kakuma, 2002) recommend mothers should aim to exclusively breast-feed their child up until the age of 6 months (Tarrant and Kearny, 2008). Current evidence proposes that babies who were breast-fed exclusively for the first 6 months of life were less likely to suffer from colic, constipation, diarrhoea, vomiting, respiratory tract infections and thrush. This finding was despite the introduction of other fluids or new foods (Harris, 2007).

The evidence supporting these claims, however, is open to interpretation. It has been proposed that there may not be enough nutrients in breast milk to fully support the needs of a growing infant, which may have the potential to cause nutritional deficiencies (Fewtrell et al, 2007). Similarly, with regard to the long-term health benefits of breast-feeding, there is no evidence to suggest that it reduces the incidence of leukaemia and cancer in later life (Turck et al, 2007).

Table 3 Guidelines for Healthy Eating During Pregnancy

Food Groups	Number of Servings Per Day
Starchy carbohydrates, including whole-meal bread, pasta, rice, breakfast cereals, chapatti and cous-cous	At least four to six portions a day Portion sizes can be: • Two slices of bread from a medium sliced loaf or one bread roll the size of the palm of the hand • 50–75 g uncooked weight of pasta per person • Three tablespoons of breakfast cereals • Half a cup of cooked rice or 125mg cooked weight
Fruit and vegetables	At least five portions per day (see *Table 1*) At least one portion of fruit should be eaten as a snack in between meals
Meat, fish and alternatives	At least two portions per day A portion of meat is 100g, which is the size of the palm of the hand, or a 125g chicken breast (raw weight) Red meat (best to help prevent anaemia), chicken, fish, tofu, seeds, pulses, eggs
Milk and dairy foods	At least three portions a day. One portion is equivalent to: • 1/3 pint of milk • One cereal bowl of milk pudding/custard • 25g or a portion of cheese (size of a matchbox)
Foods high in hydrogenated fat and sugar	Keep these to a minimum; two to three portions a week: 1/8th of a slice of cake, full fat biscuits, one to two packets of low fat crisps, confectionary

Adapted from: Pittendreigh (2005)

The relationship between exclusive breast-feeding and the development of cardiovascular disease in adult life also remains unclear (Ip et al, 2007).

Despite this, NICE (2008) still propose that breast-feeding is the safest form of nutrition that a baby can have. NICE recommend that using a combination of joint working between healthcare professionals and peer supporters, education on breast-feeding for pregnant woman, followed by proactive support, may assist in raising awareness and promotion of exclusive breast-feeding for at least 6 months—which is on the decline (Phipps, 2006).

Some babies, however, will be fed infant formula milks as opposed to being breastfed as part of their feeding regime. Despite the recommendations for exclusive breast-feeding, there is no right or wrong answer, and the decision to breast-feed or bottle-feed the child should be based on parental beliefs and lifestyle.

It has been argued that formula-fed infants have a very different growth pattern compared with breast-fed babies (Lawson, 2007). It is thought that the key reason for this is the differences in the nutrient composition of breast milk compared with formula milk (Lonnerdal, 2008). Indeed, both DHA and ARA (the two aforementioned LC-PUFA) are found naturally in breast milk. These essential fatty acids are not found in formula milk unless it is supplemented. Despite the questionable results surrounding the supplementation of formula milks with these essential fats and their effects on development, the general opinion from an expert panel is that formula milk should be supplemented with these essential fats and parents need to be made aware of their existence (Mitmesser and Jensen, 2007)

It has also been suggested that formula-fed infants are far more likely to become obese than those children who have been exclusively breast-fed (Moreno and Rodríguez, 2007). However, parents may be reassured that infant formulas, when correctly prepared and handled, are a safe alternative for mothers who do not wish to, or for whatever reason are unable to breast-feed (Lawson, 2007).

Weaning

NICE (2008) define weaning as the: 'transition from an exclusively milk diet to a diet based on solid foods.' Current Government guidelines state that weaning should be delayed and babies exclusively breast-fed until 6 months of age. It was thought that delaying weaning until this time may risk compromised growth. However, current research by Wall (2008) proposes that delaying weaning will not cause any deficit in growth in most babies, although a small percentage may be at risk. Wall also suggests that all babies should be treated as individuals when assessing their suitability for weaning.

There is now also a growing body of evidence to challenge the notion that solids should be commenced later than 6 months of age. In particular, it has been found that introducing gluten in small amounts gradually from 4–6 months, as opposed to large amounts from the age of 6 months, may reduce the risk of the child developing coeliac disease, insulin-dependent diabetes and wheat allergy (Agostoni et al, 2008). Therefore, weaning may be considered when a child is 4 months old and is showing signs that their current feeding pattern is not satisfying their appetite. In these cases, it is recommended that parents also seek advice from their GP or health visitor prior to starting their child on solids (FSA, 2007).

It is not known what constitutes an optimal diet in infancy as there appears to be relatively few studies of weaning practice in the UK (Robinson et al, 2007). Weaning is a gradual process and eventually baby will progress to eating a wide range of non-milk foods, thus ensuring that the child will receive a nutritionally adequate diet to enable healthy growth and development. It is proposed that this is best managed in stages (FSA, 2007).

The foods chosen will be dependent on the individual child's nutritional requirements with different tastes and textures being introduced within the different stages of weaning. Foods which require chewing will encourage the muscles required for speech to develop (Shepherd, 2008). Infants should be given less bottles and begin drinking from cups, which helps jaw development and can help prevent dental decay.

Table 4 Stages of Weaning, Meal Planning Ideas

Time	Stage 1: 4–6 Months	Stage 2: 6–9 Months	Stage 3: 9–12 Months	Stage 4: 12 Months Plus
	Signs of readiness for weaning: • Demanding more feeds • Not lasting 3–4 hours between feeds • Waking in the night for feed when they have been sleeping through • Ability to hold their own head up	Developmental progression for weaning: • Picking up objects between thumb and forefinger • Interest into what the family are eating • Showing signs of beginning to chew objects • Attempting to feed themselves	Developmental progression for weaning: • Drinking from a cup unsupervised • Ability to put items into a container and remove them again • Ability to hold objects between thumb and forefinger	Developmental progression for weaning: • Child can sit up fully supported • Demonstrate the signs of shuffling/crawling
Early morning	Breast milk/formula feed Total of 500–600 ml through the day	Breast milk/formula milk Total of 500–600 ml through the day with an additional fluid allowance of 120ml per kg	NA	NA
Breakfast	Infant cereal: 1–2 tablespoons mixed with formula/breast milk	Infant cereals increase to 2 tablespoons with formula/breast milk	Infant cereal: 3–4 tablespoons with a formula or follow-on milk	Breakfast cereals, e.g. wheat biscuits, porridge oats made with full-fat ordinary cow's/soya milk
Mid-morning	Milk feed	Milk feed	Boiled water or diluted fruit juices	Snack finger food, such as chopped apple/pear
Lunch	Milk feed 1–2 teaspoons of pureed vegetables, including carrot, potatoes, pumpkins or pureed fruit	Milk feed 1–2 tablespoons of mashed savoury food, including chicken, fish or red meat with a mashed potato and a mashed vegetable such as cauliflower or broccoli	Cooked chopped meat and vegetables. Emphasis on savoury foods to prevent dental caries. These can include: 3–4 tablespoons/small portion of • Pasta with a cheese sauce, • Wholemeal bread cut into either fingers • Chicken, fish, red meat • Fluid such as diluted fruit juices	Cooked chopped meat and vegetables with more texture. Small portion of: • Pasta shapes with either a cheese sauce or including chunks of vegetables in a tomato-based sauce • Chicken, fish, or a red meat with potatoes and any vegetables of choice • Fruit juice, cooled boiled water or milk
Mid-afternoon	Milk feed	Cooled boiled water or diluted fruit juices	Chopped apple, banana, pear or baby rusks/baby breadsticks	Small fingers of bread spread with a mild soft cream cheese or chopped fruit/vegetable crudités
Dinner	Milk feed 1–2 teaspoons of baby cereal with formula/breast milk OR 1–2 teaspoons of pureed vegetables, including carrot, butternut squash, potato pumpkin or 1–2 tablespoons of pureed fruit Cooled boiled water	Milk feed 1–2 tablespoons of pureed fruit, including apple, pear, apricot or banana OR Any full-fat dairy product, including fromage frais, wholemilk yoghurt (not made with eggs), wholemilk baby rice pudding	Chunky meat- or vegetable-based soup with some chopped meat or chicken, including a range of chopped vegetables in season and according to likes/dislikes	Small portion of chicken/fish/red meat served with a portion of noodles, rice, mashed potato or sweet potato Dessert of chopped pieces of seasonal fruit or a small portion of full-fat fromage frais/milk pudding
Bedtime	Milk feed	Milk feed	Breast/formula or specialist follow-on milk	Warm milk drink made with full-fat milk

Adapted from: Shaw (2007), Nestlé (2008)

There is also confusion as to what constitutes a so called 'balanced diet'. Dietary habits and lifestyle have changed radically over the past decade, and this combined with a lack of resources to help families with babies and young children to achieve nutritional status seems to complicate the matter further (Wall, 2008).

Table 4 is a guide for healthcare professionals, which summarizes when a child is showing signs that they may be ready to begin weaning, the stages of weaning and some meal planning ideas.

General Guidelines

The amount of food offered should be steadily increased by encouraging new foods at one meal then progressing to two or three meals as the baby's appetite allows. Research has shown that a new food may need to be offered between 8–15 times before it is accepted (Briefel et al, 2004). As more solid food is introduced less milk should be offered.

Feeds should continue at between 500–600 ml formula or breast milk per day through the first two stages of weaning. It is also important to note that from 7 months of age a baby taking solids also has a fluid requirement of 120ml per kg. This may be taken either as a milk feed or other suitable fluids, depending on how well the baby is feeding (Shaw, 2007).

As babies progress from formula/breast milk to ordinary milks this poses a nutritional problem. With the rising trends of obesity in children some mothers may wish to use low-fat milk. However, there is some controversy as to when low-fat milks should be introduced into the diet as there is concern that a diet low in fat may limit growth (Michaelsen et al, 2007). In addition, the literature shows that during the transition from milk foods to semi-solids, infants are at risk of developing iron deficiency anaemia. This can cause a delay in the development of the central nervous system which will have an impact on cognition/intellectual abilities later on in life (Beard, 2007).

Foods to be Avoided

Some foods should be avoided during the weaning process as they may have serious health consequences. It is not recommended to give premature babies, or babies less than 3 months of age, solid foods as this can have a profound effect on the immune system, leading to the development of allergy in later life (Calder et al, 2006). To minimize the risk of allergies, it is recommended that cooked eggs may be introduced into the diet from the age of 6 months, with whole nuts being withheld until the child is at least 5 years of age—this is for both the risks of choking as well as possible allergy (FSA, 2007).

Salt intake should be kept to a maximum of 1 g per day as any more has been shown to place a child at significant risk of developing hypertension later on in life (Brion et al, 2007).

Sugar should not be added to drinks and honey should be avoided until the child is at least 1 year of age as it is associated with the potential development of food poisoning in the form of botulism (Brook, 2007).

Conclusion

In light of the evidence presented, it is fair to say that long before fertilization and conception, parental nutrition affects a child's health and development for life (House, 2007). Achieving a healthy nutritional status in infancy by adopting good feeding habits and consumption of a wide variety of foods is vital if infants are to develop and progress into healthy children, and later on into healthy adults.

Part two of this article will seek to educate healthcare professionals as to how to promote healthy eating in childhood and adulthood, through to those aged 65 years and over. A review of the use of oral nutritional supplements in older adults is also presented.

Key Points

- Nutrition through the life-span is of fundamental importance if individuals are to achieve their full potential as they grow.
- With increasing rates of obesity and malnutrition, this tends to suggest that for whatever reason, children, adults and the older people are failing to meet their full nutritional requirements.
- Nurses are in a prime position to be able to identify individuals with acute or chronic nutritional problems and take action to rectify these.
- Safe and effective nutritional care can only be delivered if all those involved in the care of the patient are working together.

References

Agostoni C, Decsi T, Fewtrell M et al (2008). Complementary feeding: a commentary by the ESPGHAN Committee on Nutrition. *J Pediatr Gastroenterol Nutr* **46**(1): 99–110

Arterburn DE, Crane PK, Sullivan SD (2004). The coming epidemic of obesity in elderly Americans. *J Am Geriatr Soc* **52**(11): 1907–12

Arendas K, Qiu Q, Gruslin A (2008). Obesity in pregnancy: preconceptional to postpartum consequences. *J Obstet Gynaecol Can* **30**(6): 477–88

Beard J (2007). Recent evidence from human and animal studies regarding iron status and infant development. *J Nutri* **137**(2): 524S–530S

Borton C (2007). Malnutrition. Available at: http://tinyurl.com/6rgd4b (last accessed 20 Oct 2008)

Briefel RR, Reidy K, Karwe V, Jankowski L, Hendricks K (2004). Toddlers' transition to table foods: Impact on nutrient intakes and food patterns. *J Am Diet Assoc* **104**(Suppl 1): S38–44

Brion MJ, Ness AR, Davey Smith G et al (2007). Sodium intake in infancy and blood pressure at 7 years: findings from the Avon Longitudinal Study of Parents and Children. *Eur J Clin Nutr* **62**(10): 1162–9

British Heart Foundation (2007). How to keep children healthy. BHF, London. Available at: http://tinyurl.com/2nen5o (last accessed 20 Oct 2008)

Brook I (2007). Infant Botulism. *J Perinatol* **27**(3): 175–8

Calder PC, Krauss-Etschmann S, de Jong EC et al (2006) Early nutrition and immunity—progress and perspectives. *Br J Nutr* **96**(4): 774–90

Chatzi L, Torrent M. Romieu I et al (2008). Mediterranean diet in pregnancy is protective against wheeze and atopy in childhood. *Thorax* **63**(6): 507–13

Chavarro JE, Rich Edwards JW, Rosner B, Willett WC. (2007a). A prospective study of dairy food intake and anovulatory infertility. *Hum Reprod* **22**(5): 1340–7

Chavarro JE, Rich Edwards JW, Rosner BA Willett WC (2007b). Diet and lifestyle in the prevention of ovulatory disorder infertility. *Obstet Gynaecol* **110**(5): 1050–8

Chavarro JE, Toth TL, Sadio SM, Hauser R (2008). Soy foods and isoflavone intake in relation to semen quality parameters among men from an infertility clinic. *Human Reproduction* **23**(11): 2584–90

Clapp JF 3rd (2008). Long-term outcome after exercising throughout pregnancy: fitness and cardiovascular risk. *Am J Obstet Gynecol* [Epub ahead of publication]

Cottrell EC, Ozanne SE (2008). Early life programming of obesity and metabolic disease. *Physiol Behav* **94**(1): 17–28

Derbyshire E (2007). Low maternal weight gain: effects on maternal and infant health during pregnancy. *Nurs Stand* **22**(3): 43–6

Ebisch IM, Thomas CM, Peters WH, Braat DD, Steegers-Theunissen RP (2007). The importance of folic acid, zinc and antioxidants in the pathogenesis of subfertility. *Hum Reprod Update* **13**(2): 163–74

Elia M (2007). *The 'MUST' Report. Executive Summary.* BAPEN, Redditch. Available at: http://tinyurl.com/65hzy6 (last accessed 20 Oct 2008)

Emanucle MA, Emanucle NV (1998). Alcohol's effect on male reproduction. *Alcohol Health Res World* **22**(3): 195–201

European Nutrition for Health Alliance (2007). *Preventing Malnutrition of Older People in the Community: What Must Work?* ENHA, London. Available at: http://tinyurl.com/6htloc (last accessed 24 Oct 2008)

Fewtrell MS, Morgan JB, Duggan C et al (2007). Optimal duration of exclusive breastfeeding: what is the evidence to support current recommendations? *Am J Clin Nutr* **85**(2): 635S–8

Flohe L (2007). Selenium in mammalian spermatogenesis. *Biol Chemistry* **388**(10): 987–95

Food Standards Agency (2007). Starting on solid foods. FSA, London. Available at: http://tinyurl.com/6r8zlt (last accessed 24 Oct 2008)

Goyal A, Chopra M, Lwaleed BA, Birch B, Cooper AJ (2007). The effects of dietary lycopene supplementation on human seminal plasma. *BJU Int* **99**(6): 1456–60

Haggarty P (2004). Effect of placental function on fatty acid requirement during pregnancy. *Eur J Clin Nutr* **58**(12): 1559–70

Håkansson S, Källen K (2008). High maternal body mass index increases the risk of neonatal early onset group B Streptococcal disease. *Acta Paediatr* **97**(10): 1386–9

Hammoud AO, Gibson M. Peterson CM, Meikle AW, Carrell DT (2008). Impact of male obesity on infertility: a critical review of the current literature. *Fertil Steril* **90**(4): 897–904

Hanebutt FL, Demmelmair H, Schiessl B, Larque E, Koletzko B (2008). Long-chain polyunsaturated fatty acids (LC-PUFA) transfer across the placenta. *Clin Nutr* **27**(5): 685–93

Harris G (2007). A growing issue: understanding and promoting healthy growth and development. Infant and Toddler Forum, London

Harrod-Wild K (2007). Does childhood nutrition matter? *J Fam Health Care* **17**(3): 89–91

House SH (2007). Nurturing the brain nutritionally and emotionally from before conception to late adolescence. *Nutr Health* **19**(1–2): 143–61

Ip S, Chung M, Raman G et al (2007). Breastfeeding and maternal and infant health outcomes in developed countries. *Evid Rep Technol Assess (Full Rep)* Apr(153): 1–186

Kaiser L, Allen LH (2008). Position of the American Dietetic Association: nutrition and lifestyle for a healthy pregnancy outcome. *J Am Diet Assoc* **108**(3): 553–61

Kasturi SS, Tannir J, Brannigan RE (2008). The metabolic syndrome and male infertility. *J Androl* **29**(3): 251–9

Katzen-Luchenta J (2007). The declaration of nutrition, health, and intelligence for the child-to-be. *Nutr Health* **19**(1–2): 85–102

Keskes-Ammar L, Feki-Chakroun N, Rebai T et al (2003). Sperm oxidative stress and the effect of an oral vitamin E and selenium supplement on semen quality in infertile men. *Arch Androl* **49**(2): 83–94

Kramer MS, Kakuma R (2002). *The Optimal Duration of Exclusive Breastfeeding: A Systematic Review.* World Health Organization, Geneva. Available at: http://tinyurl.com/6gy53u (last accessed 3 Nov 2008)

Lawson M (2007). Contemporary aspects of infant feeding. *Paediatr Nurs* **19**(2): 39–46

Lochmuller EM, Friese K (2005) Pregnancy and sports [Article in German]. *NMW Fortschr Med* **147**(16): 28–31

Lonnerdal B (2008). Personalizing nutrient intakes of formula fed infants: breast milk as a model. *Nestle Nutr Workshop Ser Pediatr Program* **62**: 189–98

Lowdon J (2008). Getting bone health right from the start! Pregnancy, lactation and weaning. *J Fam Health Care.* **18**(4): 137–41

MacLaughlin SM, McMillen IC (2007). Impact of periconceptional undernutrition on the development of the hypothalamo-pituitary-adrenal axis: does the timing of parturition start at conception? *Curr Drug Targets* **8**(8): 880–7

Mathers JC (2007). Early nutrition: impact on epigenetics. *Forum Nutr* **60**: 42–8

McMillen IC, Maclaughlin SM, Muhlhausler BS, Gentili S, Duffield JL Morrison JL (2008). Developmental origins of adult health and disease: the role of periconceptual and foetal nutrition. *Basic Clin Pharmacol Toxicol* **102**(2): 82–9

McPherson K, Marsh K, Brown M (2007). *Tackling Obesities: Future Choices—Modelling Future Trends in Obesity and the Impact on Health.* Foresight, London. Available at: http://tinyurl.com/5wzz8s (last accessed 24 Oct 2008)

Mendola P, Messer LC, Rappazzo K (2008). Science linking environmental contaminant exposures with fertility and reproductive health impacts in the adult female. *Fertil Steril* **89**(2 Suppl): e81–94

Metwally M, Li TC, Ledger WL (2007). The impact of obesity on female reproductive function. *Obes Rev* **8**(6): 515–23

Metwally M, ledger WL, Li Tc (2008) Reproductive endocrinology and clinical aspects of obesity in woman. *Aun N Y Acad Sci* **1127**: 140–6

Michaelsen KF, Hoppe C, Lauritzen L, Molgaard C (2007). Whole cow's milk: why, what and when? *Nestle Nutr Workshop Ser Pediatr Program* **60**: 201–16

Mitmesser SH, Jensen CL (2007). Roles of long-chain polyunsaturated fatty acids in the term infant: developmental benefits. *Neonat Netw* **26**(4): 229–34

More J (2008). Promoting healthy growth through nutrition. A growing issue: understanding and promoting healthy growth and development in toddlers. Infant and Toddler Forum, London

Moreno LA, Rodríguez G (2007). Dietary risk factors for development of childhood obesity. *Curr Opin Clin Nutr Metab Care* **10**(3): 336–41

Mostafa T, Tawadrous G, Roaia MM, Amer MK, Kader RA, Aziz A (2006). Effect of smoking on seminal plasma ascorbic acid in infertile and fertile males. *Andrologia* **38**(6): 221–4

National Institute for Health and Clinical Excellence (2008). *Improving the Nutrition of Pregnant and Breastfeeding Mothers and Children in Low Income Households.* NICE, London

National Statistics (2008). Statistics on obesity, physical activity and diet: England, January 2008. National Statistics, London. Available at: http://tinyurl.com/5x8341 (last accessed 24 Oct 2008)

Nestlé (2008). Introducing solid foods and weaning stages. Nestle, Croydon. Available at: http://tinyurl.com/5manyx (last accessed 3 Nov 2008)

Oken E, Bellinger DC (2008). Fish consumption, methylmercury and child neurodevelopment. *Curr Opin Pediatr* **20**(2): 178–83

Pauli EM, Legro RS, Demers LM, Kunselman AR, Dodson WC, Lee PA (2008). Diminished paternity and gonadal function with increasing obesity in men. *Fertil Steril* **90**(2): 346–51

Phipps B (2006) Peer support for breastfeeding in the UK. *Br J Gen Pract* **56**(524): 166–7

Pittendreigh J (2005). Healthy eating during pregnancy. British Dietetic Society, Birmingham. Available at: http://tinyurl.com/6qaj3t (last accessed 3 Nov 2008)

Ricart W, Lopez J, Mozas J, Pericot A, Sancho MA, Gonzalez N et al (2008). Maternal glucose tolerance status influences the risk of macrosomia in male but not in female fetuses. *J Epidemiol Community Health* Aug 21 [Epub ahead of print]

Robinson S, Marriott L, Poole J et al (2007). Dietary feeding practice in infancy: the importance of maternal and family influences on feeding practice. *Br J Nutr* **98**(5): 1029–37

Royal College of Nursing (2007). *Malnutrition: What Nurses Working with Children and Young People Need to Know and Do.* RCN, London. Available at: http://tinyurl.com/5r4vcu (last accessed 24 Oct 2008)

Ruder EH, Hartman TJ, Blumberg J, Goldman MB (2008). Oxidative stress and antioxidants: exposure and impact on female fertility. *Hum Reprod Update* **14**(4): 345–57

Satpathy HK, Fleming A, Frey D, Barsoom M, Satpathy C, Khandalavala J (2008). Maternal obesity and pregnancy. *Postgrad Med* **120**(3): E01–9

Savitz DA, Chan RL, Herring AH, Howards PP, Hartmann KE (2008). Caffeine and miscarriage risk. *Epidemiology* **19**(1): 55–62

Shaw V (2007). Infant feeding formula: clinical guideline. UCL Institute of Child Health, London. Available at: http://tinyrul.com/5acfpa (last accessed 24 Oct 2008)

Shepherd A (2008). Paediatrics: nutrition for babies and young children. *British Journal of Healthcare Assistants* **2**(3): 132–8

Schenker S (2006). Undernutrition in the community. *Nursing in Practice* **30:** 76–80

Tarrant RC, Kearny JM (2008). Session 1: Public health nutrition. Breast-feeding practices in Ireland. *Proc Nutr Soc* **67**(4): 371–80

Turck D (2007). Later effects of breastfeeding practice: the evidence. *Nestle Nutr Workshop Ser Pediatr Program* **60:** 31–9

Turnbull B, Lanigan J, Singhal A (2007). Toddler diets in the UK: deficiencies and imbalances. 1. Risk of micronutrient deficiencies. *J Fam Health Care* **17**(5): 167–70

Vidailhet M (2007). Omega 3: is there a deficiency in young children? [Article in French]. *Arch Paediatr* **14**(1): 116–23

Viswanathan M, Siega-Riz AM, Moos MK et al (2008). Outcomes of maternal weight gain. *Evid Rep Technol Assess (Full Rep)* **May**(168): 1–123

Wall A (2008). Early childhood nutrition and growth monitoring. *Primary Health Care* **18**(7): 16–20

Weissgerber TL, Wolfe LA, Davies GA, Mottola MF (2006). Exercise in the prevention of maternal-fetal disease: a review of the literature. *Appl Physiol Nutr Metab* **31**(6): 661–74

West MC, Anderson L, McClure N, Lewis SE (2005). Dietary oestrogens and male fertility potential. *Hum Fertil (Camb)* **8**(3): 197–207

Willers SM, Wijga AH, Brunekreef B et al (2008). Maternal food consumption during pregnancy and the longitudinal development of childhood asthma. *Am J Respir Crit Care Med* **178**(2): 124–31

Williamson C (2007). Nutrition in pregnancy. BNF Briefing Paper. British Nutrition Foundation, London. Available at: http://tinyurl.com/6n8am8 (last accessed 24 Oct 2008)

Wilson RD, Johnson JA, Wyatt P, Allen V, Gagnon A, Langlois S et al (2007). Pre-conceptual vitamin/folic acid supplementation 2007; the use of folic acid in combination with a multivitamin supplement for the prevention of neural tube defects and other congenital abnormalities. *J Obstet Gynaecol Canada* **29**(12): 1023–6

Woodruff TJ, Carlson A, Schwartz JM, Giudice LC (2008). Proceedings of the Summit on Environmental Challenges to Reproductive Health and Fertility: executive summary. *Fertil Steril* **89**(2 Suppl): e1–e20

Yogev Y, Visser GH (2008). Obesity, gestational diabetes and pregnancy outcome. *Semin Fetal Neonatal Med* **Oct 14** [EPUB ahead of print]

Critical Thinking

1. What complications may occur due to poor nutrition during pregnancy?

ALISON ANNE SHEPHERD is Lecturer in Adult Nursing, De Montfort University, Leicester, and Student Community Care Practitioner, Leicester City Primary Healthcare Trust.

From *British Journal of Nursing—BJN,* 17:20, November 13–26, 2008, pp. 1261–1268. Copyright © 2008 by MA Healthcare Limited. All rights reserved. Reprinted by permission.

Nutrition in Palliative Care

Sue Acreman

Palliative care has been defined as that which:
'Affirms life and regards dying as a normal process; neither hastens nor postpones death; provides relief from pain and other distressing symptoms; integrates the psychological and spiritual aspects of patient care; offers a support system to help patients live as actively as possible until death; offers a support system to help the family cope during the patient's illness and following bereavement' (World Health Organization, 2002).

Dame Cicely Saunders described palliative care as an explicit intention to do everything possible to ease the burden (Saunders, 1984), and if the goal of palliative care is to improve the patient's quality of life, which includes the treating of symptoms, then nutrition has a clear role here.

The National Institute for Clinical Excellence (NICE) guidance, *Supportive and Palliative Care for Adults* (NICE, 2004) includes the expertise of the dietitian in the provision of specialist physical, psychological, social and spiritual care and furthermore recommends that people with advanced cancer should have their needs assessed. However, with a dearth of dietitians, nutrition is the responsibility of the entire health-care team, with nurses at the cutting edge because of their intense involvement with patient care. Accordingly, guided by the dietitian, the nurse can make sure that the patient's wishes regarding food and feeding at the end of his or her life are attended to.

The impact of the cancer and its treatment modalities exert a profound effect on nutritional status as well as compromising the elements of social interaction and sensory stimulation of food. In palliative and end-of-life care, such issues can be managed to optimize the time remaining to ensure that nutrition is not a burden but rather an enhancement during this period.

Food and Nutrition in Palliative Care

People with cancer are often malnourished, particularly when the illness is advanced (Stratton et al, 2003) and it has been suggested that as many as 20% people die from the effects of malnutrition rather than the cancer itself (Ottery, 1996). In advanced cancer, frequently-reported symptoms include those which compromise the ability to continue eating, drinking and enjoying food (Bruera and Lawlor, 1997). Nutritional care can assist in optimizing quality of life and sense of wellbeing, as well as helping to alleviate unpleasant symptoms which occur at the end of life (Acreman, 2000).

Symptom prevalence in advanced cancer (Teunissen et al 2007) is as follows:

- Anorexia (loss of appetite) 53–86%
- Weakness 60–74%
- Nausea 17–31 %
- Dry mouth 34–40 %
- Weight loss 46–86%
- Taste changes 22%.

To provide holistic care means that the health professional will need to deal with the issues and concerns that are important to the patient and their family, and often food becomes an important issue. Health professional are often asked questions such as whether patients will starve to death, if dehydration is painful and whether tube feeding would lessen the suffering. When responding to the nutritional needs and queries of the dying patient and his or her family, nurses must be aware that they are caring for the living. Food means different things to different people, but to every person food has physical, emotional and sociological importance. While Ganzini et al (2003) reported that the decision to stop eating and drinking was seen by nursing staff as a readiness to die, Fuez and Rapin (1994) found that the management of nutrition-related symptoms meant that people were able to maintain food and fluid intake up to the day of death.

Cancer and Its Effects on Nutritional Status

The presence of a tumour causes metabolic chaos resulting in increased nutritional needs at the very time when the patient is unable to meet their normal requirements. Add to this the impact of surgery, radiotherapy and chemotherapy and the result is a malnourished, weak patient. The metabolic chaos brings about altered access to protein, carbohydrates and lipids in the body.

The metabolic consequences of cancer are listed below (Stratton et al 2003):

- Altered glucose metabolism—the tumour is inefficient in the use of glucose
- Increased rate of glucose oxidation
- Increased rate of protein metabolism

- Decreased protein synthesis
- Increased protein breakdown
- Altered lipid metabolism.

Humans use meal times for social interaction and relationships. Food is used as a profession of affection as well as religious and cultural values which have deep meaning to the patient and their family.

Nutritional Impact of Terminal Ilness

Caring for the patient and family in the terminal stages of cancer is as important as treatment of the symptoms resulting from the underlying disease. Anorexia and cachexia (a wasting syndrome that causes weakness and a loss of weight) are common at the end stage of cancer. Causes include the abnormal metabolic responses brought about by the tumour and the debilitating effects of treatment. Terminal illness can alter the nutritional status of the patient in three ways:

- There is a reduction in gastrointestinal absorption and an increase in nutrient requirements brought about by physiological, metabolic and anatomical changes such as malabsorption, cachexia and increasing tumour mass
- The dying process reduces many bodily functions including gastric emptying. This results in increased satiety, decreased hunger and food intolerances
- Medication causes side-effects such as nausea, vomiting, diarrhoea and constipation. Narcotics or opioids are included here and they are the main pain-relieving agents used in this group of patients.

Consideration must be given to the psychological changes that occur at this time. The patient with advanced illness may well perceive different views of food depending on the present stage of their disease. Depression causes anorexia (Holland et al, 1977), while anger and guilt may also be present and these can also have a negative impact on dietary intake. A qualitative study among 47 patients by McClement et al (2003) reported that families were keen to do what they felt was best for the patient at this time, whether that involved encouraging food and drink in an attempt to 'fight back' against the obvious physical signs of malnutrition; letting nature take its course by shifting the focus of care away from nutrition; and a conglomeration of the two as they veer between acceptance and non-acceptance of the situation.

Aims of Nutritional Care

Overall, the goals of nutritional support in palliative care must be to minimize food-related discomfort and maximize food enjoyment. Care must be exercised, however, to ensure that any nutritional support is gentle. If eating is no longer an enjoyable experience, it is important to educate the family that perhaps aggressive feeding and over-feeding (a very common problem and one which causes conflict between the patient and family) is inappropriate and that perhaps love and care can

be demonstrated in other ways. It is very useful if the patient and family can be made aware of the impact of the disease on nutritional status. If they can understand that it is the disease reducing the appetite and not the patient being difficult; if they can realize that food can cause more discomfort than pleasure and if they can show affection by means other than feeding, the result will be some relaxation of tension.

In palliative care, nutrition should be supportive and should aim to optimize the management of nutrition-related symptoms, thus improving the sense of wellbeing felt by the patient. Additionally such support will provide reassurance to the patient and their family that changes in dietary habits are a normal response to the illness.

Assisting with the Nutritional Goals

Community nurses play an important role in helping patients to achieve their nutritional goals, below are some pointers as to how to assist:

- Assess the impact food and drink have on the patient and from there ascertain if dietary modification will improve both the physical and psychological condition
- Identify any nutritional concerns the patient and family may have
- Establish appropriate and agreed dietary goals, which must be agreed and potentially achievable
- Review and re-evaluate these dietary goals at regular intervals and implement changes where and when appropriate in response to changes in physical, psychological and emotional status.

Achieving the Nutritional Goals

- Assessment—a good nutritional assessment tool will seek to identify issues regarding nutritional status and requirements, but more importantly it should provide the opportunity for the patient to raise any physical, mechanical and psychological problems that may compromise dietary intake
- Identify patient and family concerns—it is important that we are attentive to any comments that may expose hidden fears (e.g. 'if I don't drink anything will dehydration be painful?' or 'I'm afraid that if I eat anything I will choke')
- Integrate nutrition into the plan of care—this will result from the assessment process. It is important that any nutritional goals are consistent with other medical and nursing goals as nutrition should complement other interventions. It is also important that religious and cultural traditions are taken into consideration
- Re-evaluate and review goals and implement changes where appropriate. This is very important in advanced cancer when a patient's condition changes almost daily with a consequent impact upon dietary intake. What suits on one day or week may not be appropriate the next.

Box 1
Suggestions for Improving Nutritional Intake

- Feed the patient when hungry
- Serve small portions of food
- Gently encourage–do not nag
- Set an attractive table, tray or plate
- Make much of meal times. Make them social and enjoyable. Remove bedpans, vomit bowls and other similar items from the area
- Encourage a breath of fresh air prior to the meal. Take the patient outside or open a window for a short time
- Eat outside if the weather is good enough
- The use of food supplements may or may not be appropriate here.

Source: Acreman, 2000.

It is possible that conflicting objectives may arise between the patient and the family, i.e. the patient who cannot and will not eat and the family who pushes the patient to more than he or she is willing or able. Regular review of nutritional matters will highlight these problems so that solutions can be offered. It is not easy however to achieve this. Dramatic loss of body weight and physical strength is alarming to witness and it is quite natural for family to try zealously to redress the balance. Again, it is important that the patient's wishes are upheld here and that an amicable agreement is reached regarding the provision of food and fluids.

'It is important that the patient's wishes are upheld and that an amicable agreement is reached regarding the provision of food and fluids'

Patients Who Can Eat and Want to

For this type of patient the importance of enabling them to eat cannot be over-emphasized. Medication is available to boost appetite. Alcohol can be a useful appetite stimulant provided the patient enjoys the taste and it is compatible with any medication. Other measures to employ include the use of suitable clothing to improve body image and self-esteem, attention to oral care and dentition and even a visit from the hairdresser, while fresh air and a change of scenery can improve appetite. *Box 1* provides a list of suggestions for improving nutritional intake (Acreman, 2000).

Some patients may decide to embrace an alternative or complementary diet. It is important that the health professional is not seen as condemning something the patient has taken an interest in. However, sensible advice should be offered on the importance of maintaining an adequate diet.

It is wise to consider why patients feel they need to resort to these measures–could it be that orthodox medical care has not provided suitable dietary advice in the past? This is a point worth considering.

To Feed or not to Feed

Ethical questions will be raised concerning the provision of food and fluids to a person nearing the end of their life. Questions to ask in each case include:

- What good will it accomplish for the patient?
- How much discomfort is caused by eating and drinking?
- How keen is the patient to continue eating and drinking?

The answers to such questions will be different for each patient, but they deserve honest discussion on an individual basis in order that any food related-issues can be dealt with.

There are several ethical questions that arise surrounding the nutritional support of the terminally-ill patient. While it is not possible to provide answers to them here, the professional team should be aware of them in order that supportive nutritional care can be right, good, desirable and worthy. The British Medical Association (BMA) has considered questions such as:

- Is it ever permissible to permit a patient to die from lack of nourishment?
- Is it ever permissible to discontinue nutritional support with the intent to cause a patient's death?
- Is artificial feeding ethically different from artificial respiratory support?
- Is using a nasogastric or intravenous feeding tube for feeding, a medical treatment or is it obligatory palliative care?

The BMA states that tube feeding is regarded as a medical treatment but this is not a clear-cut issue, and it is important that nutritional needs are discussed and included within the health-care plan (BMA, 1999).

While it is accepted that nutrition cannot prolong life, it should be recognized that optimal nutrition can enable and empower the patient in the following ways:

- Optimizing physically strength to fulfil last or final objectives
- To die with dignity, not of starvation
- To retain some control over the disease process–food and feeding can be a useful focus for the patient.

It is important to consider the trajectory that he or she will have endured through diagnosis, treatment and medication. An interest in feeding oneself can provide a useful focus, it often provides a sense of being in charge of something.

Another consideration is the professionals caring for the patient in the terminal stages of any treatment. Nursing staff have a clear duty of care to their patients which embraces nutrition (United Kingdom Central Council, 1996). Under the duty of care, nurses are expected to offer food and drink if a patient is able to swallow, however, the patient has the right to refuse this if he or she wishes. This is where information concerning the patient's wishes are important and nutritional needs are included within the care plan.

Conclusion

Whilst it is evident that cancer causes weakness, subsequent malnutrition makes the weak weaker. Attention to nutrition can optimize quality of life right up to the time of death. Nurses need to acknowledge that food has greater significance than the provision of nutrients and when nutrition is no longer required for physiological reasons, food is important for psychological and social reasons and as such decisions regarding its provision must be led by the patient themselves. However, the profession should be able to access education and awareness in nutrition matters at the end of life to ensure they have the skills necessary to assist the patient in this emotive area of care and need.

References

Acreman S (2000) Nutrition in palliative care. *Palliative Care Today* **9**(2): 27–8.

Breura E, Lawlor P (1997) Defining palliative care interventions. *Journal Palliative Care* **14**(2): 23–4.

British Medical Association (1999) Witholding and withdrawing life prolonging medical treatments. *British Medical Journal* **318**: 1709–10.

Feuz A, Rapin CH (1994) An observational study of the role of pain control and food adaptation of aged cancer patients during terminal care. *American Journal Dietetic Association* **94**: 767–70.

Ganzini L, Goy E, Miller L, Harvath T, Jackson A, Delorit M (2003) Nurses experiences with hospice patients who refuse food and fluids to hasten death. *New England Journal Medicine* **349**: 359–365.

Holland JCB, Rowland J, Plumb M (1977) Psychological aspects of anorexia in cancer patients. *Cancer Research* **37**: 2425–28.

McClement SE, Degner LF, Harlos MS (2003) Family beliefs regarding the nutritional care of a terminally ill relative: a qualitative study. *Journal Palliative Medicine* **6**(5): 737–48.

National Institute for Clinical Effectiveness (2004) *Guidance on cancer services—improving supportive and palliative care for adults with cancer. The Manual.* NICE, London.

Ottery F (1996) Definition of standardised nutritional assessment and intervention pathways in oncology. *Nutrition* **12**(1): S15–19.

Saunders C (1984) The philosophy of terminal care: In: *The management of terminal malignant disease* Arnold Publishers, Baltimore MD: 232–41.

Stratton RJ, Green CJ, Elia M (2003) *Disease-Related Malnutrition:An Evidence-Based Approach to Treatment.* CABI Publishing, Oxon: 392–93.

Teunissen SC, Wesker W, Kruitwagen C, de Haes H, Voest E de, Graeff A (2007) Symptom prevalence in patients with incurable cancer: a systematic review. *Journal Pain and Symptom Management* **34**(4): 94–104.

United Kingdom Central Council (1997) *Responsibility for the feeding of patients (Registrar's letter).* UKCC, London.

World Health Organization (2002) *National Cancer Control programmes: policies and managerial guidelines.* WHO, Geneva.

Critical Thinking

1. What symptoms are prevalent in advanced cancer, and how is nutrition impacted at the end of life?

2. What impact can nutrition have on a patient and his/her family during palliative care?

SUE ACREMAN is the Marie Curie Consultant Practitioner in Cancer Rehabilitation, Velindre NHS Trust.

Nutrition in the Elderly: A Basic Standard of Care and Dignity for Older People

IQBAL SINGH ET AL.

With an ageing UK population people are living longer and enjoying longer periods of retirement. The number of people over the age of 65 years in the UK is projected to rise to 20% by 2051 (Office of National Statistics, 2004). The fastest growing section of the population is those aged over 85 years—this sector is projected to reach 3.2 million in the UK by 2033. These figures also reflect trends seen across European countries. The 2008-based national population projections, EuroPop2008, predicted an increase to 30% of people age 65 years and over by 2060 (Eurostat, 2008). Similarly the number of people aged 80 years or over is projected to treble.

These demographic trends have consequences for developments in public policy and care of the elderly and also add new challenges to delivery of older people's care. Forster and Gariballa (2005) and Elia et al (2008) have highlighted the risk of malnutrition rising with age with a higher prevalence in those in receipt of care and those living in institutions. Low awareness of malnutrition among health- and social- care professionals has been addressed by publications and campaigns by several key parties. Age Concern England (2006) described the growing risk of older people being malnourished or their nutritional status getting worse while admitted to hospital. *Caring for Dignity,* a report by the Commission for Healthcare Audit and Inspection (2007), underlined the need for commitment to nutrition throughout health-care organizations.

The World Health Organization (1971) defined malnutrition as 'the cellular imbalance between the supply of nutrients and energy and the body's demand for them to ensure growth, maintenance and specific functions'. Malnutrition is consistently under-diagnosed and under-treated in both primary and secondary care. The causes of malnutrition are multifactorial. Inadequate diet quality, micro-nutrient deficiencies, chronic conditions, psychological, social and even environmental factors all contribute to under-nutrition. Older people following a major physical illness such as stroke, with other co-morbidities, those from black and minority ethnic communities and those with mental illness have a greater risk of poor nutritional status, which may be under-recognized and associated with worse outcomes. In stroke especially, poor nutrition is widely prevalent and under-recognized (Singh et al, 2004). It is a marker for increased mortality, hospital stay, morbidity and residential placement. Under-nutrition compromises the immune system, resulting in impaired wound healing and increased susceptibility to infection along with impaired physical performance.

Assessment of Nutrition

Apart from nutritional screening, which is an initial rapid evaluation method to detect significant risk of malnutrition, nutritional assessment (a more in-depth evaluation) is an integral part of comprehensive assessment and care for older people. Several screening tools exist in clinical practice with MUST (Malnutrition Universal Screening Tool) the most widely used (Malnutrition Advisory Group, 2008).

This tool is suitable for use by a range of health-care workers. It has been validated across a range of health-care settings and assesses weight status, change in weight and the presence of an acute disease likely to result in no dietary intake for more than 5 days. It categorizes subjects into low, medium or high risk of malnutrition and provides guidance on developing individualized dietary care plans. Regular nutritional assessment in stroke and regular reviews are best performed through multidisciplinary evaluation at multiple levels with patients' and relatives' involvement being an integral part of clinical practice.

Recommendations

Despite major improvements in the care of older people and general improvements in attitudes towards care of the elderly, under-recognition and management of under-nutrition remains a major challenge. To address these issues it is important that health-care organizations involved in delivering and commissioning health care are aware of the issues around under-nutrition and provide adequate training and education to all health-care staff.

All acute trusts, primary care trusts and community hospitals should include nutrition as a part of their regular clinical governance framework. There needs to be a commitment to nutrition throughout health-care organizations with a specified lead, and this should be communicated to all staff and patients. Feeding and nutrition is an integral part of dignity in older people's care. It is important to follow National Institute for Health and Clinical Excellence (National Collaborating Centre for Acute Care, 2006) guidelines in relation to nutrition and to implement a recommendation that all patients, on admission to hospital or at their first clinic appointment, should be screened for nutritional status.

As part of a comprehensive assessment, patients should undergo nutritional screening at the time of admission to care homes. Nutritional support should be considered in people who are malnourished and those who are at high risk, and especially in those following a major illness such as stroke. There are many different strategies to improve nutrition—making meals more appetising and more widely available, use of multivitamin or multimineral supplements, or oral liquid nutrition. Nutritional support should also be considered in those identified as being at risk or in a state of malnutrition and oral dietary supplements or intravenous nutritional support should be considered and provided when necessary (Mucci and Jackson, 2008).

People admitted to hospital following a stroke must have a swallow assessment so that their nutritional needs can be met. In addition to oral and intravenous support, access to the gut directly in stroke patients with dysphagia may be achieved by a nasogastric or percutaneous gastric tubes. Nasogastric tubes have a risk of extubation and aspiration, and aspiration pneumonia and oral complications of enteral feeding including diarrhea, hyperglycemia, hypercapnia, electrolyte imbalance and rebound feeding should be addressed by appropriate therapy.

Conclusions

Malnutrition in older people is multifactorial, under-recognized and often undetected by health-care professionals. Addressing nutritional needs improves recovery from illness and general wellbeing of older people, Education and training and multiagency partnership working between commissioners, providers and regulators of health and social care is key to reduction of malnutrition risks.

Key Points

- Malnutrition of older people should have no place in a modern society, yet three million people are living at risk of malnutrition in the UK.
- Older people being admitted to hospital or nursing homes should be screened and assessed for nutritional status.
- The Care Quality Commission and the General Medical Council should ensure that professionals recognize that food and help with eating is a key issue in maintaining the health and dignity of older people.
- Adequate nutritional support, both oral and parenteral, should be provided when necessary.

References

Age Concern England (2006) *Hungry to be Heard. The scandal of malnourished in older people in hospital.* Age Concern England, London.

Commission for Healthcare Audit and Inspection (2007) *Caring for Dignity. A national report on dignity in care for older people while in hospital.* Commission for Healthcare Audit and Inspection, London.

Elia M, Jones B, Russell C (2008) Malnutrition in various care settings in the UK: the 2007 Nutrition Screening Week Survey. *Clin Med* **8**: 364–5.

European Nutrition for Health Alliance (2006) *From Malnutrition to Wellnutrition: policy to practice.* A report of the European Nutrition for Health Alliance 2nd annual conference. European Nutrition for Health Alliance, London.

Eurostat (2008) *Europe in figures.* Office for Official Publications of the European Communities, Luxembourg (http://epp.eurostat. ec.europa.eu/cache/ITY_OFFPUB/KS-CD-07-001/EN/KS-CD-07-001-EN.PDF (accessed 1 September 2009).

Forster S, Gariballa S (2005) Age as a determinant of nutritional status: a cross-sectional study. *Nutr J* **27**(2): 28.

Malnutrition Advisory Group (2008) *Malnutrition Universal Screening Tool.* www.bapen.org.uk/pdfs/must/must_full.pdf (accessed 19 Novernber 2009).

Mucci E, Jackson SHD (2008) Nutritional supplementation in community dwelling elderly people. *Ann Nurr Metah* **52** (Supp l): 33–7.

National Collaborating Centre for Acute Care (2006) *Nutrition support in adults Oral nutrition support, enteral tube feeding and parenteral nutrition.* National Collaborating Centre for Acute Care, London.

Office of National Statistics (2004) *Population Trends.* http://www. statistics.gov.uk/downloads/theme_population/PTl15.pdf (accessed 19 November 2009).

Singh I, Vilches A, Narro M (2004) Nutritional support and stroke. *Br J Hosp Med* **65**(12): 721–3.

World Health Organization (1971) Joint FAO-WHO Expert Committee on Nutrition. Eighth report. Geneva, 9–18 November 1970. *World Health Organ Tech Rep Ser* **477**: 1–80.

Critical Thinking

1. What three strategies can be used to improve nutrition in the elderly population?

IQBAL SINGH Consultant Physician in Medicine for the Elderly NHS East Lancs Acorn Primary Health Care Centre Accrington BB5 1RT.

SORCHA DE BHALDRAITHE Specialty Registrar in Medicine for the Elderly and General Medicine East Lancashire Hospitals NHS Trust Blackburn. **DANIELA BONDIN** Foundation Year 2 Doctor East Lancashire Hospitals NHS Trust Blackburn. **NEETISH GOORAH** Consultant East Lancashire Hospitals NHS Trust Burnley.

From *British Journal of Hospital Medicine,* 71:1, January 2010, pp. 4–5. Copyright © 2010 by MA Healthcare Limited. All rights reserved. Reprinted by permission.

Nutrition Management of Gastric Bypass in Patients with Chronic Kidney Disease

Rachael R. Majorowicz

Growing numbers of patients with chronic kidney disease (CKD) with a BMI of 40 or greater are pursuing kidney transplantation but may be denied candidacy due to poorer patient and graft survival rates. Weight loss procedures in this population provide the prospect of kidney transplantation as well as improvement in or resolution of hypertension, diabetes mellitus, GERD, and other co-morbid conditions (Alexander et al., 2004).

Unfortunately, there is report of acute kidney injury (AKI) occurring post-gastric bypass (GBP), especially in patients with a prior history of CKD. Medical history of certain co-morbid conditions and use of certain pre-operative medications independently increase the risk of post-operative AKI, which a retrospective review determined to be "not-infrequent" following BGP surgery (Thakar, Kharat, Blanck, & Leonard, 2007). Additionally, patients with renal disease prior to GBP may be at higher risk of oxalate nephropathy. Conversely, a retrospective review by Sharma et al. (2006) concluded that "primary acute renal failure after laparoscopic GBP is an uncommon complication" (p. 389).

Despite these reports of negative outcomes, patients and healthcare providers continue to consider bariatric surgery as a means to improve kidney transplantation candidacy, since several other studies describe the benefits of GBP in patients with CKD. A retrospective review by Takata et al. (2008) concluded that laparoscopic Roux-en-Y (RYGBP) improved the candidacy for transplantation of patients. Alexander et al. (2004) and Alexander and Goodman (2007) performed RYGBP on patients with Stages 3-4 CKD, patients on hemodialysis (HD), and patients who were post-kidney transplant, and concluded that GBP may safely be recommended for patients in any stage of CKD or post-transplant.

Since there is little literature on the nutritional management of patients with CKD who undergo GBP, this article compiles both current, general guidelines with limited, available CKD-specific information. Common GBP procedures, methods of weight loss, and the nutritional implications are reviewed.

Procedures

The common GBP procedures performed on patients with CKD tend to be adjustable gastric band (AGB), sleeve gastrectomy, and standard RYGBP. The AGB and sleeve gastrectomy are restrictive weight loss procedures, in which food intake is limited by creating a smaller stomach pouch. The laparoscopic AGB (see Figure 1) generally restricts stomach capacity initially to about 15 mL, but the adjustability allows for fine-tuning as necessary. Sleeve gastrectomy is essentially the restrictive component of the duodenal switch (DS), as seen in Figure 2 (A), with the benefit of maintaining the pylorus but without any alteration of the small intestine. RYGBP (see Figure 3) restricts stomach capacity to 10 to 30 mL, bypassing the pylorus and only a short length of small intestine, but creating minimal malabsorption (McMahon et al., 2006). Multivitamin/multimineral supplements post-surgery should include 100% daily values and 1500 mg

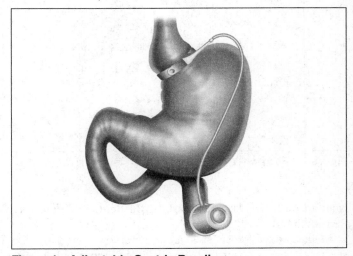

Figure 1   Adjustable Gastric Banding

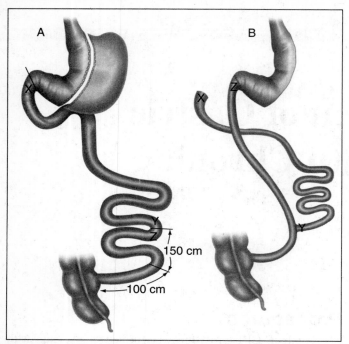

Figure 2   Sleeve Gastrectomy and Duodenal Switch

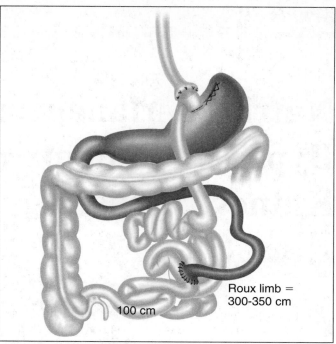

Figure 4   Very, Very Long Limb RYGBP

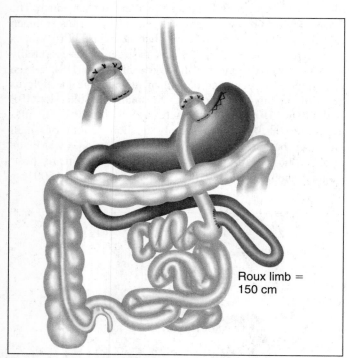

Figure 3   Roux-en-Y GBP

calcium daily (Aills, Blankenship, Buffington, Furtado, & Parrott, 2008).

Malabsorptive procedures are generally reserved for the severely obese. Longer limb RYGBP (see Figure 4) bypasses most of the stomach, the pylorus, the entire duodenum, and a greater length of the jejunum. This reroutes bile and pancreatic enzymes, limiting interaction with food to the last 100 cm of small intestine ("common channel"). The longer the limb, the greater the malabsorption, risk of bone disease, diarrhea, and oxalate nephropathy (McMahon et al., 2006). Deficiencies of iron, B12, folate, and calcium are common, and multivitamin/ multimineral supplements post-surgery should include 200% of daily values (Aills et al., 2008).

Duodenal switch creates a sleeve gastrectomy, with the pylorus and a small portion of the duodenum left intact, reducing dumping syndrome. As seen in Figure 2, (A) depicts the sleeve gastrectomy, the section of small intestine to be bypassed (XY), the remaining 250 cm of small intestine to be connected to the stomach, including the 100 cm common channel (where Y is reconnected) preceding the ileocecal valve; (B) depicts the finalized procedure, where a majority of the stomach is removed (Kendrick & Dakin, 2006). Protein and fat-soluble vitamin deficiencies are common, as well as anemia and abnormal calcium levels (Hirschfeld & Stoernell, 2004). Multivitamin/multimineral supplements post-surgery should include 200% of daily values (Aills et al., 2008).

Nutritional Complications
Preoperative

Aills et al. (2008) summarized common nutrient deficiencies prior to weight loss surgery. Pre-operative deficiency was reported in 13% to 64% of patients for B-vitamins or folate, up to 16% iron deficiency in women, and 28% for zinc. Proper identification and treatment of existing deficiencies are crucial to improved surgical outcomes and preventing post-surgical deficiencies or complications.

Table 1 Potential Nutrient Deficiencies
(Not Specific for the Renal Population Unless Stated as Such)

Nutrient	Absorption	Deficiency Post-Op	Treatment*	Laboratory Values
Thiamin (B_1)	Proximal jejunum	Post-op is rare, but occurs in all patients with GBP early if persistent vomiting Beriberi or peripheral neuropathy	Preventable with daily multivitamin 50 to 100 mg/day IV for advanced neuropathy or persistent vomiting Renal-specific multivitamin twice daily (Alexander et al., 2004)	Serum thiamin
Pyridoxine (B_6)		Rare; consider deficiency with unresolved anemia	Renal-specific multivitamin twice daily (Alexander et al., 2004)	PLP
Cobalamin (B_{12})	Decreased HCl, pepsin, and IF reduce absorption in the terminal ileum	35% occurrence Pernicious anemia	350 to 500 mcg/day oral tablet or 1,000 mcg monthly IM injection (may not be needed in patients on HD) (Alexander et al., 2004)	Serum B_{12} at least annually If symptomatic with low-normal B_{12}, watch for elevated MMA and total homocysteine
Folate	Proximal small intestine, can occur along entire small bowel with post-op adaptation	41% to 47% one-year post-op Mostly asymptomatic or subclinical; may have macrocytic anemia, neurological changes or headaches (McCann & Kelly, 2006); forgetful, irritable, or paranoid	800 to 1000 mcg/day for prevention 1000 mg/day in deficiency (not more or can mask B_{12} deficiency) Renal-specific multivitamin twice daily in patients on HD (Alexander et al., 2004)	RBC folate Normal serum and urinary MMA Homocysteine
Iron (Fe)	Most efficient in duodenum and proximal jejunum, but also reduced due to less gastric acid	20% to 49% post-op Higher risk—menstruating women and obese men, and patients less than 25 years old Microcytic anemia, fatigue, pica, pale nail beds, or spooning fingernails (McCann & Kelly, 2006)	325 mg Fe sulfate with Vitamin C for increased absorption (ADA, 2009); separate from Ca supplements by 2 hours Anemia protocols should correct for any Fe deficiency in patients on dialysis	Measure serum Fe, TIBC, ferritin, Hct, and Hgb 6 months post-op and annually.
Protein		With a DS pouch of less than 200 mL, supplementation likely needed (American Dietetic Association [ADA], 2009) In time, the colon adapts and increases absorption	For patients on HD: 1 g/kg actual weight (Alexander et al., 2004) or 1.5 g/kg ideal weight (ADA, 2009) Patients may be intolerant to meat and/or dairy products; utilize complete protein, low-sugar and low-fat, supplements as needed	Serum albumin and nPCR (in patients on HD)
Calcium (Ca)	Duodenum and proximal jejunum (facilitated by vitamin D)	Hyperparathyroidism	1,500 to 2,000 mg/day liquid or chewable Ca citrate or carbonate (ADA, 2009); divided doses In patients on HD, Ca citrate is not advised due to promoting Al^+ absorption, but Alexander et al. (2004) used 1 to 2 Ca citrate daily	Serum Ca In patients on HD: Ca $\times$ P less than 55

(continued)

Table 1 Potential Nutrient Deficiencies
(Not Specific for the Renal Population Unless Stated as Such) (*continued*)

Nutrient	Absorption	Deficiency Post-Op	Treatment*	Laboratory Values
Vitamin A**	Upper small intestine	52% in patients with RYGBP and BPD; 25% in patients with AGB Decreased vision or night blindness	10,000 IU to prevent deficiency 50,000 IU q 2 weeks to correct deficiency (retinol sources) Initiate 2 to 4 weeks post-op	Plasma retinol or Vitamin A in 6 months (McMahon et al., 2006) and annually thereafter
Vitamin D**	Jejunum and ileum	68%, with higher risk in patients who are obese Hyperparathyroidism and metabolic bone disease in the long term	2000 IU D3 (via multivitamin and Ca supplements) In deficiency, 50,000 IU ergocalciferol (NKF, 2003); in patients on HD, follow vitamin D protocol Weight-bearing exercise	Frequent phosphorus, alk phos, and Ca 25(OH)D and PTH at least in 6 months (McMahon et al., 2006)
Vitamin E**	Upper small intestine	Not prevalent	100% daily value for prevention	Plasma alpha tocopherol annually
Vitamin K**	Upper small intestine	14% one year and 68% four years post-op, especially in patients with BPD/DS	300 mcg daily; caution with coagulation therapy	Prothrombin time frequently and serum Vitamin K in 6 months (McMahon et al., 2006) and annually thereafter
Zinc	Dependent on fat absorption	Post-op: 36% to 51%, with chronic diarrhea Altered taste, impaired healing, or scaly/red skin lesions on nasolabial folds and hands (McCann & Kelly, 2006)	A zinc-containing renal-specific vitamin may be advisable for the patient on dialysis	Plasma zinc and RBC zinc (interpret according to albumin level due to being albumin-bound)

Notes. IV = intravenous; IM = intramuscular; IF = intrinsic factor; RBC = red blood cell; TIBC = total iron binding capacity; Hgb = hemoglobin; Hct = hematocrit; alk phos = alkaline phosphatase; Al⁺ = aluminum; Ca = calcium; PLP = pyridoxal-5'-phosphate; MMA = methylmalonic acid; PTH = parathyroid hormone.
*Multivitamin and individual supplements should initially be provided in liquid or chewable forms, progressing to whole tablets/capsules as tolerated.
**Can be provided in water-soluble form. Encourage a very low-fat diet to decrease loose stools (Hirschfeld & Stoernell, 2004).
Source: Aills et al. (2008) (except where otherwise indicated).

Initial

In the first post-surgical year, dumping syndrome and vomiting can be common following malabsorptive procedures. Dumping syndrome occurs in RYGBP due to bypass of the pylorus, resulting in the inability to regulate gastric emptying of simple carbohydrates into the intestine. Symptoms include lightheadedness, sweating, nausea, weakness, and sometimes diarrhea (McMahon et al., 2006). Vomiting and dehydration are common with smaller pouch sizes. Reinforce with the patient to avoid high fat or sugar foods, limit portions to appropriate pouch size, chew well, and eat slowly. Generally, 6 to 8 cups of fluid per day are recommended for patients with adequate kidney function. There are no guidelines available for patients with CKD, so close monitoring for signs of dehydration is necessary, as is frequently adjusting the estimated dry weight for patients on dialysis. Patients can drink water, skim/low-fat milk, or unsweetened fruit juices as tolerated and as laboratory results allow.

Long-Term

Numerous long-term complications can arise following malabsorptive procedures, especially those with longer limb lengths. See Table 1 for common nutrient deficiencies and suggestions for management. There are no guidelines regarding the frequency of monitoring laboratory values of patients with CKD post-GBP, but some general recommendations are included in Table 1. Due to the limited literature on patients with CKD who have GBP, especially patients on dialysis, it is advisable to monitor electrolytes frequently or until initial complications subside, laboratory results stabilize, and the rate of weight loss declines. In addition, patients should be observed for changes in subjective global assessment, dry weight should be adjusted frequently in patients on hemodialysis, calorie or protein supplement doses should be adjusted regularly, and diet recommendations should be modified as needed. If possible, the nephrology team should work closely with the bariatric care team to best manage the needs of the patient.

Summary

Frequent monitoring by a dietitian can ensure desirable progression of the post-surgery diet, adequate nutritional composition, management of food intolerances, and ongoing education/reinforcement of the post-surgical nutritional needs. Additionally, it is critical for dietitians to assess the type of GBP because increased nutritional risks result with longer limb lengths and will require closer monitoring. With close follow up and adherence to recommendations, patients with CKD who undergo GBP can reduce the risk of post-surgical, nutrition complications.

References

Aills, L., Blankenship, J., Buffington, C., Furtado, M., & Parrott, J. (2008). ASMBS allied health nutritional guidelines for the surgical weight loss patient. *Surgery for Obesity and Related Diseases, 4*(5, Suppl.), S73–S108.

Alexander, J.W., & Goodman, H. (2007). Gastric bypass in chronic renal failure and renal transplant. *Nutrition in Clinical Practice, 22*(1),16–21.

Alexander, J.W., Goodman, H.R., Gersin, K., Cardi, M., Austin, J., Goel, S. . . . Woodle, E.S. (2004). Gastric bypass in morbidly obese patients with chronic renal failure and kidney transplant. *Transplantation, 78*(3), 469–474.

American Dietetic Association (ADA). (2009). Bariatric surgery. *Nutrition Care Manual.* Retrieved from http://nutritioncare-manual.org

Hirshfield, L., & Stoernell, C. (2004). Nutritional considerations in bariatric surgery. *Plastic Surgical Nursing, 24*(3), 102–106.

Kendrick, M.L., & Dakin G.F. (2006). Surgical approaches to obesity. Supplement on bariatric surgery in extreme obesity. *Mayo Clinic Proceedings. 81*(10), S18–S24. Retrieved from www.mayoclinicproceedings.com/content/81/10_Suppl/S18.full.pdf

McCann, L., & Kelly, M.P. (2006). *Comprehensive nutrient assessment: From lab results to physical findings.* Presented at the 2006 NKF Clinical Meetings, Chicago, IL, April 21, 2006.

McMahon, M.M., Sarr, M.S., Clark, M.M., Gall, M.M., Knoetgen, J., Service, F.J. . . . Hurley, D.L. (2006). Clinical management after bariatric surgery: Value of a multidisciplinary approach. Supplement on bariatric surgery in extreme obesity. *Mayo Clinic Proceedings, 81*(10), S34–S45. Retrieved from www.mayoclinicproceedings.com/content/81/10_Suppl/S34.full.pdf

Sharma, S., McCauley, J., Cottam, D., Mattar, S., Holover, S., Dallal, R. . . . Schauer, P. (2006). Acute changes in renal function after laparoscopic gastric surgery for morbid obesity. *Surgery for Obesity and Related Diseases, 2*(3), 389–392.

Takata, M., Campos, G., Ciovica, R., Rabl, C., Rogers, S., Cello, J. . . . Posselt, A. (2008). Laparoscopic bariatric surgery improves candidacy in morbidly obese patients awaiting transplantation. *Surgery for Obesity and Related Diseases, 4*(2), 159–164.

Thakar, C., Kharat, V., Blanck, S., & Leonard, A. (2007). Acute kidney injury after gastric bypass surgery. *Clinical Journal of the American Society of Nephrology, 2*(3), 426–430.

Additional Readings

Nasr, S., D'Agati, V., Said, S., Stokes, M., Largoza, M., Radhakrishnan, J., & Markowitz, G. (2008). Oxalate nephropathy complicating Roux-en-Y gastric bypass: An underrecognized cause of irreversible renal failure. *Clinical Journal of the American Society of Nephrology, 3*(6), 1676–1683.

National Kidney Foundation (NKF). (2003). *K/DOQI clinical practice guidelines for bone metabolism and disease in chronic kidney disease.* Retrieved from www.kidney.org/professionals/KDOQI/guidelines_bone/index.htm

Critical Thinking

1. What are some of the post-op deficiencies a patient may experience after a gastric bypass?

RACHAEL R. MAJOROWICZ, RD, LD, is a Clinical Dietitian, Mayo Clinic Dialysis Services, Rochester, MN.

From *Nephrology Nursing Journal,* vol. 37, no. 2, March/April 2010, pp. 171–175. Copyright © 2010 by Anthony J. Jannetti, Inc. Reprinted with permission of the publisher, the American Nephrology Nurses' Association (ANNA), East Holly Avenue, Box 56, Pitman, NJ 08071-0056; 856-256-2320; FAX 856-589-7463; E-mail: nephrologynursing@ajj.com. www.annanurse.org

UNIT 7

Men in Nursing

Unit Selections

Learning Outcomes

After reading this unit, you should be able to:

• Describe the changes in marketing over the last few years that have been designed to encourage more men to enter nursing.

• Discuss the impact male nurses are now making on the nursing profession.

• Explain why men in nursing were hired to work in unfavorable areas within the community.

• Describe how the persistent images of feminine subservience deter both male and female students from entering the profession.

Student Website

www.mhhe.com/cls

Internet References

The American Assembly for Men in Nursing
   http://aamn.org
Men in Nursing History
   www.geocities.com/Athens/Forum/6011
Texas Health Resources
   www.minoritynurse.com

The first nurses throughout the world were men. In approximately 250 B.C. the first nursing school was established in India. Only men were admitted because it was believed that only men were pure enough to be nurses. During the Black Plague in Europe, a group of men formed one of the first hospitals to care for the victims. During the American Civil War, both sides had military men serving as nurses, although historic accounts focus on the Union nurse volunteers, who were predominately female. The Confederate Army identified thirty men per regiment to care for the wounded. The Union Army also had men serving as nurses within their ranks—the poet Walt Whitman was one of these volunteers. In the early 1900s, men were excluded from nursing in the military and did not resume this function until the early 1950s, after the Korean War. Today, about 6 percent of the nursing workforce in the United States is made up of men. However, in the military, at least 35 percent of the nursing forces in each of the three branches are composed of men. With the tremendous shortage of nurses, men are again being encouraged to join the nursing profession. Ways that could help relieve the shortage would be to retain nurses, both male and female, who are already in the profession, recruit more males into the profession, and present nursing as a gender-neutral career to male high school students: Nursing is not a gender, it is a profession. This unit presents information and issues related to males in the profession. Encouraging more men to enter the field requires education and a mindset change.

Although only 6 percent of working nurses are men, 13 percent of nursing students are males, which is encouraging. Nursing educators, many of whom are women, must guard against using stereotypes and gender-specific terms that could alienate male students. The misconceptions that male nurses are "wannabe" doctors who couldn't make it in medical school, or are predominately gay must be eradicated. Most males go into nursing for the same reasons that females enter the profession; they want to provide comfort and aid their patients in recovery. As more men enter the field, the profession and female colleagues will benefit, because men tend to be more assertive and will speak-up on issues related to salary and work conditions.

During this unit, you will read about the shortage of qualified Registered Nurses and the history of men's evolution into

© David Fischer/Getty Images

nursing practice. Male nurses now practice in specialty areas. The promotion of diversity in nursing is encouraged as we support and recognize men as nurses and eliminate the term "male nurse" when referring to men who practice as nurses. Men also played a pivotal role in community nursing. They were often appointed to work in areas of the community that were deemed too dangerous for female nurses to visit.

The common theme through these statements is that nursing is a rewarding career that provides challenges and opportunities. This message needs to be delivered to men who are considering career choices. They need to know that they are welcomed in the nursing profession. Men in the profession wish to stand proudly next to their female counterparts, to provide care for the sick and injured, and to be respected for their contributions. The profession of nursing will only be able to meet the health-care needs of the global community if the best and the brightest of both sexes are welcomed.

Men in Nursing Today

Erika Icon

When many people think of men as nurses, the first image that pops into their head is Greg Focker from Meet the Parents. Focker's patients confuse him for a doctor and his future father-in-law ribs him for being a nurse. But as the nursing shortage escalates, the field is opening up and welcoming men to join. Today, you will find them alongside women in doctors' offices, emergency transport helicopters, forensic police divisions, and hospitals.

Even though nursing may be considered "woman's work," the first nurses were actually men. The first nursing school started in India around 250BC, when only men were considered "pure" enough to become nurses. The Army has a long history of male nurses, as do other branches of the armed services. 25–30 percent of our military nursing population is male.

A Growing Trend

There are more men in nursing than ever before. They account for 6% of the nursing population and this figure grows yearly. According to the National Survey of Registered Nurses, 13 percent of the students in nursing schools are men. Men are attracted to the profession for the same reason women are. Nursing is consistently ranked as one of the most respected and trusted professions in the nation. It is a high-demand career that offers a lot of flexibility. Nurses feel rewarded when they are saving lives and helping people.

The nursing shortage is driving up salaries. In 1992, the average salary of RN was $37,738. Today, the average nationwide salary of a Staff RN is $46,258 and upwards. That's a 23 percent increase in pay; very few professions have seen that kind of salary increase. Nurses can often get a healthy sign-on bonus or/and perks such as new cars or student loan pay-offs.

Do Men Make Good Nurses?

Men who enjoy working with people and have good communication skills tend to enjoy their roles as nurses. Male nurses are more likely than their female counterparts to be assertive and speak up. Men are team players and understand the value of teamwork. Studies also show that men exude a certain level of confidence that can calm a patient's anxiety.

Yes, there are prejudices. Many male nurses encounter a gay stereotype, which statistically is not the case. Or, people assume a male nurse dropped out when the path to medical school became too tough. Again, statistically not true. Most male nurses seek out the profession specifically. Some nurses, doctors and patients feel like male nurses don't belong in the delivery room, on OB/GYN floors and women's health wards. Yet, as more men become nurses, this situation should change.

Bucking Stereotypes

According to the Journal of Health Affairs, male nurses are twice as likely to leave the profession. 7.5 percent of new male nurses left the profession in their first 4 years versus 4.1 percent of women. Why are men bolting from the profession? Some female nurses don't always play nice to their male nurse colleagues. Even doctors can be indifferent to men who are nurses. Some patients are uncomfortable; older women are the most likely to express surprise or disapproval. For a male nurse, encountering a constant barrage of negative attitudes from coworkers and patients can be demoralizing. It's hard to buck stereotypes. As a society, we're removing the 'he' connotation from the word doctor, but not the 'she' from nurse.

Making Strides

In the last 10 years there have been great strides made for men in nursing. Spend some time on the Internet and you will find recruitment ads and banners aimed specifically at men. Many hospitals are encouraging men to join their staff.

In the past, ads featured only women as caregivers, but that's been turned on its head. In 2002, Johnson & Johnson kicked off campaigns that featured men in half their ads and one-third of the nurses profiled on their site are men. Hospitals are getting creative with ads specifically geared at encouraging young men—Gen X, Y and Z—to join their nursing staff.

Of course, nursing colleges are catching on to the trend too. The Oregon Center for Nursing features the headline asking, "Are You Man Enough to Be a Nurse?" The ad features nine men in different garb such as a sport coat and tie, martial arts gear, scrubs or a white coat and even toting a snowboard.

Mentoring programs have been formed to help men through nursing school and beyond. Linda Panter, RN, MSN, FNP and Assistant Professor at Alfred State College in Alfred, NY, started the Adopt-A-Nurse Program. After they're on the job as nurses, they become mentors, a kind of "pay it forward" idea. The program will be extended to other schools in the future. There's even a magazine or two specifically geared at men. Pick up a copy of Male Nurse Magazine.

Slowly, but gradually, men are getting the respect that they deserve. With gender roles changing over the last 30 years, more men are finding it socially acceptable to become nurses. In 1966, men made up only one percent of the nursing population, but today they make up almost 6 percent. That's a 500 percent increase! We can't lose sight of what nursing is—it's caring for others. Good nurses come in all shapes, sizes, and sexes. Men are definitely an important part of this growing field.

Critical Thinking

1. What changes in marketing over the last few years have been designed to encourage more men to enter nursing? How does this relate to gender inequity within the profession?

ERIKA ICON is a Los Angeles-based writer.

Men in Nursing: Addressing the Nursing Workforce Shortage and Our History

Jennifer Bonair and Nayna Philipsen

As we all face a shortage of qualified RNs in the work-force, one of the proposals that we hear to solve this problem is to attract more men to the profession. The shortage is partly the result of about half of our population (males) feeling shut out. Men comprise only about 5.4% of the total nursing population now. Even if these men were recruited at the same rate as now, and retained at the same rate as women nurses, their representation would grow in the future. That is because the average male nurse is younger than the average female RN. The average age of the RN population in the United States is 45.2 years, and only 9.1% of all RNs are under the age of 30. About 38% of male RNs are under 40 compared with 31% of female RNs. Twenty-one percent of male RNs are 50 years of age or older, compared with 34% of female RNs.

This is an urgent issue. In January 2006, one out of three nurses under the age of 30 in the United States plans to leave their current employment within one year, and the workforce is clearly aging (Blais, Hayes, Kozier, & Erb, 2006). In 2007 the percentage of male graduates in nursing did not increase compared to 2006, but held steady at 12 percent. If the total population of men in nursing was the same as women there would be no nursing workforce shortage.

Many of us fail to realize that in recruiting more men to nursing, we are also dealing with the past. As William Faulkner said, "The past is not dead. In fact, it is not even past." We have to identify and address that past to reshape nursing in the future as a profession open to all qualified and motivated future nurses.

Like teaching, nursing was primarily a male profession throughout early history, before it became "maternal" around the beginning of the twentieth century. Then a pattern of gender segregation and subsequent stereotyping began.

In the Parable of the Good Samaritan, Jesus mentions a male innkeeper being paid to nurse an injured man. Nursing schools were for men. In 1783, James Derham, a slave, earned money to buy his freedom by working as a nurse. However, Florence Nightingale's reforms made it clear that nursing was a "natural" profession for women, who were viewed as naturally caring and nurturing, the requirements for a good nurse. This was a reflection of the comfort with social stereotyping and gender

segregation of her era in Victorian England. In the 1900's, men became nurses at their own social peril. Male nurses were increasingly discriminated against by society, military, and female nurses. There was open and active gender discrimination within the profession. Twentieth century nursing schools incorporated an all-female residence dorm life. Men were considered minority and second-class citizens, and they were paid half the salary of female nurses.

The American Nursing Association is one of the oldest nursing organizations. In 1911, the Nurse's Associated Alumnae (female) was changed to the American Nursing Association. The American Nursing Association (ANA) offers memberships to nurses in all 50 states. It promotes voluntary certification for specialty and advanced practice. It supplies data and research and engages in public policies and political education. The ANA also determines and publishes standards for nursing professionals, as well as nursing practices for generalized and specialty areas.

The ANA sets the standards of practice in nursing. Section 1-5 of the Code of Ethics of the ANA mentions the relations of nurses with colleagues and others. This includes respecting all individuals, maintaining a compassionate and caring relationship with colleagues, and precludes any and all prejudicial actions.

Male nurses fought for equality in the twentieth century. They were up against some of the same ignorance and stereotypes that they face today. Why would a man want to be a nurse? How would a man feel about taking orders from a woman if she were his supervising nurse? Men wrote letters to the Surgeon General, and finally, in 1949 and with the help of the ANA, they were formally recognized as equals in nursing society. Male nurses flooded the military and entered the nursing profession in general. They are now 35% of the army corps.

Males are now attracted to the competitive field of nursing, due to its high status and high pay. Male nurses are now working in competitive areas such as: Intensive Care Unit, Emergency Room, and Flight Nurse. Today, the average male nurse is higher educated and higher paid than female nurses. This is a reflection of other inequalities in our society, but it also reflects a genuine contribution to the quality of care from the men in nursing.

The salvation of nursing and of quality patient care lies in eliminating gender bias and discrimination in nursing, and increasing our diverse population, especially men, in the workforce. What can nursing do to accelerate this trend? Suggestions include: 1. Our nurse educators must cease teaching an ethnocentric history of our profession, for example, referring to Linda Richards as the first "trained" nurse without mentioning the Alexian Brothers nursing school which opened in 1866 for men; 2. We should all eliminate use of the term "male nurse" and just refer to all nurses as nurses; 3. Identify and eliminate references to the stereotype of nursing as a feminine profession in our daily practice, and protest when it is done in the media; 4. Schools and professional associations should focus on recruiting men into the profession; and 5. Fight the isolation of men that comes with being a minority by supporting peer groups and mentoring for our male nursing students and nurses.

Some actions to promote equity and diversity in nursing and health care in our communities, including advocacy for men as professional nurses, are actions that an individual nurse must take, such as recognizing references to the stereotype of the female nurse at the workplace, and objecting when we hear a comment such as, "Male nurses are good because they can lift patients." Other actions require us to work together as an authoritative group, for example, by joining organizations such as ANA and the Maryland Nurses Association (MNA). This is the way that individual nurses can contribute to actions such as working with representatives of the media to promote a nondiscriminatory image of nursing, or with the Maryland legislature to support workforce diversity. The MNA Center for Ethics and Human Rights is an example of a Maryland group that is ideally located to address the nursing workforce shortage while they promote equality and human rights.

References

American Nurses Association Guide. Retrieved December 4, 2008, from www.righthealth.com/health/america-nurses-association.

History of male nurses. Retrieved December, 4, 2008, from http://history.amedd.army.mil/anewebsite/articles/malenurses.htm

American Nurses Association. Retrieved December 4, 2008, from www.nursingworld.org.

Blais, K., Hayes, J.S. Kozier, B. & Erb, G. (2006). Professional nursing practice: Concepts and perspectives (5th ed.) Upper Saddle River, NJ: Pearson Education, Inc.

Chung, V. Men in nursing, Minority Nurse, (2001) Retrieved on March 15, 2009, from www.minoritynurse.com/minority-nursing-statistics.

History of the Alexian Brother. (2007). Retrieved on March 20, 2009, from www.alexianbrothers.org/index.php?submenu=Heritage&src=gendocs&ref=ICP_History&category=ICP.

National League for Nursing. Nursing education research: Annual survey of schools of nursing academic year 2006–2007: Executive summary. Retrieved on March 30, 2009. from www.nln.org/research/slides/exec_summary.htm.

Critical Thinking

1. What impact are male nurses making on the nursing profession?

HELENE FULD School of Nursing, Coppin State University.

The Male Community Nurse

As district nursing celebrates its 150th anniversary, Alison Blackman assesses how the role of men in the profession has evolved.

ALISON BLACKMAN

'And there is no trade or employment but the young man following it may become a hero.' So wrote Walt Whitman. On hearing that his brother had been wounded in the American Civil War (1861–1865), Whitman went immediately to care for him. His brother's wounds were not life threatening, but Whitman stayed to discover more about the experience of soldiers in the field hospitals. He subsequently worked as a volunteer nurse, feeding the wounded and treating gangrenous wounds. The experience changed Whitman's life forever and inspired his greatest poetry.

Nursing was a male-dominated profession from at least 250BC until the 19th century and the era of Florence Nightingale. Miss Nightingale had strong views on women in nursing and felt that only women possessed the traits of a good nurse, such as caring, nurturing and empathy. The training schools that were set up based on her philosophy refused entry to men.

In 1898, an article in *The Nursing Record and Hospital World* stated that there were no male nurses in England because there was no advantage in having them. It said there was no demand for male nurses and therefore no need to supply them.

Male orderlies supported the work of nurses by helping to care for the 'unmanageable' and 'delirious'. They were also in charge of cooking, gardening and general upkeep of the hospital. In 1936, the *British Journal of Nursing* published an article that expressed concerns about the shortage of nurses. It was agreed subsequently that ex-service nursing orderlies, both men and women, would be eligible to train as nurses. The training would last for 12 months with state exams after three and nine months.

By 1947, the *British Journal of Nursing* reported that male nurses were now allowed to take up senior posts as matrons or assistant matrons and to be paid the same salary as female nurses. This article also highlighted the fact that the Queen's Institute of District Nursing had begun to train men and there was already one practising in Leicester.

No Prejudice

DJ Gillett began his Queen's Nurse training in 1947 along with four other men as part of a national policy experiment to recruit men as district nurses. He had reservations about whether he would be accepted by female colleagues and patients, but he found no evidence of prejudice. His patients were amazed that there were male district nurses and wanted to know what to call him and more about his role.

Mr Gillett remembers that he was often mistaken for an insurance salesman, as he did not wear a uniform and only put his white coat on when performing his duties. These 'early pioneers', as they were known, were only allowed to visit male patients and therefore would cover much more ground during the day than their female colleagues. Mr Gillett felt that the experiment of male district nurses was successful and believed that men would continue to become more prominent in the profession.

Doug Morison had wanted to be a nurse from the time he was brought up by his aunt, who was a matron in New Zealand. This did not please his father, who felt women were nurses, whereas men became doctors. In the 1950s, New Zealand still did not recruit male nurses, so Mr Morison left for the UK to train at London's Mile End Hospital (now the Royal London). On his first day, the matron said to him: 'Do you realise you have walked in to a den of women and there are some things I will not be able to save you from?'

However, Mr Morison never felt any animosity from staff or patients and recalls that some men preferred to receive their care from a male nurse. In fact, being a male nurse on the district had its perks. At the time there were two men working on the district. There were also five London female superintendents, who were known to be competitive.

One afternoon, Mr Morison and the other male nurse were summoned to matron Agnes Evans's office and told there was a black tie ball that night and that both of them were to attend with her. She assured them that their patient visits were covered. As Ms Evans was about to make her grand entrance to the ballroom, she hooked up her arms with both men, one on either side. Being a ball for nurses, the room was full of women, so the two male nurses found themselves in great demand. As Ms Evans circulated, she told the other superintendents: 'Don't worry ladies—as the night progresses I will lend you one of mine.'

Dangerous Duties

Robert Mounce also found his gender advantageous at times in nursing, as all the women in his district wanted to be visited by the handsome male nurse and were known to apply a couple of coats of lipstick before he arrived. However, he recalls that some patients could not understand the concept of a male nurse and believed him to have failed as a doctor. His school headmaster also reacted with disbelief, saying: 'You are throwing your life away, boy. I suggest you leave quietly and tell no one else of your plans.'

Mr Mounce's duties on the district were to care for all the men, including those who frequented the local cider house and were deemed too dangerous for the female nurses to visit. He remembers going to dress their infected wounds and attempt some health promotion, but instead regularly being on the receiving end of a clenched fist. Mr Mounce recalls that men in nursing at that time were referred to as 'humpers and plumbers' as they would carry out all male catheterisations and would do all the bathing and lifting that was deemed too much for their female counterparts. If Mr Mounce had no patients he would stock cupboards and deliver linen to patients' homes, commonly being mistaken for the delivery man.

By the 1980s, he was allowed to visit female patients to give injections but only if they were to be administered in the arm. He retired five years ago but still looks back fondly on his career as a district nurse.

Andrew Harrison trained as a district nurse in 1999 after working in A&E and intensive care. During his training there were 20 women on the course and four men. He has not experienced any animosity from staff or patients, although he recalls that some of his male nursing colleagues have had difficult times. However, he feels he has been expected to take on more physical roles because of his gender, including restraining patients' guard dogs and, on one occasion, fighting with a flock of angry geese. Mr Harrison remains passionate about district nursing today.

Tony McCoy has also experienced nothing but positive responses from patients and colleagues. He left school with few qualifications but wanted to join the ambulance service, so he decided to gain some experience as a care assistant. In 1999, his manager seconded him to undertake nurse training and he qualified in 2002. In 2005, he completed his district nurse training with first class honours and now supports a team of community nurses.

A Nurse First

Mr McCoy believes good nursing care is not an issue of gender. he feels he is a nurse first and a man second. He also believes that communication is the key to a good relationship with patients. Before a new visit Mr McCoy always contacts the patient by phone to ask whether it would be all right for him to visit—he has rarely been refused because of his gender. In fact, as he explains, he works in small villages and his patients tell their friends, who then ask him to dress their wounds or help manage their medication. He often gets stopped in the street by strangers asking for his help.

The NHS Information Centre for Health and Social Care published figures in March 2008 showing gender differences in nursing between 1997 and 2007. It showed that 18.5 per cent of nurses in 2007 were men. The percentage of male district nurses was lower still, at only 4.3 per cent. Men made up 17 per cent of nurse managers, 33.7 per cent of mental health nurses and 27 per cent of learning disability nurses.

In the past ten years there has been a slight increase in men entering community nursing and a decrease in men entering community psychiatry (Drennan and Davis 2008). In 2006–07, however, applications for nursing courses from men decreased by 48 per cent (King's College London 2009).

As we shift nursing care from hospitals to the community and the roles of nurses advance alongside more attractive pay structures, we may see an increase in men attracted to nursing. Recruitment advertising needs to target men more effectively. We have come a long way in the past 150 years but still have some way to go if we are to see real gender equality in nursing. There were many changes to the nursing workforce throughout the 19th and 20th centuries, with nursing going from a male-dominated profession to female-dominated. Only time will tell where the 21st century will take us in relation to men in nursing.

We must not lose sight, however, of the fact that nursing is not about the proportion of males versus females, but about committed, dedicated nurses providing a high standard of care.

References

Drennan V, Davis K (2008) *Trends Over Ten Years in the Primary Care and Community Nurse Workforce in England.* St Georges University of London/Kingston University, London.

Information Centre for Health and Social Care (2008) *Workforce statistics.* www.ic.nhs.uk/statistics-and-data-collections/workforce/nhs-staff-numbers/nhs-staff-1998—2008-non-medical (Last accessed: May 20 2009.)

King's College London (2009) *Who Wants to be a Nurse?* National Nursing Research Unit, King's College, London.

The British Journal of Nursing (1936) The training of the male nurse. August, 91.

The British Journal of Nursing (1947) The Queen's Institute of District Nursing. August, 92.

The Nursing Record and Hospital World (1898) nursing echoes. June 25. 517–518.

Critical Thinking

1. Why were men in nursing hired to work in unfavorable areas within the community?

ALISON BLACKMAN is professional officer, The Queen's Nursing Institute, London.

The State of the Profession
"Code White: Nurse Needed"

March 1, 2005—Today *The State* newspaper, of Columbia, South Carolina, ran the final installment of a massive, three-part special report by Linda H. Lamb about the nursing shortage, "Code White: Nurse Needed." The report addresses the causes of and potential solutions to the shortage, and it has many excellent elements, notably extensive examinations of the problems with nursing's public image, issues related to men in nursing, and aspects of the training of new nurses. Perhaps the most glaring problem is the report's failure to mention what many believe is the primary immediate cause of the current shortage, namely the managed care-driven hospital budget cuts of the 1990's which led to the dangerous nurse short-staffing that has driven many nurses from the bedside. The piece gives the impression that any short-staffing is merely an effect of the shortage, rather than a leading cause of it. In addition, a short sidebar on the growing use of foreign nurses in the U.S. fails to mention the devastating effect such migration is having on the health systems of many developing nations.

The special report consists of three long articles published February 27–March 1. Each of the three is accompanied by relevant sidebars, "nurse vignettes," which are three short quotes from nurses or nursing students that echo the basic themes of the article, and a collection of photos. The first article, "Aging population drives crisis," explains how "South Carolina is coping with the nursing shortage—and why it will get worse." The piece explains some of the basic parameters of the problem, including the aging population, the aging nursing workforce, and the devastating nursing faculty shortage, which has led to the rejection or waitlisting of many qualified nursing school candidates. The piece deserves credit for citing recent research linking higher nurse staffing and better educated nurses with better patient outcomes, and for explaining the magnet hospital concept. The piece is also commendably nurse-driven, using nurse sources almost exclusively, though one section features quotes from a local ED physician about the need for more experienced nurses on "his staff": "I don't have the experience [on the staff] that I want to have." Perhaps this person actually is the administrative manager of the ED, but if a nurse held that position (as many do), we wonder if a piece would feature a quote from him or her about whether she had adequate physician experience on staff. This physician is also identified as "Dr. Ron Fuerst," but no quoted nurse receives the doctoral label, even though at least three have doctorates: South

Carolina Board of Nursing head Sylvia Whiting, Medical University of South Carolina (MUSC) nursing dean Gail Stuart, and leading Vanderbilt nursing shortage expert Peter Buerhaus, on whom the piece seems to rely more heavily than any other single source. Consistent with this failure to fully appreciate the autonomous and intellectual components of nursing, the lengthy piece never really explains what it is that nurses do to save lives and improve patient outcomes.

On the whole this part of the report suggests that the shortage is not a crisis now, but may become one in a few years when the baby boomers retire, citing statistics that show an increase in the number of nurses in the last few years and lower hospital vacancy rates. Of course, vacancy rates are a function of positions the hospitals decide are necessary, and may or may not reflect real patient needs. To its credit, the piece cites Buerhaus for the idea that the current shortage is different from prior ones because of its deep demographic bases and its duration: four times longer than any previous shortage with no apparent end in sight. But the piece fails to identify hospital staffing and other budget cuts driven by the increasing influence of managed care and declining reimbursement rates in the 1990's as a primary cause of the shortage, suggesting that a lack of nurses is a threat to patient safety, but never that nurse short-staffing is the result of anything but a simple lack of nurses. There is no mention of the fierce ongoing battles over mandatory nurse staffing ratios in California and Massachusetts, nor of the many local hospital labor disputes that have centered on nurses' objections to short-staffing. Many nurses feel the nursing shortage is a crisis right now, here and around the world, and that it has claimed countless lives. This is not an obscure point; it has figured heavily in a number of national articles and books on the shortage in the last few years. Even if the piece countered the assertions of nursing activists with multiple hospital administrators arguing that hospital and insurer decision-making is no factor in the shortage, the failure to even raise the issue is a grievous one. The piece does include quotes from nursing leaders decrying the lack of sufficient action by the South Carolina legislature to address the shortage, and it notes that surrounding states seem to be doing more, for instance creating nursing centers to coordinate efforts to improve working conditions and recruiting.

The first part of the report concludes with what seem to be brief previews of the later two parts. The first of these sections introduces the problem of the low rates of men and minorities in nursing, citing the many other career options for those in these groups (and for "bright female students"), and the difficulty of overcoming prevailing stereotypes. The piece notes that a local hospital's human resources director "hates" Ben Stiller's "warped and bumbling male nurse" character in the 2000 film "Meet the Parents." In fact, that is not a fair description of the smart, caring and resourceful Stiller character, who overcomes just the kind of stereotypes the piece is discussing, but it illustrates the understandable sensitivity to such stereotypes in the current social environment. The final section notes that time, energy and money is being spent to address the shortage, but it's not clear yet what is working; it closes with Buerhaus arguing that the single most important thing states can do now is to address the critical faculty shortage.

Perhaps the most noteworthy and commendable aspect of the special report is that the second part, "Image problems are a subtle, persistent factor," focuses on the role of nursing's public image in the shortage. It's fairly unusual for pieces on the shortage to spend much time on this, even though in our view the deep and widespread public undervaluation of nursing is the bedrock problem underlying most of the more immediate causes, including short-staffing and the demographic factors. Lamb effectively frames the image discussion with a scene from the popular NBC drama "ER" in which resident physician Greg Pratt rejects a care idea from nurse Abby Lockhart, saying, "I fly the plane. You serve the coffee." The piece accurately summarizes the message as being that the physician is the male "sovereign" and that the nurse is the female "handmaiden," including a quote from MUSC dean Gail Stuart that "that attitude is quite pervasive." The second part of the report ends with a note that hospitals should hope nurses do not respond to such attitudes as Lockhart did, since "[t]ired of Pratt's condescending bluster, Lockhart went on to medical school," and "[n]ow, she's a doctor." In fact, the episode with Pratt's "coffee" comment aired in April 2002 and the episode in which Lockhart decided to return to medical school aired in October 2003; there was no such direct causation. But Lockhart's decision did appear to be driven at least in part by feelings of disrespect from physicians and the public. Of course, nurses are still at least 50 times more likely to pursue graduate education in nursing than medicine, or (sadly) to simply leave the bedside. "ER"'s presentation of the "coffee" comment was intended to show that Pratt was arrogant and green—the show was not *consciously* endorsing the comment—but much of "ER" actually does support the myth that physicians provide all significant care, and the episode, very typically, contained no effective rebuttal of the substance of Pratt's remarks. It is surprising and unfortunate that the second part of the report, despite using "ER" as its central example, does not really examine the impact of the media itself in fostering and reinforcing the social attitudes it discusses.

However, the second part does provide a great deal of valuable material. It describes some of the efforts hospitals have made to curb "I fly the plane" physician arrogance, including emphasizing teamwork, monitoring physician conduct, and supporting nurses' advanced training. Yet the piece rightly notes that the persistent "image of feminine subservience" continues to deter today's male and female students. The piece profiles veteran male nurses who have battled through the stereotypes,

notably that they must be gay. It deserves credit for including a candid comment by one that he spent a lot of time "overcompensating" by doing "manly, adventure-type things," and for noting that some male nurses feel that female nurses actually resent them. The piece explains why the male nurses it profiles have stuck with nursing, but does not sugarcoat the difficulties. One complains that, while there are plenty of female physicians on "ER," male nurses are rare. (In fairness, the percentage of female physicians in real life is much higher than that of male nurses, though more male nurses do work in the ED.) The piece then circles back to the importance of good relationships with physicians, and the difficulties caused by physician disruptive behavior, which remains a problem. A local OB/GYN describes physician mentoring programs his hospital has implemented to curb such behavior, and a medical school dean stresses that fostering a team approach in his students enhances their critical ability to collaborate with nurses. His message: "A good nurse can save you hundreds of times a day." What he means is, a good nurse can save patients hundreds of times a day, but we'll take it. Finally, the piece addresses the nursing uniform issue, noting that the proliferation of colorful scrubs has made nurses harder to identify, and that some feel the newer uniforms seem less professional. Here again, we learn that "Dr. Ron Fuerst" "has had to draw the line at male nurses in do-rags." The second part closes by suggesting that "what may do the most to improve nurses' image is their own commitment to professionalism," citing hospitals where nurses play a major role in generating patient safety ideas and are encouraged to obtain advanced certifications. It might have also noted the importance of nurses—and the media—speaking up about just what nurses do to improve patient outcomes; there is little in this piece to tell people exactly what that is.

The third part of the special report, "Hospitals, educators getting creative," examines efforts to resolve the shortage. It focuses on hospital "efforts to create a culture where nurses feel valued," which includes "creativity, collaboration and cash," as employers focus on "competitive pay and perks, flexible schedules, reducing paperwork, enhancing patient safety, and buying special lifting equipment to spare nurses' backs." The piece also discusses hospital mentoring, educational innovations, the merits of shift bidding in reducing vacancy rates and the use of agency nurses, and outreach to potential new nurses like middle school boys. It discusses the benefits of "nurturing" hospital nurses through mentoring, the use of patient simulators to streamline clinical education, and the efforts of South Carolina hospitals to attain magnet status, which is designed to reward hospitals that "empower and respect nurses;" apparently none of the hospitals have magnet status yet. Commendably, the piece explains magnet status in some detail, noting that it is associated with lower death rates, shorter stays, and better outcomes generally. Although the piece fails to discuss in any depth the growing ranks of master's-prepared advanced practice nurses—a potential career path that may attract many to nursing—it does note that it's a great time to be a nursing student, as potential employers lavish students with attention, perks and offers. But here again, the failure to examine the short-staffing issue is glaring. Some would contend that when these bright-eyed students hit the continuing reality of short-staffing, there is a significant risk that, perks or no perks, they will not be at the bedside long—assuming they even realize the extent to which they are unable to provide the critical care their patients need, since they may never have had the chance to see such care given. The piece discusses potential legislative assistance for nursing education, including quotes from a state legislator who co-owns an assisted living facility and believes that ultimately "the market will solve the problem" through rising salaries. The piece could have benefited from a response to this view, particularly from someone involved in efforts to ensure adequate staffing. Despite this, the third part of the special report concludes with an excellent quote from local health economist Lynn Bailey, who stresses that the shortage is a patient-safety issue: "Patients, fundamentally, go to the hospital for nursing care . . . When you give highly qualified, experienced nurses too many patients to take care of, mistakes happen, and patients die."

A short sidebar, "Hospitals turn to foreign-born nurses," accompanies the third part of the report. Also written by Ms. Lamb, this item notes that although foreign nurses are not yet a huge factor in South Carolina, local hospitals have recruited a number of Filipino nurses. The piece focuses on two such nurses. These two note that despite some bigotry and other obstacles, they have assimilated and are satisfied with their work, as well as the far higher salaries that drew them in the first place. A local nursing director notes that the Filipino nurses have done well on the national licensing exam. The piece does not explore the potential difficulties some have identified in care from some foreign-born nurses due to differences in training and social factors that may not be measured on exams, such as the ability and inclination to perform U.S. nurses' vital patient advocacy duties. More broadly, the piece fails to mention that the migration of nurses to wealthy nations with shortages has dealt severe blows to a number of developing nations with health systems that are already fragile, including the Philippines. Nor does it discuss the ethical implications of active recruiting of such nurses, which has become a significant issue internationally, though it is rarely mentioned in U.S. articles.

The photos accompanying the pieces show many of the nurses and places discussed in the report. They provide a pretty good portrait of nursing diversity and of the training of new nurses by senior ones. However, one photo shows an NP-in-training in an OR setting where, as the caption describes it, "anesthesiologist Dr. Thomas Warren . . . walks over to instruct the nurse anesthnatist [sic]," who is not shown. Obviously, we don't know what was really going on here, but it's pretty clear that most readers will get the idea that anesthesiologists tell nurse anesthetists what to do in OR settings. In fact, certified nurse anesthetists are highly autonomous professionals whose care has been shown to be at least as good as that of anesthesiologists, though in most states they must nominally work under the "supervision" of anesthesiologists. If the paper was going to mention the nurse anesthetist at all, it should at least have given him or her a name, rather than leaving the reader with Dr. Warren instructing some nameless nurse. Spelling "anesthetist" correctly would have been a bonus.

On the whole, although there are significant flaws in *The State*'s special report, the paper and Ms. Lamb deserve a great deal of credit for making an unusually serious and comprehensive effort to address the crisis in nursing.

See the three-part series "Code White: Nurse Needed" by Linda H. Lamb: Part I "Aging population drives crisis" on February 27, 2005; Part II "Hospitals, educators getting creative" on February 28, 2005; and Part III "Hospitals turn to foreign-born nurses" in the March 1, 2005 edition of the *The State*. Also see the *Nurse Vignettes*.

Critical Thinking

1. How do persistent images of feminine subservience deter both male and female students from entering the profession?

From *Center for Nursing Advocacy*, March 1, 2005. Copyright © 2005 by Center for Nursing Advocacy Inc. Reprinted by permission.

UNIT 8
Nursing Education

Unit Selections

Learning Outcomes

After reading this unit, you should be able to:

- Discuss two benefits of incorporating simulation into the nursing curriculum.

- Identify the reason for including reflective self-assessment activities during simulation experience.

- List the positive outcomes of peer-mentoring.

- List four reasons for the fairness and sensitivity review process.

- Describe the challenges of conducting test item analysis.

Student Website

www.mhhe.com/cls

Internet References

The Commission on Collegiate Nursing Education
www.aacn.nche.edu/accreditation
The National League for Nursing
www.nln.org
The National Student Nurses' Association
www.nsna.org

Nursing education began in America in 1872 at the New England Hospital for Women and Children. A year later, Linda Richards graduated as the first American-trained nurse (see Article 7 Unit I). Initially nurse-training programs were based in hospitals; classes were taught by physicians and nurses employed by the hospital; students lived on the hospital grounds and worked side-by-side with the hospital nurses. In the late 1960s to early 1970s, many hospital-based diploma programs joined with community colleges and more academic courses were added to the curriculum. By the end of the 20th century, only a few hospital-based diploma programs remained.

Today, the overwhelming majority of nursing programs are in public colleges. Associate degree RN programs are often offered at community colleges and BSN programs are at most state colleges and universities. Additionally, a number of private colleges have nursing departments. Practical/vocational nursing programs (PN/VN) and Associate Degree in Nursing programs (ADN) are taught in high schools, vocational technical schools, community colleges, and proprietary schools.

With the advancement of nursing education, many nursing programs have chosen to use the opportunity to be innovative and incorporate simulation into their nursing curriculum. This groundbreaking opportunity has placed many nursing programs on the cutting edge of technology. Simulation experience provides students with a safe environment to make errors and apply the knowledge ascertained in the classroom. Unlike the traditional classroom setting, simulation allows the learner to function in an environment that is as close as possible to a real-life situation and provides the opportunity for the learner to think spontaneously and actively rather than passively (Billing & Halstead, 2005, p.195). This technology rich environment provides many life-like scenarios that promote and stimulate critical thinking and teamwork and improve clinical judgment. The incorporation of this pioneering strategy has helped to facilitate a seamless transition for nursing students into their roles as novice nurses.

As nursing education continues to evolve, nurse educators are tasked with utilizing effective strategies that identify students who maybe at-risk for academic failure and provide the learners with a student-centered environment that focuses on peer-to-peer mentorship and academic support so that students can achieve academic success. In order for this to occur, academic institutions must provide nurse educators with the proper training and support so that students can achieve successful outcomes. Several methods of providing students support not only increases the likelihood of academic success but also helps to decrease nursing program attrition rates.

The face of nursing has changed not only in gender and age but also in culture and ethnicity. Students are more diverse, may have responsibilities other than school, they may be employed or have extensive family responsibilities. Nursing program directors and faculty must be prepared to meet the students' diverse needs. Just as the student population has become culturally diverse, so has the patient population. It is common to view others' beliefs in the context of our own, but in health care it is important to meet the needs of patients while taking into account their basic beliefs.

A practicum experience in which nurse educators mentor nursing students to enhance the teaching-learning process and to assess students' complex learning needs helps to develop students' practice at various levels of the clinical setting.

© Jose Luis Pelaez/Getty Images

In order to practice as a nurse, every student who graduates from a nursing program must take and pass the National Council Licensure Examination (NCLEX). The review discussed in this unit looks at the steps that are taken to ensure stereotyping on the NCLEX does not occur and that language, symbols, words, phrases, or examples that are sexist, racist, potentially offensive, inappropriate, or negative toward any group are avoided.

One of the goals of nurse educators is to ensure that nurse graduates are prepared and pass the NCLEX on the first attempt. The authors of this article review item analysis and how to measure central tendencies. The ability of faculty to understand an item analysis and to utilize the components when revising tests can help to decrease the frustration many faculty members often experience.

The articles within this unit will further elaborate on these topics.

References

Billings, D. M., Halstead, J. A. (2005). Teaching in nursing. A guide for faculty (2nd ed.). St. Louis, MO: Elsevier Saunders.

An Evaluation of Simulated Clinical Practice for Adult Branch Students

STEPHEN PRESCOTT AND JOANNE GARSIDE

Higher Education Institutions and healthcare organisations strive to develop effective teaching and learning strategies for students and staff. Simulated clinical practice is one strategy that is being developed in pre and post-registration healthcare education. Considerable funds are being invested in the development of sophisticated clinical simulation rooms to help create a variety of healthcare environments—from general ward and complex critical care areas to replica community healthcare settings.

Technological developments, changes in healthcare placement provision and a shift in what is considered ethical in practising essential clinical skills have, over the past few years, led to an increased use of simulation in professional education and training in healthcare (Alinier et al 2006).

Billings and Halstead (2005) defined simulation as: 'A near representation of an actual life event; may be presented by using computer software, role play, case studies or games that represent reality and actively involve learners in applying the content of the lesson.' Simulation is the promotion of understanding through doing.

During 2005, a review of existing theory and skills sessions was undertaken in the authors' institution to implement a move away from a task-orientated programme to a more holistic and problem-based approach to simulation. The review concluded that a scenario-based approach should be used, not only to enhance the delivery of theory, but also to provide an alternative approach to allow students to incorporate theory into practice in a safe clinical environment. For example, theory sessions on the anatomy and physiology of circulation and the nursing care of a patient with cardiac disease would be followed by a simulation session involving scenarios about the assessment and nursing care of a patient with a cardiac illness.

This article shares experiences from the delivery and evaluation of simulation implementation for a group of second-year diploma nursing students.

Literature Review

The use of simulation as a teaching strategy allows the learner to practise, repeatedly if necessary, in a safe environment (Cioffi 2001). Simulation offers learners an opportunity to practise safely before they perform the procedure in clinical practice, without exposing patients to any risk. The learning experience through simulation depends on the reality of the scenario and the environment (Parr and Sweeney 2006). Care should be taken when writing scenarios to ensure that they are not embellished for additional effect, but reflect the reality of clinical practice. Similar care should be taken to prepare the simulation environment, with appropriate, fully working equipment and consumables in a manner, that as closely as possible, reflects the clinical setting. This can involve substantial expense in terms of time and money.

Simulation also offers the opportunity for a 'time out' to be called, either by facilitators or students (Bland and Sutton 2006), for example, to ask how to use a particular piece of equipment, to ask about the rationale for a particular treatment, or to review progress made so far or refocus the team. Time outs are not always practical in the clinical setting, but are possible in simulation.

Simulation can involve a range of simulated patients and can take many forms. Examples include the use of simple task or procedure trainers designed to develop psychomotor skills such as tracheal intubation, intravenous (IV) cannulation, and defibrillation (Perkins 2007) through a range of low, intermediate, and high-fidelity manikins (Alinier et al 2006, Perkins 2007) or through the use of actors (Bland and Sutton 2006). The term fidelity refers to the extent to which the manikin reflects reality (McCallum 2007), and as technology has developed over the years so has the 'life-like' nature of some of the manikins. Some high-fidelity manikins have been programmed to mimic human physiology and will respond 'appropriately' to a given treatment. For example, administration of IV fluids will correct the signs of hypovolaemia (Perkins 2007). Intermediate-fidelity manikins often require the facilitator to alter physiological parameters in real time, in response to patient deterioration or treatment. Some software programmes include scenarios or give facilitators the ability to programme their own, although this can be a time-consuming process.

Simulation not only allows individual students to practise a particular clinical skill, but can also be used to promote team working. As Medley and Horne (2005) suggested, pre-registration nursing has only recently begun to realise the potential of simulation in undergraduate programmes.

Simulation can be seen as a different teaching strategy (Lammers 2007), but should not only be viewed as a stage between delivery of theory (often in the form of a lecture) and clinical practice. In other words, simulation does not only allow theory to be integrated into practice, but also offers an opportunity to deliver theory in conjunction with the delivery of practice. During simulation sessions, the practise (doing) is undertaken; additionally the learner is offered the opportunity to revisit any theory that has been delivered to integrate theory and practice.

In this study, the simulation sessions were repeated to allow the large cohort to be split into small groups. At each session the students, around 15 students per group, were further divided into three groups of five. One group was given a scenario and became the team that

---

Box 1
Scenario

Clinical History and Setting

Simon is 57 years old. He is admitted to your ward via the emergency department. He arrived in the emergency department about four hours ago, complaining of central chest pain. The emergency department nurse hands an electrocardiogram (ECG) to you, which she says was done in the emergency department and is 'normal'. The nurse informs you that Simon has angina. His only medication is a glyceryl trinitrate (GTN) spray which he uses as required. He tries getting himself off the emergency department trolley as he 'does not want to be a nuisance'. Simon's wife, Louise, is at work. Nursing staff have tried to reach her, but have not been successful.

Clinical Course

Simon looks pale and anxious. He informs you that the pain is much better than it was earlier. He tells you that when he came to the emergency department the pain was much worse than anything he had ever experienced. He would like a cup of tea as he has not had anything to eat or drink since he left home. An intravenous cannula has been inserted in the back of his right hand.

Initial Assessment

Airway (A) = Clear. Talking in full sentences. Oxygen mask removed in the emergency department
Breathing (B) = Respiratory rate 20 beats per minute (bpm). Oxygen saturation 96%
Circulation (C) = Pulse 80 bpm, regular (sinus rhythm), blood pressure 148/90mmHg
Disability (D) = Conscious and alert
Exposure (E) = Nil of note

Key Points to Observe

Prepare bed area appropriately
Welcome Simon onto the ward
Receive handover from emergency department nurse
Attach monitor
Record initial observations
Bleep appropriate medical staff
Offer psychological support
Check that relatives have been contacted

Clinical Progression

One hour after settling on the ward, a first-year nursing student informs you that Simon is experiencing severe pain in his chest. When you arrive at the bed he is cold and clammy.
A = Clear
B = Respiratory rate 24bpm. Oxygen saturation 90%
C = Heart rate 110bpm, sinus tachycardia, blood pressure 160/100mmHg
D = Alert–complaining of severe central chest pain
E = Cold and clammy

Key Points to Observe

Rapid assessment
Reassurance and psychological support
High-flow oxygen
Fast bleep appropriate medical staff
GTN spray
Other analgesia (on arrival of medical staff–as prescribed)
ECG–different from admission (shows acute inferior changes)
Possible transfer to critical care unit
Consider other treatment options
Inform relatives of action taken

---

assessed and implemented the required care to the simulated patient. The remaining two groups observed the care delivered and were encouraged to critique positively the performance of the first group. An example of a scenario used is outlined in Box 1.

The adult branch cohort consists of 60–80 students and runs twice a year (September and January). Smaller groups of about 15 students were established and each smaller group attended a two-hour session. Each student attended ten simulation sessions on various subjects relating to the theory that had been covered during the academic year.

Before the first session, the students were introduced to the concept of simulation, and were shown SimMan®, the simulated manikin used for the majority of the sessions. SimMan® is an advanced, instructor-driven, full-size patient simulator (Perkins 2007). SimMan® generates realistic heart, breath, bowel, and blood pressure sounds. Vital signs are displayed on an accompanying monitor, with the displayed range of parameters adjustable depending on the requirements of the students.

During each scenario, students were encouraged to work as a team and systematically assess and care for a given patient with a variety of health needs, including physical, emotional, and social problems. Students were required to identify the key physiological abnormalities and treatment options, to know their own limits, and when to get

appropriate help. During a two-hour session each student had the opportunity to participate as a member of a team and to observe and critique two different scenarios undertaken by the other two teams. Parr and Sweeney (2006) stated that the wearing of uniforms by students and facilitators encouraged professional thinking. Students and facilitators in this evaluation did not wear uniforms, although this is under review. Active participation by all students during the scenarios was encouraged, with the facilitator emphasising that although they were being 'watched' they were not being tested. Students were encouraged to ask questions, and time was allowed for the team to discuss potential solutions before the facilitators intervened.

In each scenario one of the facilitators stood at the head of the bed allowing students' focus to remain on the patient (Parr and Sweeney 2006). The sessions were usually led by academic staff, but practice staff were involved in some sessions to share their expertise and for personal development.

The scenarios were designed to be as realistic as possible, using the appropriate equipment and documentation. The students were not expected to undertake roles or demonstrate knowledge outside their expected scope of practice. Additional facilitators undertook the role of any medical and other health professional staff required.

Box 2
Expectations of Students During Simulation Sessions

- Relax and enjoy the sessions. Simulation training involves a type of role play, which some find a little uncomfortable. Remember, we are not trying to embarrass or undermine you in any way.
- Participate. You need to 'do'.
- Work as a team.
- Treat the manikin as you would a patient. This is not easy, but will develop as we perform more simulations. The

manikin can never replace a real patient, but if you can get beyond it being a lump of plastic, you will learn more.
- Appreciate that inevitably there will be slight differences between trusts in paperwork and occasionally in the treatment options and protocols used.
- Be prepared to critique yourself and others.
- Be prepared to be critiqued.
- Dress appropriately.

Feedback on students' performance is a vital component of simulation (Perkins 2007), therefore each scenario was followed by a period of feedback and debriefing. First, members of the group who were assessing and caring for the patient were asked to critique their performance following the format outlined by Mackway-Jones and Walker (1998). Students were asked what went well in their performance and what they were particularly pleased about. They were also asked what they would like to improve if they were to repeat the simulation. Second, members of the other groups critiqued the assessing group. Nurses' ability to critique their practice positively and the practice of others is a vital skill, and these sessions gave students the opportunity to develop this skill further. The facilitators were then able to critique and evaluate what had happened, reinforcing what had already been discussed by students, and to add any further comments. The feedback was followed by the opportunity to revisit various aspects of theory and to address any questions the students had in relation to the scenario. Box 2 summarises the facilitators' general expectations of the students during the simulation sessions.

Aim

To evaluate simulation strategies used for adult branch nursing students in one university.

Method

Students were given a questionnaire using a numeric method of analysis. In addition, to broaden the type of data collected, sections for open comments were included in the questionnaire.

The quantitative section of the questionnaire asked a variety of questions, from biographical characteristics to questions using an adapted Likert scale where students were asked to quantify their level of knowledge and understanding, skills and confidence before and after their simulation experiences. The quantitative section of the questionnaire was electronically read to reveal the statistics for analysis. The open-ended section offered the students large text boxes in which they were asked to describe their expectations, their learning experiences and any areas they felt could be improved.

The authors had undertaken many of the developments on simulation, therefore care was exercised to protect against bias by obtaining support and advice from an academic supervisor. It was recognised that it was important to have a comprehensive understanding of the subject area, such as the circumstances surrounding simulation and the rationale for it, which the researchers possessed.

A pilot consultation of the questionnaire was undertaken with senior lecturers and students from a different cohort and was adapted

accordingly. The sample for the evaluation consisted of a cohort of second-year diploma nursing students.

The information sheet, consent form and questionnaire were given to all students in the cohort at the end of delivery of the module and the simulation sessions. Forty-five out of 60 nursing students chose to complete the questionnaire. These students were 18 months into their three-year course. They had completed the theory element and were about to embark on their second practice placement in year two. All students in this group were in the diploma in higher education of nursing studies (adult branch) programme.

Ethical considerations Ethical approval for the evaluation process was obtained from the university's research ethics panel. The students were given an information sheet and were required to complete a consent form, which included a guarantee of confidentiality and reassurance that pseudonyms would be used when direct quotes were cited. Students could withdraw from the evaluation at any time with no repercussions.

Results

Thirty-two (71%) respondents identified that they had not experienced simulation before. All had, however, undertaken task-orientated skills sessions in the common foundation programme but were new to the scenario/problem-based style of simulation.

Before the simulation experiences, just over half (23, 51%) the respondents identified their level of understanding as 'good' but reported that they did not feel confident. More than one quarter (12, 27%) of students described their understanding as poor. After the simulation experiences the majority (26, 58%) of respondents described their level of skill as competent and confident.

The findings from the qualitative data have been categorised into five themes: simulation as a learning method; theory to practice; confidence building; individual support; and feelings.

Simulation as a Learning Method

Although the evaluation did not provide any quantifiable answers about the effectiveness of simulation, it did provide a positive evaluation of respondents' perceptions of simulation as a learning method:

'I really enjoy these sessions and feel there should be more as it is often easier to learn when the lesson is practical' (Respondent A).
'The sessions are a really good way of simulating a ward environment. I believe the sessions are more useful than lecture sessions' (Respondent B).
One student commented that simulation provided an opportunity to '. . .piece together problems and question why things are occurring' (Respondent C). Others commented on the team

working required in the scenarios, with one suggesting that 'team working is a priority; working together is essential' (Respondent D) and another saying 'team work is the foundation of all areas of practice' (Respondent E).

Theory to Practice

The key aims of the simulated sessions were to strengthen the theory sessions delivered in the module and to support the development of practical skills in a safe, simulated practical environment. Simulation cannot replace the practice experience for nursing students but it was hoped that it would help prepare the students for their practice placements. To explore whether these aims were being met, respondents were asked if they felt better prepared for practice, to which they all gave a positive reply with 23 (51%) students agreeing and 22 (49%) strongly agreeing.

When asked to describe what they had learned from the simulation experiences in the module, many identified the link between theory and practice:

'I personally find it a lot easier to understand theory when in a practical situation; it makes things easier to understand' (Respondent F).

'[I have learnt] how to put theory into practice and . . . to practise procedures in a safe environment' (Respondent G).

'[I have learnt] to think about the possible diagnosis behind the symptoms a patient may present with' (Respondent C).

When respondents were asked if their understanding had improved, 44 (98%) gave a positive response, with 24 (53%) strongly agreeing, 20 (44%) agreeing and one (2%) not responding.

Confidence Building

A key area commented on by many respondents was confidence building. When asked about their confidence levels after the simulation, 22 (49%) agreed and 22 (49%) strongly agreed that their confidence had increased. One (2%) did not respond. Comments included:

'They [the simulation sessions] made me feel much more confident about putting these skills into real practice before we go to clinical placements' (Respondent H).

'It has made me more confident when working on a ward and I feel conscious of the rationale behind what I'm doing' (Respondent B).

'Thank you for the sessions, they put me out of my comfort zone, but I feel good to have done the session and it has helped my confidence massively' (Respondent I).

Although many commented that initially the sessions were frightening, they reported that as the module progressed and they gained more experience in the simulation environments, they felt significantly more confident.

Individual Support

One of the problems with teaching in the lecture format is the issue of whether all students comprehend the subject being taught and are learning. Working with smaller numbers of students in the simulation exercise gave students the opportunity to ask questions and lecturers the opportunity to identify students who were struggling to understand the subject matter and therefore repeat or reinforce some areas. The following statement supports this:

'I feel that they [the simulation sessions] prepare you for the real thing. They give you the confidence to answer questions, even if you are unsure' (Respondent J).

Feelings

The single negative aspect that emerged from the evaluation came from a small number of respondents (4, 9%) who were concerned about being watched by peers:

'The other groups learn by watching but it is off-putting when people are watching you' (Respondent K).

'It made me more nervous having people watching me, I think it would be better for me if they weren't' (Respondent L).

'Standing in front of the group makes the experience more nerve-racking and I felt my nerves held me back from being able to participate fully and confidently' (Respondent M).

'The only thing I dislike is that it is quite embarrassing having to get up in front of a group to do activities alone' (Respondent N).

Discussion

A limitation of the evaluation was that although the feedback was valuable, the opinions were subjective. This meant that the results cannot be assumed to be an accurate measure of the learning that occurred and cannot be generalised to other settings or other student cohorts. The results of the evaluation were encouraging and suggested that simulation is a valuable learning method in nursing education. The results have informed the development of a larger project on simulation in the school.

Simulation is resource intensive compared to the traditional lecture, in relation to staff and equipment. At least two facilitators are generally required at each session. As only small groups of students can be accommodated at each session, to allow the large numbers of pre-registration students, the sessions need to be repeated several times. In addition, much of the technical equipment is expensive. Therefore, the development of simulation relies heavily on the enthusiasm of the team and support from management.

The concern raised by some of the respondents about being watched by their peers will be considered with future cohorts. However, the practicalities of ensuring simulation for large numbers of students are unlikely to permit major changes in this respect. Since completing the evaluation, developments have included the following:

- Students are summatively assessed using clinical simulation scenarios; this accounts for 50% of the module assessment.
- The evaluation has been extended to all pre and post-registration courses/cohorts that use simulation as a teaching method.

Implications for Practice

- Simulation can play a valuable part in supporting theory and practice elements of nurse education.
- In the current healthcare environment, where there has been a shift in the availability and accessibility of appropriate clinical placements, simulation provides a vital learning strategy for professional training.
- Nursing students find they are more confident and retain material after using simulation manikins.

- Further investment has been made in additional technical equipment.

Future developments include:
- Development of inter-professional learning will be incorporated into the scenarios, for example the role of the physiotherapist will be undertaken by a physiotherapist or physiotherapy student.

Conclusion

Respondents reported that simulation was generally a positive experience that allowed them to understand the concepts involved, endorsing and developing the theory covered.

Simulation will never replace traditional methods of delivering theory, such as lectures or seminars, and neither will it replace good quality, practice experience for nursing students. The results of this evaluation, however, showed that respondents appreciated and valued simulation as an additional teaching strategy. Many commented on how much their confidence had improved. This not only involved a feeling of increased confidence in the simulated setting, but also more importantly, increased confidence in dealing with similar situations in clinical practice.

References

Alinier G, Hunt B, Gordon R, Harwood C. (2006). Effectiveness of intermediate-fidelity simulation training technology in undergraduate nursing education. *Journal of Advanced Nursing.* 54, 3, 359–369.

Billings DM, Halstead JA (2005) *Teaching in Nursing: A Guide for Faculty.* Second edition. Elsevier, St Louis MO.

Bland A, Sutton A (2006) Using simulation to prepare students for their qualified role. *Nursing Times.* 102, 22, 30–32.

Cioffi J (2001) Clinical simulations: development and validation. *Nurse Education Today.* 21, 6, 477–486.

Lammers RL (2007) Simulation: the new teaching tool. *Annals of Emergency Medicine.* 49, 4, 505–507.

Mackway-Jones K, Walker M (1998) *Pocket Guide to Teaching for Medical Instructors.* BMJ Books, London.

McCallum J (2007) The debate in favour of using simulation education in pre-registration adult nursing. *Nurse Education Today.* 27, 8, 825–831.

Medley CF, Horne C (2005) Using simulation technology for undergraduate nursing education. *Journal of Nursing Education.* 44, 1, 31–34.

Parr MB, Sweeney NM (2006) Use of human patient simulation in an undergraduate critical care course. *Critical Care Nursing Quarterly.* 29, 3, 188–198.

Perkins GD (2007) Simulation in resuscitation training. *Resuscitation.* 73, 2, 202–211.

Critical Thinking

1. What are two benefits of incorporating simulation into the nursing curriculum?

STEPHEN PRESCOTT and **JOANNE GARSIDE** are senior lecturers, University of Huddersfield, Huddersfield. Email: s.f.prescott@hud.ac.uk.

Nursing Students' Self-Assessment of Their Simulation Experiences

MARY L. CATO, KATHIE LASATER AND ALYCIA ISABELLA PEEPLES

One Goal of Simulation is the Development of Clinical Judgment, or the Ability to Think Like a Nurse. Tanner (2006) analyzed and described this complex process in a model of clinical judgment, defining clinical judgment as "an interpretation or conclusion about a patient's needs, concerns or health problems, and/or the decision to take action (or not), to use or modify standard approaches, or to improvise new ones as deemed appropriate by the patient's response" (p. 204).

- Simulation offers excellent opportunities for students and faculty to examine students' clinical judgment and help develop their thinking. While some students interact in a simulation scenario, others watch the action. Then, a full group debriefing follows, involving faculty and peers.

- While this strategy is designed to foster students' development of clinical judgment, it has been the authors' goal to make the connection between simulation and clinical judgment more transparent. This article describes how simulation faculty introduced and helped students use the Lasater Clinical Judgment Rubric, a theoretical and empirically grounded assessment tool for simulation, as a personal, reflective, self-assessment tool (Lasater, 2007a).

- Students and faculty now have a language in common as well as a means for assessing students' progress in their clinical judgment development and goal setting. The self-assessments are additional evidence of clinical thinking for faculty who supervise students in the clinical setting to better support student learning, development, and evaluation.

The Dilemma

Nursing students have traditionally spent clinical time in various health care facilities, providing care to patients under the supervision of a clinical instructor and/or preceptor. Clinical faculty may find it difficult to evaluate students' clinical judgment skills when they are supervising multiple students in the same setting. With the availability of increasingly realistic patient simulators and faced with a shortage of clinical sites as well as faculty, nursing programs are utilizing new avenues of clinical learning, such as manikin-based simulation.

Since 2004, junior nursing students in a clinical adult acute care course have participated in manikin-based simulation as an integrated clinical learning activity. The amount of simulation per course has varied, from two to eight four-hour sessions each term, depending upon experiences available in clinical sites, course objectives, and learning needs. Students in the course attend their four-hour simulation sessions in groups of 12. Each session consists of four scenarios, or patient cases, during which a team of three students provides care for the patient while the remaining nine students observe via live video broadcast from a debriefing room. Two faculty, serving as facilitators, remain consistent throughout the course. They develop the scenarios, run the equipment, direct the case, and facilitate the debriefing. The scenarios are written to complement the content the students are learning in lectures and skills labs; for instance, after a lecture on the endocrine system and an intravenous fluid and pump lab, a scenario event may be based upon a diabetic patient on an insulin drip who requires a focused assessment and a particular action.

Debriefing that follows each scenario focuses on the team's care of the patient, including safe practice, priority setting, continuous assessment, communication, resource management, and leadership. Faculty facilitate debriefings with specific course competencies in mind and lead the group, comprised of scenario participants and observers, toward discussion of clinical judgments in each case.

Despite having extensive discussions surrounding the cases, students have requested more personal evaluations and feedback regarding their actions in simulation (Lasater, 2007b). The dedicated simulation faculty also needed a better way to provide evidence of clinical thinking and provide feedback to faculty who supervised students in other clinical settings. Thus, a tool was required to help students reflect on their experiences in simulation, assess their progress in developing their clinical judgment, and provide feedback and guidance toward the attainment of higher levels of thinking.

Literature Review

Reflective journaling is often described in the nursing literature (Craft, 2005; Lasater & Nielsen, 2009; Murphy, 2004; Nielsen, Stragnell, & Jester, 2007), but written self-assessment strategies rarely appear. However, self-assessment or self report strategies

Table 1 Lasater Clinical Judgment Rubric Scoring Sheet

Student Name: Observation Date / Time: Scenario #:

Clinical Judgment   **Observation Notes**

Components Noticing:
- Focused Observation: E   A   D   B
- Recognizing Deviations from Expected Patterns: E   A   D   B
- Information Seeking: E   A   D   B

Interpreting:
- Prioritizing Data: E   A   D   B
- Making Sense of Data: E   A   D   B

Responding:
- Calm, Confident Manner: E   A   D   B
- Clear Communication: E   A   D   B
- Well-Planned Intervention/Flexibility: E   A   D   B
- Being Skillful: E   A   D   B

Reflecting:
- Evaluation/Self-Analysis: E   A   D   B
- Commitment to Improvement: E   A   D   B

Summary Comments:

in education are routinely evaluated and reported. One review in particular strongly suggested that self-directedness in students is enhanced when students have opportunities and structures by which to self-assess, that is, set goals and monitor their progress toward them (Nicol & Macfarlane-Dick, 2006). The authors cited a study by Boud (1986) that concluded that students should be engaged in identifying and working with standards that will pertain to their work and use those standards to assess their work. These findings are congruent with the use of a rubric for formative self-assessment.

Likewise, Fitzpatrick (2006) found that self-assessment had a positive impact on students' personal and professional development, especially their sense of autonomy and thinking skills. These conclusions were further supported by a study that examined enhanced self-directed learning as an outcome of self-assessment (Maclellan & Soden, 2006). In addition, self-assessment, which is closely related to reflection (Nicol & Macfarlane-Dick, 2006; White, Crooks, & Melton, 2002), can help students develop plans for personal growth, an important professional skill. When shared, self-assessment offers faculty the opportunity to observe students' thinking (Davies, 2002). Marienau (1999) concluded that self-assessment furthers learning from experience, more effective functioning, and commitment to competency, self-agency, and authority. Manikin-based simulation certainly offers the potential for such learning.

The Reflective Self-Assessment Activity

The Clinical Judgment Rubric (Lasater, 2007a) was selected as a means to provide students with more personalized feedback. The rubric's language and developmental progression

has greatly enhanced the sharing of thoughts and ideas among groups of students as well as simulation and clinical faculty. Students use it for personal self-assessment and as a feedback mechanism for themselves and for supervising clinical faculty.

In instituting use of the rubric, a clinical judgment scoring sheet was posted on an online learning management system along with a description of the behaviors at each level (Exemplary, Accomplished, Developing, Beginning) for the four phases of Tanners Clinical Judgment Model (2006) (Noticing, Interpreting, Responding, and Reflecting). (See Table 1.) The rubric provided a framework for students for organizing their thoughts about managing various patient situations. Students completed a scoring sheet, based on their own self-reflection of their practice in simulation, using specific examples of the scenarios in which they participated, as well as their contributions in debriefing.

When the rubric was first used, students completed their self-assessments twice, at midterm and at the end of the term. Simulation faculty read them and gave online feedback on the students' thoughts. The completed documents were then available to the students and clinical faculty. Although student performance in simulation was not graded, clinical faculty found the information useful as further confirmation of students' progress in the clinical areas. In addition, the process of reflecting on clinical judgments by students fostered their development and expertise (Lasater, 2007b).

Due to the number of students in the course (48 per term), certain measures were necessary to provide individualized feedback. These included consistency of simulation faculty throughout the term, to facilitate continuous observation of students in the simulation environment; familiarity of simulation faculty with the course competencies, the scenarios, and their objectives; and contact with clinical faculty regarding specific student issues. Table 2 offers a sample of students'

Table 2 Examples of Student Self-Assessment of Clinical Judgment Development through Simulation

	Clinical Judgment Dimension	Self-Assessment	Evidence for Student Assessment
Student A	Effective Noticing • Focused Observation • Recognizing Deviations from Expected Patterns • Information Seeking	Beginning	"With the patient, I sometimes get 'lost' in the assessment. More specifically, I know what I see but don't know what to do with the information. When I was doing the assessment for Juanita, my mind simply went blank about what pertinent questions to ask of her relative to her COPD exacerbation. I guess I was 'expecting' her to allow me to proceed with an uninterrupted head-to-toe assessment (which clearly was not a reasonable expectation). She kept asking me questions and was unable to breathe easily. This, of course, threw me off so that my assessment was not finished appropriately, and I got sidetracked with ensuring that she was getting enough O_2. Essentially, I think I just got overwhelmed by the situation and my lack of understanding of the big picture. I also neglected to carry out information-seeking opportunities with my teammates in order to troubleshoot more effectively."
Student B	Effective Interpreting • Making Sense of Data • Prioritizing Data	Developing	"I learned so much from this scenario about priorities because I had them completely wrong. The priority here was getting the consent form signed so that he could go to surgery—his pain was important, but I delayed him getting treatment for his condition by giving him morphine and making it impossible for him to sign the consent form. I was so focused on the skill of drawing up the medication that I completely lost sight of the big picture. I realized later that I am not doing the patient or my team members a favor by tuning out of the situation. I really tried to participate and actively interpret data in the next two simulations based on what I learned. In this aspect only, I felt really proud of myself and excited that I was able to assess and make sense out of what was going on with my patients."
Student C	Effective Responding • Calm, Confident Manner • Clear Communication • Well-Planned Intervention/Flexibility • Being Skillful	Developing	"Being in the leader's role in this scenario, I realized that in a stressful situation I get very tunnel visioned and start to feel that it is solely my responsibility to manage what is going on. I remember thinking that I needed help. Once I got the phone orders, my thinking became very narrow. I told my team that we needed to give her meds, but I became completely stressed and disorganized. Later, I realized that when someone is having an anaphylactic reaction, the calm, confident, and effective leader would have opened up the door and yelled for help instead of thinking that it was solely up to me to administer 4 medications. I know now that part of being calm and confident is knowing that you are not really all alone—you have lots of help and you can make huge mistakes when you don't use the resources that are available to you."
Student D	Effective Reflecting • Evaluation/Self-Analysis • Commitment to Improvement	Accomplished	I spend a great deal of time outside of the simulation reflecting on my performance, what I did poorly, what I felt confident about, and how to improve my performance and knowledge each time. At times, I am overly critical; I have high expectations of myself and get frustrated. But I also know that becoming a good nurse is an ongoing learning process. There will always be things to learn and situations that challenge me. I recognize that I also need to push and challenge myself in the learning process so that I can constantly grow in my skill proficiency and confidence. I find the debriefings very helpful for my learning in that I can verbalize my perceived shortcomings and learn from them as a result of the group's responses and observations.

self-assessments about their performance in simulation using the Lasater Clinical Judgment Rubric.

When the rubric was first used, students and faculty expressed satisfaction with the feedback process. However, simulation faculty found two times per term to be impractical for thorough, individualized review and comments on each student's self-assessment. To make the process more manageable and continue to provide high quality of personalized feedback, students now do self-assessments once per term in selected courses.

Significance for Nursing Education

The ability of students to engage in self-reflection and share their clinical thinking about their practice is evident in their submissions. The majority of students show an ability to think deeply about the situations they encounter in simulation, analyze the patient events and their responses, and apply their experiences to their broader knowledge of nursing and the clinical judgment required to practice safely and effectively.

The tool has proved to be a useful method for clinical and simulation faculty to review as they plan learning activities for students; requiring specific examples offers additional insight into students' thinking. Often, the students' self-assessments parallel the observations of the supervising clinical faculty. Reading the score sheets offers evidence of students' thinking that enables faculty to select appropriate patients, provide supervision at the clinical sites, and observe students more closely when necessary.

The Lasater Clinical Judgment Rubric (2007a) offers students the language needed to describe their progress. The four phases of Tanner's Clinical Judgment Model provide a framework for students to organize their thoughts about managing various patient situations. Very specific examples from the scenarios are described, indicating that students have encountered valid learning experiences in simulation. Their reflections are often deeper and more significant than what they discussed in debriefing, demonstrating that their reflective thinking about the scenarios continues, days and weeks beyond the actual simulation experience.

References

Boud, D. (1986). *Implementing student self-assessment.* Sydney, Australia: Higher Education Research and Development Society of Australia.

Craft, M. (2005). Reflective writing and nursing education. *Journal of Nursing Education, 44*(2), 53–57.

Davies, P. (2002). Using student reflective self-assessment for awarding degree classifications. *Innovations in Education and Teaching International, 39*(4), 307–319.

Fitzpatrick, J. (2006). An evaluative case study of the dilemmas experienced in designing a self-assessment strategy for community nursing students. *Assessment & Evaluation in Higher Education, 31*(1), 37–53.

Lasater, K. (2007a). Clinical judgment development: Using simulation to create an assessment rubric. *Journal of Nursing Education, 46*(11), 496–503.

Lasater, K. (2007b). High fidelity simulation and the development of clinical judgment: Student experiences. *Journal of Nursing Education, 46*(6), 269–276.

Lasater, K., & Nielsen, A. (2009). Reflective journaling for clinical judgment development. *Journal of Nursing Education, 48*(1), 40–44.

Maclellan, E., & Soden, R. (2006). Facilitating self-regulation in higher education through self-report. *Learning Environments Research, 9,* 95–110.

Marienau, C. (1999). Self-assessment at work: Outcomes of adult learners' reflections on practice. *Adult Education Quarterly, 49*(3), 135–146.

Murphy, J. I. (2004). Using focused reflection and articulation to promote clinical reasoning. *Nursing Education Perspectives, 25,* 226–231.

Nicol, D.J., & Macfarlane-Dick, D. (2006). Formative assessment and self-regulated learning: A model and seven principles of good feedback practice. *Studies in Higher Education, 31*(2), 199–218.

Nielsen, A., Stragnell, S., & Jester, P. (2007). Guide for reflection using the clinical judgment model. *Journal of Nursing Education, 46*(11), 513–516).

Seropian, M.A., Brown, K., Gavilanes, J. S., & Driggers. B. (2004). High fidelity simulation: Not just a manikin. *Journal of Nursing Education, 43*(4), 164–169,

Tanner, C.A. (2005). What have we learned about critical thinking in nursing? *Journal of Nursing Education, 44*(2), 47–48.

Tanner, C.A. (2006). Thinking like a nurse: A research-based model of clinical judgment. *Journal of Nursing Education, 45*(6), 204–211.

White, D. R., Crooks, S. M., & Melton, J. K. (2002). Design dynamics of a leadership assessment academy: Principal self-assessment using research and technology. *Journal of Personnel Evaluation in Education, 16*(1), 45–61.

Critical Thinking

1. Why should reflective self-assessment activities be included during simulation experience?

MARY L. CATO, MSN, RN, an instructor and lead simulation specialist, Simulation and Clinical Learning Center, Oregon Health & Science University in Portland, is one of the expert authors of the NLN/Laerdal Simulation Innovation Resource Center (SIRC). **KATHIE LASATER, EdD, RN, ANEF,** is an assistant professor and served in 2007–2008 as interim statewide director of simulation learning, Oregon Health & Science University. **ALYEIA ISABELLA PEEPLES, BSN, RN,** is an intensive care unit nurse and simulation specialist, Oregon Health & Science University Hospital. For more information, write to Ms. Cato at catom@ohsu.edu.

Mentoring as a Teaching-Learning Strategy in Nursing

Current nursing and faculty shortages necessitate development of strategies that prepare all students to function in their roles immediately after graduation. This study used a practicum experience through which nurse educator students mentored nursing students to enhance the teaching and learning of both groups. Study methods, evaluation, and results are discussed.

MARGUERITE RILEY AND ARLEEN D. FEARING

As the nursing and faculty shortages persist, a need exists to develop unique teaching and learning strategies to facilitate students' transition into their professional roles immediately following graduation. The development and use of student-centered practicum experiences while students are still under the guidance of faculty is an effective strategy to provide real situations they are likely to encounter as graduates.

The nursing faculty role includes application of numerous teaching-learning strategies to address students' complex learning needs. This typically involves teaching large and small groups in both classroom and clinical settings. Nurse educator students are instructed on teaching-learning strategies, and during their teaching practicum course they are encouraged to utilize a variety of effective strategies in their classroom and clinical practice teaching. An area that often lacks emphasis is how to assist students having difficulty with the nursing content on a one-to-one basis. Strategies for individual, at-risk students or nontraditional students should be varied, including academic support to meet their specific learning needs and help them experience academic success in the nursing program (Jeffreys, 2001; Price & Balogh, 2001).

The purpose of this descriptive study using student-centered learning theory was to examine the effectiveness of using a nurse educator graduate student in an undergraduate nursing student mentoring program. Overall expectations were to provide a true life practicum experience for the nurse educator students, and enhance the teaching and learning that could be transferred to their professional roles after graduation for both groups.

Literature Review

A review of the nursing literature using the key words *mentoring undergraduate/graduate nursing* and *practicum* in the Cumulative Index to Nursing and Allied Health Literature (CINAHL) database (1997–2007) revealed a number of studies generally related to mentoring undergraduate students using faculty members, alumni, and various levels of student peers. These studies (Fredricks & Wegner, 2003; Neary, 2000; Sprengel & Job, 2004) primarily examined formal mentoring programs to promote clinical development and increase student retention. Research literature was limited in the area of using graduate students as mentors for undergraduate students. Only one study explored the use of graduate (MSN) students as mentors for BSN students (Lloyd & Bristol, 2006). No studies were located involving graduate nurse educator students as mentors for undergraduate students. Barker (2006) discussed mentoring of advanced practice nurse (APRN) students. In addition, most of the studies explored mentoring as a learning process for undergraduate students (Kostovich & Thurn, 2006; Morrison-Beedy, Aronowitz, Dyne, & Mkandawire, 2001; Scott, 2005; Sword, Byrne, Drummond-Young, Harmer, & Rush, 2002). None of the studies examined mentoring as a student-centered teaching-learning strategy to benefit both the mentors and the mentees. This literature review reports on the studies found related to formal mentoring programs in nursing which includes faculty and alumni as mentors, student peer-to-peer mentoring, and graduate student mentors.

Faculty and Alumni Mentors

Kostovich and Thurn (2006) conducted a qualitative study at a liberal arts university school of nursing to investigate the faculty's perceptions of doing group mentoring with students in nursing courses. Faculty mentors were asked to volunteer; eight participated. Students enrolled in a 1-hour mentoring course led by one of the eight faculty mentors for four consecutive semesters. The researchers explored group mentoring along with the process of faculty becoming mentors for nursing students. The results of their study showed some faculty "role ambiguity" but overall definite "personal and professional satisfaction" for both faculty and students (p. 12).

Morrison-Beedy and colleagues (2001) described the use of experienced faculty members to mentor students and junior faculty in the research process. The purpose of the study was to extend the concept and practice of mentoring beyond its traditional focus on clinical training to the realm of nursing research. Students and junior faculty participated in a research project that allowed the principal investigator to delegate many different tasks as team members learned aspects of the research process. No original research was involved, only reporting studies by others; authors concluded that good mentoring could extend to all partners involved in the research process.

Ryan and Brewer (1997) described a formal mentoring program for undergraduate BSN students utilizing faculty as mentors. The authors described how a mentorship program and a professional role development course were integrated into a BSN program. This seminar-type program (graded *satisfactory/unsatisfactory*) involved mentors who had fulltime faculty positions and a minimum of 2 years teaching experience in the BSN program, and were willing to serve as mentors for up to 10 students for 2 years. Even though students gave positive comments regarding the weekly program, they expressed some concern about the commitment of time required for an ungraded course.

Another mentor program design found in the literature involved assigning alumni members as mentors for undergraduate BSN students (Sword et al., 2002). A Canadian nursing school conducted a mentoring program in which baccalaureate nursing students were mentored by alumni from the same nursing program. Unlike preceptorship, which generally involves clinical supervision and performance evaluation, this mentorship was focused more on sharing and nurturing to promote personal and professional growth. Student involvement was voluntary. The use of alumni as mentors was considered an innovative approach which allowed the students to benefit from their mentors' experiences within nursing and the shared experience of graduating from the same undergraduate program. Because the alumni mentors had no formal evaluation role for the students' course grade, greater reciprocal relationships were possible. Benefits for students included increased understanding of the roles and responsibilities of nursing. Mentors also provided students with career development information and employment references as well as other learning opportunities. The mentors reported a sense of satisfaction from working with students and "gained an increased awareness of trends and issues in nursing education" (Sword et al., 2002, p. 430).

Price and Balogh (2001) discussed nurse alumni mentoring of at-risk students in an effort to reduce attrition. Nursing graduates who met GPA and other standards volunteered, were selected as mentors, and matched to at-risk students who signed a contract agreeing to meet objectives for the one semester program. Even though abiding by the terms of the contract was problematical for both mentors and students, 21 of 24 students completed the term. A high majority of the mentors and students indicated this mentoring program met their needs. Conflicting work and school schedules interfered with the mentoring contract and led to meetings being missed or postponed.

Student Peer-to-Peer Mentoring

Jeffreys (2001) described and evaluated aspects of an enrichment program for students, with study groups led by peer mentor/tutor students. The author noted many students entering nursing programs are nontraditional students (older, employed, parents) who often are at-risk; a program of enrichment was designed "encompassing the various stages of the educational process" (p. 143). Upper-level nursing course students or those in the RN-BS nursing program who performed at high academic standards in prerequisite nursing courses and clinical skills, and who had excellent communication skills, were selected and trained to be peer/mentor tutors. Participants in the enrichment program achieved positive academic outcomes, and the author concluded such support strategies should be encouraged and developed. However, Jeffreys also found retention is influenced more often by environmental variables than academic variables.

A peer-to-peer mentoring program to teach collegiality was developed in response to the shortage of student advisors and the increased faculty workloads caused by budget limitations (Scott, 2005). Students were encouraged to use email, face-to-face exchange, or the telephone as mentoring vehicles. Senior students could experience nurturing, leading, and advising of young student nurses, while the junior students entering the program received benefits of a support system. Scott concluded this type of mentoring teaches the value of collegiality; that lesson may carry into the practice environment and reduce the possibility of new graduate burnout.

A similar program by Sprengel and Job (2004) involved second-year nursing students who served as mentors for first-year students within their initial clinical setting. Each first-year student was assigned a second-year student mentor who was enrolled in the medical-surgical course. Specific role preparation was given to both levels of students. The mentors worked with their mentees pertaining to clinical preparations, client care, and expectation for future clinical courses. This mentoring took place for 4 hours once during the semester because that was the only clinical they received in their 2-hour "Fundamentals of Nursing" course. Even though only one 4-hour session was involved, those acting as mentors generally believed the experience was positive and a boost to self-confidence; the mentees were impressed by the knowledge and clinical skills of the mentors.

Fredricks and Wegner (2003) described a program in which senior nursing students mentored freshman students enrolled in a human anatomy and physiology course. Each freshman student shadowed a senior nursing student on a critical care nursing clinical rotation for exposure to critical thinking skills that are necessary to apply theory to practice. As a result of their experience, the freshmen tended to emphasize the importance of anatomy and physiology in nursing coursework, and the seniors validated nursing knowledge gained during their 4-year nursing program.

Graduate Student Mentors

Lloyd and Bristol (2006) offered the only study of graduate students (MSN) mentoring undergraduate (BSN) students. This pilot matched the MSN students as mentors with BSN student mentees in a community clinical practicum. Faculty and clinic staff developed the mentoring network as a team to implement health education programs for the clients. Ten students participated along with one faculty member from the MSN and BSN programs. Two BSN students were matched with each MSN student to plan effective client teaching for clients with asthma, hypertension, and diabetes mellitus. All students completed a Likert survey regarding the mentorship and collaboration processes. The survey included six items relating to the perceived effectiveness of the mentoring program in the community practicum, with responses ranging from 1 (*strongly disagree*) to 5 (*strongly agree*). The collaboration four-item survey used the same Likert scale. BSN students gave scores of 4.4–5.0 points on the Likert scale for the mentoring portion, and 4.0–4.8 for the collaboration survey. MSN student evaluations were 4.6–5.0 for the mentoring portion and 4.4–5.0 for the collaboration process. These positive results demonstrated the effectiveness of the process for this project for both levels of students in the community clinical practicum.

Mentoring of the APRN was discussed by Barker (2006) as a means to support growth and develop success in the advanced practice role. Her review of studies in advanced practice nursing and other disciplines, including vocational behavior, management, psychology, guidance counseling, and ethics, concluded a successful mentoring relationship includes understanding the nature of mentoring, monitoring the progress of the relationship, realistic expectations, and positive compatibility of the mentor and mentee. Pitfalls identified in the review included poor communication, improperly identified limits, and inappropriate objectives. The author concluded properly structured mentoring relationships tend to improve professional growth, productivity, and competence.

In summary, this literature review demonstrated the use of mentoring as a means to develop student practice at various levels in the clinical setting. This finding is consistent with the nursing tradition of mentoring used in clinical to promote professional practice (Lloyd & Bristol, 2006; Morrison-Beedy et al., 2001; Ryan & Brewer, 1997). In most of these studies, mentoring was a teaching strategy with the main goal to support students in their clinical courses (Fredricks & Wegner, 2003; Neary, 2000; Scott, 2005; Sprengel & Job, 2004). Therefore, the purpose of the current study was to explore the effectiveness of a formal mentoring program in which nurse educator graduate students mentored undergraduate students. Effectiveness was determined by students' academic outcomes and completion of an evaluation survey at the end of the semester. For this study, formal mentoring was a planned teaching-learning situation that included expectations of mentor and mentee as well as time to participate in the assigned relationship. In addition, the process was monitored and evaluated.

Study Methods
Design

This descriptive study examined the effectiveness of using a nurse educator graduate student to mentor an undergraduate nursing student. The mentoring program was developed and implemented in the nurse educator teaching practicum course.

With the cooperation and assistance of the practicum preceptor, each nurse educator graduate student arranged to mentor an undergraduate nursing student who was having academic difficulty. Mentoring was done in face-to-face sessions which could be supplemented by online communication via email.

Sample and Setting

A convenience sample of 18 nurse educator graduate students was obtained from a university school of nursing. All of the nurse educator students from the Southern Illinois area were enrolled in the teaching practicum course, which is the final course for the nurse educator master's specialization.

The 18 nurse educator graduate students chose an undergraduate nursing student with academic problems to mentor for one semester. The graduate student's practicum site preceptor helped identify at-risk undergraduate students who might benefit from this mentoring. Fourteen of the practicum sites were at ADN programs, two were BSN programs, and two were LPN programs. The level of undergraduate students varied based on the preceptor's teaching assignment.

Ethical Considerations

Institutional review board approval for the study was obtained from the university's committee. An explanatory cover letter was developed to describe the program and sent to each of the graduate and the undergraduate students. Written consent to participate was obtained from all graduate and undergraduate students who chose to participate. All responses, reports, and evaluations gathered during the program were handled in a confidential manner by the researchers. Permission to use the VARK questionnaire in this study was obtained from Neil D. Fleming, designer of the inventory.

Implementation and Assessments

In the first meeting between the graduate student and assigned undergraduate student, the nurse educator student assessed the undergraduate student's strengths and opportunities based on a discussion of study skills habits and academic history.

The graduate student administered the Learning Style Assessment using the VARK Survey (Visual, Aural, Read/write and Kinesthetic) (Fleming, 2001). The VARK survey, designed for adults age 18 and older, consists of 13 items that identify a person's preferences for gathering, organizing, and thinking about information. VARK is the learning style modality of instructional preference. Other modalities are personality characteristics, information processing, and social interaction. In the *visual* mode, the learner prefers information in charts, graphs, hierarchies, circles, pictures, media, videos, and websites. In the *aural* mode, learning occurs best when information is heard or spoken, such as with lectures, group discussions, seminars, tutorials, and talking with other students. In the *read/write* mode, the learner prefers information displayed in text and printed words. Learners using the *kinesthetic* mode prefer experience and practice that is connected to reality (Fleming, 2001).

In considering reliability, consistency of scores over time is not an expectation of the VARK. An individual's preferences for learning are predicted to change over time based on experience and trends toward multimodalities with aging. Content validity of the VARK is reflected using multiple studies indicating a matching of preferences with a person's perceptions and learning strategies. The VARK does not have predictive validity as its design is not diagnostic or predictive. However, some studies have shown that the preferences identified do predict successful study methods. Learning is facilitated when students and teachers have similar preferences (Fleming, 2001).

As determined at the initial mentoring meeting, identified academic and study skills strengths and weaknesses and the student VARK survey results related to learning preference provided the basis for the graduate student to develop an academic plan with individualized outcomes for the assigned undergraduate student. With the approval of the practicum course faculty and the graduate student's preceptor, the plan was presented to the undergraduate student and implemented with his or her acceptance. Meetings usually were face-to-face, but email also occurred often. The graduate student submitted progress reports and a summary evaluation of student performance to the course faculty at designated intervals during the semester.

At the end of the semester, the undergraduate students completed a 12-item Likert scale evaluation from 1 (*strongly agree*), to 5 (*strongly disagree*) about the effectiveness of the mentoring. The tool assessed communication, level of assistance, accessibility, feedback and response time, support, and attitude. Three open-ended questions addressed the most helpful strategies, areas for improvement, and any additional comments. This formal mentoring program was a graded assignment for the graduate students, providing an actual one-on-one teaching-learning experience.

Communication

The graduate student and undergraduate student met face-to-face for the initial assessment and administration of the VARK. E-mail was used for clarification, communication, and questions. Subsequent meetings were arranged jointly; many of the graduate students met weekly with their undergraduate student. Graduate students used emails to send academic plans and progress reports to course faculty members. Faculty members were available to answer questions via email, or on the Web-based practicum course site. They also provided feedback on the academic plans and progress reports via email. Preceptors were available at the practicum site to discuss concerns.

Results
Vark Survey Results

The following single modal learning preferences were found for the undergraduate students: one aural, one visual/read write, one visual/aural, and four kinesthetic. Eleven students were *multimodal,* with no one channel predominant; two or more,

up to all four, sensory modes are preferred by the learner, who is able to adjust to a variety of teaching strategies used by an instructor. Also, they may be able to adapt to other students' modes when working in peer groups. This process is known as *matching.* Even though all modes can be used by a learner, one mode may be preferred more strongly.

Areas Identified for Mentoring

Eight major areas for mentoring were identified by the graduate students in their assessment plans. Many students cited writing skills as a weakness. Undergraduate students also reported correct use of American Psychological Association referencing format and care plan or care map development as problematic. For senior students, résumé development was identified as an area for improvement. In the area of study skills, the graduate students identified undergraduate student needs in management of large reading assignments; test-taking skills, especially for standardized multiple-choice questions; and time-management skills. Another area in which students frequently requested assistance was preparation for clinical assignments. A number of graduate students worked on specific skill acquisition, such as venipuncture or intramuscular injection, with their undergraduate student. Also, graduate students assisted the undergraduates to understand specific content from their courses. Each graduate student developed objectives and teaching strategies, as suggested in their nurse educator courses, in the academic plan to address the areas identified.

Teaching Strategies Used by Graduate Students

The graduate students developed teaching strategies based on the VARK results, as well as discussions of learning needs and preferences with the undergraduate student. A variety of strategies was utilized for the multimodal students with an emphasis on what the individual undergraduate needed. For single-preference students, the strategies best suited for the identified mode were included; however, many graduate students also included strategies to help strengthen the student's other modes of learning to assist with adaptation to a variety of learning situations.

End-of-Semester Evaluation Results

Sixteen of the 18 undergraduate students completed the evaluation tool; 89% of the scored items were in the *strongly agree* or *agree* categories. One item on each of two evaluations was scored as *disagree* or *not enough information* to answer. The rest of the items on these evaluations were scored in the *strongly agree* or *agree* categories. Several undergraduate students who had been in danger of failing the course stated they passed in part because of the extra help from their mentors. Specific areas of assistance mentioned by the undergraduate students were critical thinking case studies and multiple-choice questions developed by the mentor, learning their study strengths and weaknesses, time management and organization skills, individualized support, and encouragement. No negative comments were received.

Graduate Students' Evaluation of the Mentoring Experience

Many graduate students commented that online communication via email facilitated implementation of the academic plan they developed. Many used teaching strategies based on the VARK results and reported these helped meet the undergraduate student's learning needs. The graduate students reported progress on the goals of the academic plan and also indicated a belief that study skills, critical thinking, understanding of content, clinical time management skills, confidence levels, and overall clinical performance improved as a result of the mentoring project. A number of graduate students stated the experience with one-on-one mentoring increased their self-confidence in the ability to apply the nurse educator content in the practicum setting. Many reported in narrative comments that prerequisite nurse educator courses prepared them for this project.

Several of the graduate students expressed concern about undergraduate students who missed appointments for mentoring, were late, or were unprepared for the mentoring session by not completing assignments. Some believed their students' busy schedule was the reason. These graduate students expressed some frustration and disappointment because they had invested their time and energy into the project. Even with these problems, common to nursing faculty at all levels, the graduate students reported they had made a positive difference in the undergraduate students' academic situation and level of achievement for the semester.

Implications for Nursing

This study demonstrated that an effective mentoring relationship can be developed to assist with academic achievement and clinical performance of nursing students. Online tools such as email can facilitate communication in the mentoring relationship and lead to positive outcomes for all levels of students. Use of an assessment tool such as the VARK can help mentors develop effective teaching and learning strategies for the student.

Future research should assess the effectiveness of a totally online mentoring program. Also, a longer study could assess further the impact of the mentoring relationship on both parties. Matching the mentor and student according to VARK preferences could be done to determine if student learning and communication with the mentor are enhanced. Mentoring of a new graduate in the health care setting including a VARK assessment and an individual academic plan can assist with NCLEX-RN® preparation. This type of mentoring program also could be used in orientation and inservice or continuing education programs for new graduates in a variety of health care settings. For staff development planning, a VARK assessment could be completed with nurses on each unit. The unit educator then could plan programs to complement the staff learning styles. As

identified in the literature review, a need exists for continued research on mentoring in nursing.

References

Barker, E. (2006). Mentoring: A complex relationship. *Journal of the American Academy of Nurse Practitioners, 18,* 56–61.

Fleming, N. (2001). *Teaching and learning styles—VARK strategies.* Christchurch, New Zealand: Author.

Fredricks, K., & Wagner, W. (2003). Clinical relevance of anatomy and physiology: A senior/freshman mentoring experience. *Nurse Educator, 28,* 197–199.

Jeffreys, M. (2001). Evaluating enrichment program study groups: Academic outcomes, psychological outcomes and variables influencing retention. *Nurse Educator, 26,* 142–149.

Kostovich, C., & Thurn, K. (2006). Connecting: Perceptions of becoming a faculty mentor. *International Journal of Nursing Education Scholarship, 3,* 1–15.

Lloyd, S., & Bristol, S. (2006). Modeling mentorship and collaboration for BSN and MSN students in a community clinical practicum. *Journal of Nursing Education, 45*(4), 129–132.

Morrison-Beedy, D., Aronowitz, R., Dyne, J., & Mkandawire, L. (2001). Mentoring students and junior faculty in faculty research: A win-win scenario. *Journal of Professional Nursing, 17*(6), 291–296.

Neary, M. (2000). Supporting students' learning and professional development through the process of continuous assessment and mentorship. *Nurse Education Today, 20,* 463–474.

Price, C.R., & Balogh, J. (2001). Using alumni to mentor nursing students at risk. *Nurse Educator, 26*(5), 209–211.

Ryan, D., & Brewer, K. (1997). Mentorship and professional role development in undergraduate nursing education. *Nurse Educator, 22*(6), 20–24.

Scott, E. (2005). Peer-to-peer mentoring: Teaching collegiality. *Nurse Educator, 20,* 52–55.

Sprengel, A., & Job, L. (2004). Reducing student anxiety by using clinical peer mentors with beginning nursing students. *Nurse Educator, 29,* 246–250.

Sword, W., Byrne, C., Drummond-Young, M., Harmer, M., & Rush, J. (2002). Nursing alumni as student mentors: Nurturing professional growth. *Nurse Education Today, 22,* 427–432.

Critical Thinking

1. What are the positive outcomes of peer-mentoring?

MARGUERITE RILEY, PhD, RN, is an Associate Professor, Southern Illinois University, Edwardsville School of Nursing, Edwardsville, IL. **ARLEEN D. FEARING,** EdD, RN, is an Associate Professor Emerita, Southern Illinois University, Edwardsville School of Nursing, Edwardsville, IL.

Author note—the authors and all *MEDSURG Nursing* Editorial Board members reported no actual or potential conflict of interest in relation to this continuing nursing education article.

From *MEDSURG Nursing,* July/August 2009, pp. 228–233. Copyright © 2009 by Academy of Medical-Surgical Nurses (AMSN). Reprinted by permission of Anthony J. Jannetti, Inc. www.medsurgnursing.net

NCLEX Fairness and Sensitivity Review

The National Council Licensing Examination is an examination that is used for the purpose of licensing registered nurses and practical/vocational nurses. To help ensure that the examination questions (items) are fair and unbiased, all items undergo a fairness and sensitivity review as part of the item development process. This article focuses on the fairness and sensitivity review.

ANNE WENDT, PhD, RN, CAE, LORRAINE KENNY, MS, RN, AND MICHELLE RILEY, DNP, RN

Each year, nursing faculty interface with more culturally diverse student populations and entry-level nurses care for more diverse clients. According to the 2001 US Census Bureau, minority groups will continue to grow.[1] Understanding a client's culture and its relationship to healthcare is critical for all healthcare personnel, whether practicing in a clinical, educational, or research setting.[2] Faculty are in a unique position to administer course examinations in nursing programs that reflect unbiased and culturally sensitive questions (items).

The National Council Licensing Examination (NCLEX) is a high-stakes examination that assesses the knowledge, skills, and abilities of entry-level nurses to protect the public from unsafe practitioners. The item development process used to prepare NCLEX content includes multiple steps to ensure that the questions (items) on the examination are fair and valid for all candidates who take the licensing examination. It is essential that content on the NCLEX is presented in a fair and sensitive manner without bias, so a reliable and valid measurement of nursing ability can be made.

Each item on the NCLEX is reviewed for fairness and sensitivity. This fairness and sensitivity review consists of an analysis of the examination items to eliminate offensive material to allow the candidates to demonstrate their competence fairly. This fairness review is best accomplished with written guidelines that ensure that all aspects of fairness are considered and the reviews are objective.[3] The purpose of a fairness review is to identify and remove any construct-irrelevant factors in items. By removing construct-irrelevant factors, candidates can respond to items in ways that allow appropriate inferences to be made about their knowledge, skills, and abilities. Thus, the fairness review is intended to identify and eliminate aspects of test items that might hinder accurate measurement of a candidate's performance on an examination. It is important to note that the driving force behind a fairness review is to ensure *validity* of the examination, not political correctness.[3]

As part of the item development process, fairness and sensitivity review guidelines were implemented in the 1980s by Educational Testing Service.[4] Educational Testing Service guidelines and procedures promoted a general responsiveness not only to the cultural diversity issues in the United States but also to the changing roles and attitudes toward groups in the society, diversity of background, cultural tradition, and viewpoints found both in the United States and the international test-taking populations. In the sensitivity review, items are evaluated to ensure that "stereotyping and language, symbols, words, phrases or examples that are sexist, racist, potentially offensive, inappropriate, or negative toward any group" are avoided.[4] For the purpose of the NCLEX, fairness and sensitivity reviews are not to be confused with NCLEX-Differential Item Functioning process, which is a separate evaluation of items based on statistical information and expert judgment for potential bias.[5]

Fairness and Sensitivity Review Process

To ensure that fairness and sensitivity are embodied within the NCLEX, the guidelines are given to all participants involved in the development of content for the NCLEX. In addition, after all items have completed the item writing and review stages, they are brought forward to a Sensitivity Review Panel. This panel is created to represent the diverse groups who take the NCLEX in groupings such as ethnicity, culture, sex, age, disabilities, and English as a second language. Using their unique perspectives and expertise, the panel evaluates items to ensure that the items are fair and unbiased. The ultimate goal is to have no group stereotyped, offended, defamed, or patronized in any manner in NCLEX items.

Panel members are not required to have a background in nursing because their focus is not related to the nursing content of the items. However, a nurse serves on each panel to assist in defining specific content or providing explanations about content that may otherwise appear to panelists as potential sensitivity issues. Each panelist is oriented to the sensitivity process at the start of the review session. Once educated in the process, panelists review examination items, asking for clarification as needed. After individual review of the items, the members discuss their findings with the larger group. Any item deemed by the group to contain an issue of sensitivity is "flagged" or identified for further review and forwarded on to the NCLEX Examination Committee for final decision as to the item's disposition.

As stated above, there are specific guidelines used to judge if an item is potentially unfair or insensitive to any group or population. All examination items are reviewed against the following categories, which are applied to both the NCLEX-RN and the NCLEX-PN:

- Inappropriate terminology
- Stereotypes
- Underlying assumptions

- Ethnocentrism and Elitism
- Tone of language/Language
- Inflammatory material
- Age

Using these categories to judge the fairness and sensitivity of an item fosters the recognition not only of the roles and contributions of ethnic and minority groups but also individuals with disabilities and the continually changing roles of women in society.[4]

Inappropriate Terminology

As the world becomes more diverse, content in an item must remain clear, concise, and easy to interpret for all candidates who take the NCLEX. Terminology, accents, and use of "medical" or "faculty" jargon can all affect a clear understanding of the item and problem being posed.[6] To have NCLEX items that are fair, balanced, and neutral, each item must be reviewed for inappropriate terminology. This means that the client in an item is not characterized by his/her disability, disease, religion, or ethnicity, as in ". . . the blind client" or the ". . . alcoholic male with . . . ."[4] Rather, the preferred phrasing would be "the client with a diagnosis of . . ." Also, item writers should avoid the generic use of "he," as in "the nurse became the spokesman . . ." as well as use of specific pronouns such as "he" or "she" unless it is germane to the content.[7] Whenever possible, use the terms *client, spouse,* or *partner* and *child* in lieu of pronouns. Additional inappropriate terminology occurs with the use of idiomatic expressions such as *kick the bucket, running late, submarine or hoagie* (referring to sandwiches) *pop, sod, sushi,* or *cop,* which may be unfamiliar to some candidates.

In writing an examination item, there are several other common terms that can be identified as inappropriate. Item writers should not modify the description of the nurse, such as "new nurse" or "older preceptor," unless it has a specific impact on the question. In addition, using the marital status of the client's mother/father as a "single parent" when not relevant to the content being tested may actually be distracting to the candidate and will not foster a better demonstration of the requisite nursing knowledge.

Stereotypes

The next guideline identifies any actual (or perceived) words that may be viewed as stereotypical in an item. A stereotype is a standardized mental picture, opinion, or attitude that attributes certain features to a group or individual related to race, ethnicity, sexual orientation, sex, age, disability, or religion.[4]

When developing items, it is essential that faculty avoid any statements about a group that may imply that everyone in the group shares a specific characteristic, or any suggestion that one group is culturally or genetically inferior or superior to another. The ability of a candidate to answer an item correctly could be impacted negatively if false or untrue generalizations about a group or groups of people are presented in the item.[8] Failure to remove such content or wording from an item may inadvertently perpetuate a false view and could cause the examinee to answer the item incorrectly or could indirectly impact a candidate's ability to remain objective and neutral when answering the item.

Examples of stereotypes to avoid in items would include the portrayal of specific sexes in a negative manner, the insinuation that all persons in an ethnic group possess certain characteristics, and the view that all persons with disabilities are unproductive or that older persons are senile, incompetent, or dependent.[4] Some examples of stereotyping include the following: (1) suggesting that women are portrayed as overly concerned with their appearance or more intuitive than men

are; (2) African Americans characterized as living only in depressed, urban areas; and (3) people with disabilities distinguished as heroic or victims because of their disability. These are all offensive and unfair for any person viewing an item with such content.[4]

Additional examples of stereotyping that are more subtle to detect occur when minority groups are described as sharing the same basic culture or when clients who are older than 65 years are categorized as elderly. One simple way to evaluate whether an item or phrase in an item is potentially offensive is to substitute the group being discussed with one that the item writer/reviewer relates to strongly. If the passage becomes offensive with this substitution, it probably includes stereotyping.[7]

The inclusion of the income level of a group (eg, poor Hispanic clients) in an item may distract candidates when attempting to choose the correct answer. This may transmit an attitude or generalization about the ethnic group identified in the item. In addition, identifying the ethnicity of clients being taught by the nurse may convey a sense that all persons in this ethnic group lack a basic understanding of health practices. Using specific idioms or slang related to an ethnic group conveys the notion that all people in this ethnic group use the phrase when communicating, which is an unfair generalization. Respecting cultural practices is an important part of the holistic understanding of client situations encountered, but care must be taken so stereotypes are not perpetuated and all clients are cared for in a fair and sensitive manner.

Underlying Assumptions

Underlying assumptions may be considered extremely subtle stereotypes.[4] Underlying assumptions may include subtle ideas such as all elderly are dependent, men with AIDS are homosexual, or anorexia is a disease of only teenage girls. These assumptions require that the candidate see the world with the same perspective as the item writer. It is illogical to insist that people share the same world view, and although it may be unintentional, it is still an unacceptable and unfair assumption. In addition, allowing an assumption because it occurs in an incorrect answer option does not make it acceptable. While not knowing which answer is correct, the candidate may have a personal assumption confirmed by choosing an answer that reinforces the negative stereotype.[4]

Faculty who are developing items should pay particular attention to how an item is written and if the item is supporting an unfair or insensitive underlying assumption about a group or population. Using the underlying assumption, guideline review the following item and identify any underlying assumptions:

When caring for a Buddhist client who speaks little English, the nurse should consider which of the following to be most important?

- Being aware of her own self, values, and biases
- Studying the language and religious practice of the minority group
- Being sympathetic or nondiscriminating
- Working among or close to the minority-group members

In this item and its answer options, there are several underlying assumptions. The religious affiliation of the client and the fact that the client speaks little English are being subtly linked in the item and may not be needed. This information makes the item less neutral and objective, because nurses work with clients from various religious backgrounds who speak many languages. In addition, a nurse would need to be cognizant of his/her own values and beliefs when caring for any client, not just a Buddhist or one with limited English proficiency. The fourth option may perpetuate the assumption that successful nurses work closely only with minority group members.

In addition, the nurse is referred to as "her" in option 1 and the sex is not needed. Overall, the item should be viewed and written from a neutral, unbiased perspective so that performance levels of examinees are not negatively impacted because of a false underlying assumption.

Ethnocentrism and Elitism

Ethnocentrism is where assumptions made about aspects of Western culture are mistaken for universal norms or contribute to misunderstandings of a particular group.[4] Such assumptions may include pain beliefs, death practices, or gestures of disrespect, as well as religious values and traditions. A particular example of this would be using the term *Eskimo* to describe inhabitants of the Arctic region. This term originated in Western culture and became a universal norm, despite the fact that people in these regions call themselves Innuvialuit.[4]

Although the NCLEX does test religious beliefs and practices as they relate to medical care, it would be important to note that nonreligious items should not require Judeo-Christian values and traditions to answer the item appropriately.[4] For example, including the word *bible* in an item may support the religious beliefs or values of one group over another. Also, saying "God bless you" after a client sneezes or offering a bible to a client who is terminally ill can be offensive to clients who do not practice Christianity.

Elitism refers to vocabulary related to socioeconomic status that is unfair to include in an item because candidates and clients are from varied backgrounds and have individualized cultural beliefs and practices. An elitist, patronizing, sarcastic, derogatory, or inflammatory tone is always unacceptable. Elitist terms such as *penthouse, polo,* or *estate* are words that tend to be associated with a specific socioeconomic class and are therefore not considered neutral words to include in test items. When developing an item to assess nursing knowledge, faculty should review items to be certain that ethnocentric or elitist language is not used.

Tone of Language

The tone of a test item reveals the item writer's attitude toward the subject and/or the reader.[4] As stated before, an elitist tone is one that uses terms more likely to be familiar to a particular socioeconomic, ethnic, or geographic group. Elitist terms may include *junk bond, yacht,* or *regatta,* whereas geographic or ethnic term examples may include *soul food, stickball,* or *purse.* Another example of the tone of language that should not be included in an item is one that is condescending to any person or group. Condescending tones may include *lady lawyer, the little woman,* or a *coed* in that these terms imply that women are not included in the mainstream of society. The tone of language, although familiar or acceptable to a particular ethnic group or sex, is not appropriate to include in a test that is written for all groups of people.

Patronizing language, sometimes viewed as complimentary, is also a form of elitist tone. Phrases such as "Native Americans are closer to the earth" or "Buddhism is very meditative or peaceful" are examples of universal or cultural norms assumed by the item writer that may be considered patronizing. Tone of language that uses terms that are overly familiar or condescending (eg, *sweetie* or *honey*) may be not only insulting but also distracting to candidates.

Inflammatory Material

Avoidance of inflammatory material in items is important so that a negative emotional impact will not be perceived by any population or subgroup. If items need to test understanding of material that may bring about negative and subjective, emotionally laden feelings, they should be handled in a fair, balanced, objective, and conscientious manner.[4] In the nursing profession, controversial issues may be difficult to avoid and necessary to test on examinations. The reason for including such issues on a test is their relevance to the course objective or purpose of the test. When information on topics such as abortion, AIDS, or teenage sexuality are required to meet the objectives of the test, it is important to ensure that ethnic, racial, and sex groups are represented proportionately and appropriately. Review the sample item below to identify inflammatory material in the item:

A nurse is caring for a client who had an amniocentesis and has just been informed that the fetus has Down syndrome. The client has unfortunately indicated a desire to terminate the pregnancy. Which of the following actions should the nurse take?

- Tell the client that permission from the spouse must be obtained for an abortion.
- Remain available within the client's room to talk with the client.
- Ask a coworker who has a child with the same diagnosis to talk with the client.
- Provide the client with the telephone number of an agency that researches birth defects.

In this item, the content is appropriate to test; however, the value judgment made with the term *unfortunately* adds to the sensitivity of the diagnosis. Down syndrome and reproductive choices are examples of requisite knowledge for entry-level nursing practice. There is a need to cover controversial and highly sensitive content in nursing curricula, and care should be taken to avoid the inclusion of personal thoughts or feelings in the item, such as *unfortunately*. Using neutral and unbiased words is one way to remain sensitive and fair when presenting content.

In summary, using the categories detailed above to create a fair and sensitive examination, requires the following:

- Ensure that language is clear and concise.
- Review items to determine words or phrases that may be offensive to individuals or groups.
- Avoid using adjectives and adverbs.
- Determine if information included is needed to clarify health issues or medical diagnoses.
- Ensure that value judgments and underlying assumptions are not required to answer the item.

Conclusion

The NCLEX is administered nationally and internationally to candidates from various ethnic backgrounds and cultures. Faculty are in a unique position to administer course examinations in nursing programs that reflect unbiased and culturally sensitive items. By reviewing items for categories of fairness and sensitivity as discussed here, faculty can avoid interjecting bias into test.

The fairness and sensitivity review process is a formal step used in the item development process for NCLEX. Fairness and sensitivity review provides each candidate an equal opportunity to answer all items presented without adding potential bias into the examination. Nursing faculty could benefit from developing and implementing a formal process to review content on their examinations for fairness and sensitivity using the examples provided above. Through the use of fairness and sensitivity review process, contributions of various groups and cultures are supported and respected. Faculty may learn more about writing items in the NCLEX style and may participate in our item development program by accessing the National Council of State Boards of Nursing website at www.ncsbn.org.

References

1. Broome B. Culture and diversity issues: culture 101. *Urol Nurs.* 2006;26(6):486–489.

2. Glazner LK. Cultural diversity for health professions. *Work.* 2006;26:297–302.

3. Zieky M. Fairness reviews in assessment. In: Downing S, Haladyna T, eds. *Handbook of Test Development.* Mahwah, NJ: Erlbaum; 2006:359–376.

4. ETS Fairness Review Steering Committee. *Overview: ETS Fairness Review.* Princeton, NJ: Educational Testing Service; 1998.

5. Wendt A, Worchester P. The National Council Licensure Examination/differential item functioning. *J Nurs Educ.* 2000;39(4):185–187.

6. Davidhizar R, Shearer R. When your nursing student is culturally diverse. *Health Care Manag.* 2005;24(4): 356–363.

7. Ramsey PA. Sensitivity review: the ETS experience as a case study. In: Holland PW, Wainer H, eds. *Differential Item Functioning.* Hillsdale, NJ: Erlbaum; 1993:367–388.

8. Educational Testing Service. Fairness review guidelines. 2003. Available at www.ets.org/Media/About_ETS/pdf/overview.pdf. Accessed February 18, 2009.

Critical Thinking

1. What are four reasons for the fairness and sensitivity review process?

How to Read and Really Use an Item Analysis

A frequent challenge for nursing faculty is to write a test that effectively evaluates learning and prepares students to be successful on the NCLEX-RN examination. Use of item analysis is an approach often used to provide an objective evaluation of examinations. Interpreting these analyses, however, can be frustrating. The authors provide an explanation of the various components of an item analysis, how to make an analysis useful for faculty, and how to use the components of an item analysis in revising tests.

THAYER W. MCGAHEE, PHD, RN AND JULIA BALL, PHD, RN

Writing the perfect examination is an idealized and unrealistic goal for faculty, but one for which to continually strive. Faculty often spend hours writing and rewriting test questions each semester, only to find some new problem or issue with each version of the test. Test banks are frequently used in generating questions, but items are sometimes found to be poorly written,[1] so the frustration continues. Faculty often think they have written an excellent test, but without an objective item analysis it is really impossible to have any assurance of its evaluative competence. Using a computer-generated item analysis can be extremely useful to faculty in their constant endeavor to write the perfect test.

A well-written test serves to confirm that students are appropriately challenged, have a good grasp of the material that was taught, and are prepared to progress. Although many alternative forms of teaching and evaluation are available and frequently used in nursing curricula, the use of multiple-choice tests is the most common mode of objectively evaluating course work. This format is also used for initial licensure examinations, so creating good tests helps to prepare students to successfully take the NCLEX-RN licensing examination. Because licensure is the ultimate goal for a graduating nursing student, faculty have an obligation to prepare them well for this.

It has become common practice for nursing faculty to have an item analysis performed after administering an examination. A standard item analysis report yields a wealth of information, but to many faculty, the data are not meaningful, and therefore, the report is not used. When used appropriately, an item analysis can guide the faculty in revising and improving tests. There is, however, a dearth in the nursing literature on this subject.

What Is an Item Analysis?

Slight variations exist in the statistics used in an item analysis, depending on the software used, but the general elements of the analysis are the same. Item analysis is a process of statistically examining both the test questions and the students' answers to assess the quality of the questions and the test as a whole.[2] The analysis assists in determining the extent to which individual test items contribute to the overall reliability, or internal consistency, of the test. The basic elements of an item analysis include measures of central tendency (mean, median, mode, standard deviation), correct group responses, response frequencies, nondistracters, point biserial, and a reliability coefficient. Looking at the item analysis in this order will provide a clear process to follow and enable faculty to systematically examine tests.

Measures of Central Tendency

Measures of central tendency are among the most basic of all the statistical results specified in an item analysis. The *mean* is simply the average of all individual student scores for a particular test or examination. The *median* is the number at which 50% of all scores on that test fall below. The *range* is the difference between the highest and lowest scores. *Standard deviation* is the measurement of variability. In other words, it is the measure of dispersion of student scores or how much on average scores vary around the mean. Although these statistics are easily understood by most, their interpretation may be somewhat skewed in a population of nursing students. For instance, in upper-level nursing courses, it is not expected that a percentage of students will fail the test or the course. In fact, the further along in the nursing curriculum the students are, the greater the expectation that students will pass. Therefore, while a mean of 80 might look attractive to a professor in a lower-level general education class, it may not be desired for a nursing test. If a mean such as this is obtained, it does not necessarily mean that the test is too hard or the students are not capable. Further investigation is required. It could be that there are 1 or 2 very low grades that are outliers and have skewed the mean.

Item Difficulty/Item Discrimination

In addition to measures of central tendency, other rather simple descriptive elements are in an item analysis. These include the correct group responses, response frequencies, and nondistracters (Table 1).

The correct group responses category is divided into 3 columns. One column gives the total percentage of students who answer an item correctly. This is a basic indicator of item difficulty. The greater the percentage of students answering a question correctly, generally, the easier that question is. By contrast, if a question has a zero percentage of students answering it correctly, that item does not contribute to distinguishing between individual differences among the students and is a question that needs to be revised. If a question has more than 50% incorrect responses, the faculty needs to examine the item and determine whether it needs to be revised or deleted or if there was a coding error.

The next column indicates the percentage of the upper third of the students answering an item correctly. The last of these 3 columns shows the percentage of the lower third of the students answering an item correctly. These last 2 columns of the correct group responses can be very useful. They tell you how the students making the highest grades and the lowest grades on this examination did on a particular question. This analysis is the first step in what is often called *item discrimination*. It would be expected that the upper third would do the best on all test items. However, if the reverse is true and students in the lower third do best on the item, it means that the question was not a good one or was worded poorly, thus misleading or discriminating against the students in the upper group. Any question that favors the lower third of the students and not the upper third needs to be revised. In addition, if an analysis shows low percentages in the upper third of the test takers selecting the correct answer, it warrants looking at those questions and determining if they need to be revised. A professor may have taught the content and the students just missed it, or it could be that the content explanation was not clear. In any event, if the upper third missed a test item, the professor needs to go back and reteach the material. The best test questions discriminate between those students who do well on the examination and those who do not.

The response frequencies are merely a tally of how many students responded to each of the possible answers to a particular question. It also usually indicates with an asterisk the correct response to that question. This is particularly useful when trying to understand why the majority of a class misses a question. For example, if no students answered an item correctly, it may be that the answer sheet was keyed incorrectly. Teacher error in keying is always a possibility. The response frequencies also help to illustrate which distracters are most challenging.

The nondistracters are the potential answers that none of the students chose as correct. When there are nondistracters, it means that the number of plausible answers is more limited than intended. In other words, if there are 4 potential answers in a multiple choice question, and 2 of those are nondistracters, the students were in essence answering a question with only 2 potential answers. Nondistracters are often too easy. All alternative answers should be plausible. It is important to remember that when writing questions for nursing examinations, they should mimic NCLEX-RN questions as closely as possible, so silly or obviously incorrect potential responses should not be used.

Point Biserial

The point biserial is the second and more complete calculation of test item discrimination and is used to judge item quality. It tells how much predictive power a test item has and whether the students who would be expected to answer a question correctly are actually doing so.[2] The point biserial is a correlational calculation determined by the dichotomous variable of student responses to a particular test question (1 = right or 0 = wrong) and the continuous variable of their total score on the overall test. In other words, it is a correlation between item score and total score. This coefficient is an interaction between item discrimination and item difficulty. It is the measurement that illustrates how well an item separates, or differentiates, between those students who answer an item correctly or incorrectly, and have a high or low test score, respectively.[3] This number can range from -1 to $= 1$. Very easy or very difficult test items will have little discrimination. Items of moderate difficulty (60%–80% answering correctly) generally are more discriminating.

The point biserial is designed to reflect the degree to which an item and the examination as a whole are measuring a single attribute topic and will be lower for examinations that measure a wider range of content. The higher the point biserial, the better that examination item is at discriminating among students on the basis of how well they really know the material.[4] A positive biserial indicates that those scoring higher on the test were more likely to answer that question correctly. If the students in the lower third answer an item correctly more frequently than the upper third of the students, the point biserial will have a negative value. This usually indicates that that test item is flawed and should be revised. A low value usually means that the question was too easy. There are no universal guidelines as to what point biserial value is most desirable on a nursing examination, but there are common ranges considered to be acceptable. As a general rule, anything below 0.20 is considered a poor question and in need of revision;

Table 1   ParSCORE Analyses for two 50-Item Examinations

Standard Item Analysis Report On Exam3 Version A

Course #: ANRS 312

Course Title: Pathophysiology

Day/Time:

Instructor:

Description:

Term/Year: Fall 2008

Total Possible Points:	50.00	Median Score:	41.00	Highest Score:	48.00
Standard Deviation:	5.64	Mean Score:	.39.92	Lowest Score:	26.00
Student in this group:	36	Reliability Coefficient (KR20):	0.81		
Student Records Based On:	All Students				

| No. | Correct Group Responses | | | Point Biserial | Correct Answer | Response Frequencies-`indicates correct answer | | | | | | | | | Non Distractor |
	Total	Upper 27%	Lower 27%			A	B	C	D						
1	97.22%	100.00%	90.00%	0.42	B	1	`35	0	0						CD
2	86.11%	100.00%	80.00%	0.21	B	3	`31	1	1						
3	86.11%	100.00%	70.00%	0.54	A	`31	2	2	1						
4	77.78%	100.00%	60.00%	0.30	D	6	2	0	`28						C
5	66.67%	60.00%	80.00%	− 0.16	C	11	0	`24	1						B
6	52.78%	80.00%	50.00%	0.17	D	9	7	1	`19						
7	88.89%	90.00%	70.00%	0.37	A	`32	2	2	0						D
8	88.89%	100.00%	60.00%	0.67	A	`32	3	1	0						D
9	100.00%	100.00%	100.00%	0.00	B	0	`36	0	0						ACD
10	83.33%	100.00%	50.00%	0.52	D	2	1	3	`30						

Standard Item Analysis Report On Exam3 Version A

Course #: ANRS 312

Course Title: Pathophysiology

Day/Time:

Instructor:

Description:

Term/Year: spr 2008

Total Possible Points:	50.00	Median Score:	41.20	Highest Score:	48.00
Standard Deviation:	3.29	Mean Score:	41.10	Lowest Score:	33.00
Student in this group:	39	Reliability Coefficient (KR20):	0.45		
Student Records Based On:	All Students				

| No. | Correct Group Responses | | | Point Biserial | Correct Answer | Response Frequencies - `indicates correct answer | | | | | | | | | Non Distractor |
	Total	Upper 27%	Lower 27%			A	B	C	D						
1	97.44%	100.00%	90.91%	0.30	B	0	`38	1	0						AD
2	92.31%	100.00%	90.91%	0.24	B	0	`37	0	3						AC
3	92.31%	90.91%	90.91%	0.10	A	`36	0	3	0						BD
4	84.62%	81.82%	81.82%	0.06	D	6	0	0	`33						BC
5	53.85%	63.64%	54.55%	0.00	C	12	1	`21	5						
6	58.97%	81.82%	45.45%	0.42	D	11	2	3	`23						
7	84.62%	90.91%	72.73%	0.32	A	`33	6	0	0						CD
8	97.44%	100.00%	100.00%	0.01	A	`38	0	1	0						BD
9	100.00%	100.00%	100.00%	0.00	B	0	`39	0	0						ACD
10	82.05%	72.73%	81.82%	- 0.05	D	5	2	0	`32						C

items with a value between 0.20 and 0.30 are considered fair and could be improved upon, and items between 0.40 and 0.70 are considered good.[2,4,5] However, each question should always be evaluated in terms of the purpose of the test and the purpose of the individual question. For example, there may be a question that is so critical to the knowledge base of the students that the professor desires and expects 100% of the students to answer it correctly. In that case, a point biserial of 0 may be the goal.

Reliability Coefficient

The overall reliability of an examination is analyzed using a reliability coefficient. It may be reported as a Cronbach α or a Kuder-Richardson Formula 20 (KR-20) coefficient. This is a measure of the stability or consistency among the test scores or the internal consistency of a test.[4] The higher the reliability coefficient, the more likely a test will produce consistent scores when administered to similar groups. It is designed to measure how well a test measures a single cognitive factor. A low reliability coefficient may be reflected when a test covers multiple topics. The KR-20 is used for tests that have right and wrong answers. Cronbach α can be used for instruments that have right, wrong, and no right-wrong answer, such as in an attitude survey. For this reason, the KR-20 is most often used in education.

The KR-20 index ranges from 0 to 1, and reflects 4 different things: (1) the total number of test questions, (2) the proportion of the responses to an item that are correct, (3) the proportion of responses to an item that are incorrect, and (4) the variance for that set of scores.[6] Low reliability may mean that the test questions are unrelated to each other in terms of who answered them correctly and that the test scores reflect peculiarities of the test items more than the students' knowledge of the subject. The most common cause of a low reliability score is that the questions are too easy. Other reasons for a low reliability include an excessive number of very difficult items, unclear or poorly written items that do not discriminate, or test items that do not test a unified body of content.[7] A high reliability coefficient indicates that the individual questions on a test tended to pull together to measure 1 topic, and the students who did well overall were likely to answer each question correctly.

Practically speaking, a reliability coefficient of greater than 0.50 can be considered a good coefficient for a nursing examination because most nursing examinations cover multiple concepts and topics. Even if a test does cover multiple areas of content, such as cardiovascular and respiratory systems, the central construct for most examinations is still nursing knowledge, so the KR-20 is useful in analyzing the reliability of the test items relative to this construct. There are several ways to improve test reliability: administer longer tests, have a more heterogeneous group of students, or attempt to change the questions and thus the item difficulty to where 70% to 80% of the students answer test items correctly. However, none of these may be a viable option in nursing courses.

Examples of Test Analyses

Table 1 is an example of 2 different item analysis reports for 50-item examinations. Only the first 10 items are used for illustration. The first example has a high KR-20 of 0.81, indicating strong internal consistency. The mean of this examination is 79.84. The passing grade for this particular course is 80. There was 1 outlier of a grade of 52, which brought the mean down. Point biserials indicate that items 1 to 4 are good questions that discriminate well between the upper third of those scoring on this examination and the lower third. Item 5 is an example of the lower third scoring better than the upper, thus rendering a negative biserial. When this item was examined, it was found to be a knowledge-level question, so it is possible that the upper third of the class overanalyzed the question and the lower third did better because they simply took it at face value. Because the entire class answered item 9 correctly, the biserial is 0. The second example in Table 1 has a lower KR-20, but a higher mean. There is also more variability in the correct group responses.

Table 2 is an example of an item analysis of a 10-item quiz. As would be expected, most of the upper third of the students did very well on this quiz. Eight of the point biserials are above 0.20, indicating good discrimination between the upper and lower thirds of the class.

Now That I Can Read and Interpret an Item Analysis, So What?

The general interpretation of an item analysis can be more difficult when used with nursing students in the upperlevel courses of the curriculum. The typical normal distribution of grades in a bell curve that might be expected in a freshman-level course should not be seen in this population of students. Nursing students in a baccalaureate nursing program should show a positive skewed distribution because they have already completed approximately 3 or 4 semesters of general education and science courses before they are accepted into a nursing program. The nursing major is a rigorous one, and the criteria for acceptance into nursing programs are more stringent than those of many other majors. There is nothing "average" about nursing students. This makes interpreting and using an item analysis even more complex because the basic rules of interpretation may not be valid for this population.

In nursing schools, everything has to be examined in context. A faculty member may administer an examination and have an item analysis that indicates that the test is an excellent one. It may have a reliability coefficient of 0.85, but a mean score of 76. If the cutoff for passing at a particular school is 80 and the mean score on an examination is 76, it means that the average score is not a passing one. This may be an indicator that the material for this examination was not grasped well by the students as a whole and needs to be reinforced or taught in a different way. It may also mean that the students simply did not prepare adequately. A good way to distinguish

Table 2 ParSCORE Analysis for a 10-Item Quiz

Standard Item Analysis Report On Quiz3 Version A

Course #:   ANRS 312
Course Title: Pathophysiology
Day/Time:

Instructor:
Description:
Term/Year: fall 2007

Total Possible Points:	10.00	Median Score:	8.17	Highest Score:	10.00
Standard Deviation:	1.37	Mean Score:	8.17	Lowest Score:	4.00
Student in this group:	30	Reliability Coefficient (KR20):	0.40		
Student Records Based On:	All Students				

No.	Correct Group Responses			Point Biserial	Correct Answer	Response Frequencies - * indicates correct answer								Non Distractor
	Total	Upper 27%	Lower 27%			A	B	C	D					
1	88.67%	100.00%	62.50%	0.55	A	*26	4	0	0					CD
2	86.67%	100.00%	62.50%	0.55	D	2	1	1	*26					
3	86.67%	100.00%	62.50%	0.41	C	1	1	*26	2					
4	93.33%	100.00%	75.00%	0.52	A	*28	1	1	0					D
5	93.33%	100.00%	75.00%	0.23	B	2	*28	0	0					CD
6	40.00%	87.50%	12.50%	0.55	B	17	*12	1	0					D
7	73.33%	87.50%	75.00%	0.35	A	*22	8	0	0					CD
8	100.00%	100.00%	100.00%	0.00	C	0	0	*30	0					ABD
9	70.00%	100.00%	37.50%	0.45	D	3	0	6	*21					B
10	86.67%	100.00%	87.50%	0.12	A	*26	0	4	0					BD

between these 2 causes is to administer the same or very similar test each semester and compare the analyses over time. Item analysis interpretations are usually normed with typical grading scales, so each nursing school, each course, and each examination must be considered when deciding what is acceptable on the item analysis and what indicates a need for a change in a particular examination. Caution should always be used when interpreting statistics based on a small sample size because the results may simply be random chance.

Evaluating examinations is difficult. Faculty members tend to like questions they write and want to think it is strictly the fault of the students if they miss questions. An item analysis is an objective measure to scrutinize examinations more critically and determine which questions really do need to be revised, what material needs to be revisited, and which questions need to be kept. If an item analysis indicates a poor question, it does not mean that the question must be discarded and everyone given credit for the item on that examination. It is most useful to look at the item analysis along with the test blueprint to see the whole picture. Looking at the whole picture is helpful to see where to focus to develop new questions. It is also helpful to have colleagues look at tests and give feedback. An objective outsider will be able to discern awkward wording or poorly written questions.

Statistics from an item analysis are useful in understanding student performance on that examination, but the purpose of the examination should always remain paramount. There are both theoretical/conceptual and practical reasons for writing examination items and designing the test as a whole.[8] Results of an item analysis should always be carefully examined and not used as the sole determinant for rewriting or revising a test. Data from the analyses are always going to be influenced by the number of students taking the examination, the type of students, the variances in teaching, and the inevitable errors attributable to chance. Table 3 gives a brief summary of suggestions of when to revise and when not to revise questions.

When using a software package to do item analysis, it is possible to add results of a current examination to prior examinations to increase the sample size. This is helpful only if examination questions have not been changed. If the questions have been changed, it is helpful to compare the analyses to determine if the changes made the examination better.

Once a faculty member has learned how to read and interpret an item analysis, the analysis can be very helpful both in refining tests and also in indicating what may need strengthening or deleting in the teaching of the course. The analysis helps to find flaws or errors in a test so it can

Table 3   What to Do with Test Questions

Upper Third	Lower Third	Biserial	Nondistracters Present	Action
Correct	Incorrect	+	−	None
Correct	Incorrect	+	+	Revise nondistracters
Incorrect	Correct	−	−	Revise test item
Incorrect	Correct	−	+	Revise test item and nondistracters
Incorrect	Incorrect	+ or −	+ or −	Reteach content

be adjusted before grades are posted. For instance, it may indicate that there are 2 right answers and both should be accepted or that a question was keyed incorrectly and the tests must be rescored. The analysis is also useful in determining which questions are too difficult or too easy so that those questions can be reworded or revised to be more appropriately challenging. When questions are found to have high levels of difficulty, it may be necessary for the material to be stressed more carefully in class or more fully explained. Identifying the common misconceptions of the students by looking at the frequencies of the distracters also helps to determine material that needs to be further clarified in class.

Summary

Writing the perfect examination is a lifelong challenge, but the goal should be continual improvement. Interpreting examination results by use of an item analysis yields a wealth of information that is useful in both improving test items and in improving teaching. Because one of the goals of every nursing school is to have its students pass the NCLEX-RN examination on the first try, the time and effort given to interpreting examinations and using this information are invaluable.

References

1. Masters JC, Hulsmeyer BS, Pike MaryE, Leichty K, Tiller MT, Verst AL. Assessment of multiple-choice questions in selected test banks accompanying text books used in nursing education. *J Nurs Educ.* 2001;40(1):25–32.

2. Matlock-Hetzel S. Basic concepts in item and test analysis. Paper presented at Annual Meeting of the Southwest Educational Research Association; January 23–25, 1997; Austin, Texas.

3. McGill exam results. Available at http://www.mcgill.ca/ncs/products/exams/. Accessed January 14, 2009.

4. Oermann MH, Gaberson KB. *Evaluation and Testing in Nursing Education.* 2nd ed. New York, NY: Springer Publishing Co; 2006:173–176.

5. Introduction to Item Analysis. Scoring Office: Academic Technology Services. Available at https://www.msu.edu/dept/soweb/itanhand.html. Accessed January 14, 2009.

6. Bodner GM. Statistical analysis of multiple-choice exams. *J Chem Educ.* 1980;57(3):188–190.

7. Kehoe J. Basic item analysis for multiple-choice tests. Practical assessment, research & evaluation. 1995;4(10). Available at http://PAREonline.net/getvn.asp?v = 48&n = 10. Accessed January 14, 2009.

8. Brown JD. Questions and answers about language testing statistics. *Shiken: JALT Testing & Evaluation SIG Newsletter.* 2001;5(3):12–15.

Critical Thinking

1. What are the challenges of conducting test item analysis?

THAYER W. MCGAHEE, PhD, RN, is a Assistant Professor, College of Nursing, University of South Carolina Aiken. **JULIA BALL**, PhD, RN, is a Dean and Associate Professor, College of Nursing, University of South Carolina Aiken.

From *Nurse Educator*, July/August 2009, pp. 166–171. Copyright © 2009 by Lippincott, Williams & Wilkins/Wolters Kluwer Health. Reprinted by permission via Rightslink.

UNIT 9

The Profession and Professionalism

Unit Selections

Learning Outcomes

After reading this unit, you should be able to:

- Discuss how nurses can combat the age-old problem of lack of respect.

- Describe the four generations of nursing today.

- Discuss how hospital restructuring in the 1990s resulted in the lay-off of thousands of nurses. Relate this to the nursing shortage facing the profession today.

- List the three essential attributes of professional nursing practice.

- Identify several things an applicant can do to prepare for a job interview for a position in nursing.

- Describe some of the emotional and psychological issues experienced by nurses today.

Student Website
www.mhhe.com/cls

Internet References

The American Nurses Association
www.nursingworld.org/
National Center for Health Workforce Analysis
http://bhpr.hrsa.gov/
The National Council State Boards of Nursing
www.ncsbn.org/
University of North Carolina Chapel Hill
http://nursing-research-editor.com

Nursing has been referred to as a calling, a ministry, a career, and a job. No matter what term is used, nursing needs to be considered a profession and its members need to carry out their duties in a professional manner. A profession is an occupation or career that requires considerable training and specialized study and is also the body of knowledge that qualifies persons in an occupation or field. The primary definition of professionalism is that membership in a profession carries with it a set of internalized values that will be reflected in the way work is carried out and in the individual's adherence to ethical standards. As an example, medical professionalism would require that doctors prescribe the most appropriate treatment for their patients rather than one that would yield higher fees or would meet an externally imposed objective of cost control. Nursing has worked for the last 150 years to become a profession in its own right. For decades, nurses were considered to be physician's helpers. They were trained in the medical model and carried out the physician's orders without question. Eventually, nursing developed its own theories and diagnoses; the profession has come of age. The nurse is still expected to follow the physician's orders, but today the nurse is also expected to critically think and advocate for the patient. The nurse is even expected to evaluate care and treatment and question the physician when it is in the best interest of the patient. The nursing workforce today is made up of four generations; nurses 59 and older to recent graduates in their early twenties. Differences need to be recognized and strengths maximized if all the nursing generations are to have a peaceful coexistence.

This unit looks at some of the issues facing the profession today. The dwindling supply of new nurses to meet the healthcare needs of the aging population can no longer be ignored. The nurse must have more knowledge and skills to care for the patient holistically—body, mind, and spirit. Technical skills must be continually upgraded to make use of the new, highly specialized equipment. Every day 5 to 10 drugs are introduced to the marketplace and the nurse must understand the side effects, interactions, and uses of each. Nurses must also have the ability to teach the patient or patient's family how to administer drugs or treatments. Because hospital stays are shorter, family members are called upon to provide follow-up care at home. The nurse must be able to prepare them for this task. Although the majority

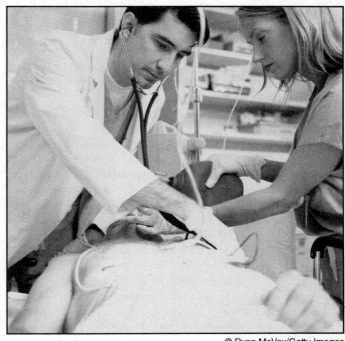

© Ryan McVay/Getty Images

of nurses are still employed in the hospital setting, they must be equipped to provide care in a variety of settings. Hospitals are facing other crises besides the nursing shortage and nurses must be aware of how these other issues may impact their ability to do their job. Hospital restructuring can have a negative physical and emotional effect on the nursing workforce. The stress of being short-staffed, or working with other nurses who are suffering from burnout, can have a significant impact on the nurse. Self care is important if the nurse is going to carry out the responsibilities of the profession and behave in a professional manner. Burnout should never be an option or an excuse for the nurse. If burnout is evident, a change should be instigated. This unit provides tips for job interviewing that can be valuable for a new graduate, a nurse returning to the workforce, or someone who is ready for a change.

Nurses Step to the Front

In hamlets and high-tech hospitals, nurses are taking on bigger roles.

SAMANTHA LEVINE AND ANGIE C. MAREK

When white-haired Harry Curry shuffles into the Minnie Hamilton Health Care Center in rural Grantsville, W.Va., he says he'll see only "his doctor." That's his name for Teresa Ritchie, the nurse practitioner who looks after the 71-year-old at this tiny complex tucked in the Appalachian hills. And it's not really a misnomer. The veteran nurse takes on everything from minor surgery to emergency room crises. Ritchie has admitting privileges at area hospitals, still unusual for a nurse, and can prescribe medication with just a doctor's checkoff. Her autonomy surprises her as much as anyone. "When I started out, nurses were not told we could think for ourselves," says the West Virginia native, who delivers care with a dollop of down-home gossip and finds ways to give her predominantly low-income patients free medicine. "We just did what a doctor planned out for us."

But those times are long gone. Many of the country's more than 2 million nurses are taking on jobs that were once the purview of physicians, like administering chemotherapy and running their own primary-care practices. They are carving new niches in fields such as genetics and computerized patient records, where nurses were once hard to find, and bringing philosophies oriented toward health promotion and problem prevention to geriatric care and case management. "When we are allowed to think outside the box, there is a lot we can do," says Jane Barlow, a University of North Carolina nurse who is developing a disability screening and intervention system for children in her home state. "In every situation, there is more that nurses can do if they feel empowered."

The seeds of nurses' liberation from doctors' white coattails were sown in the 1960s. That's when a nationwide shortage of primary-care physicians, especially in rural and inner-city areas, pushed many nurses into advanced roles. "Nurses were doing things that most people thought just physicians were doing, like seeing patients and recommending medication," says Lynne Vigesaa, who in 1972 became the first nurse practitioner in the state of Washington and later helped write the state's regulations defining and governing that role. Through the 1980s, the idea of nurses' doing more than just assisting doctors gained acceptance as patients began seeking out nurses—who seemed to have more time for them—and resistance from physicians' organizations eased. States began formalizing nurses' expanded roles. "It's pretty amazing, when you think about it," says Vigesaa, who now manages a dermatology clinic outside Seattle.

Filling a void. The advent of managed care also opened doors. The law defining health maintenance organizations was passed in 1973. But the idea—prepaid plans that enroll members and arrange their care from a designated network of doctors—brought on serious complications. The plans often tightly control how, when, and why doctors offer medical services. The reimbursements doctors get from the plans are prearranged but haven't all risen in step with the ballooning cost of healthcare. That problem now extends beyond HMO s to all insurance plans, doctors complain, which means they must see ever increasing numbers of patients to remain profitable. Those busy schedules have created voids in patient care—and nurses are filling them. "When patients call doctors for advice these days, many times nurses are the ones at the other end of the line," says Patricia Rowell, a senior policy fellow for the American Nurses Association.

Over time, as the hours and roles of medical residents and interns changed, many nurses with master's degrees who specialized in fields such as pediatrics were called upon to perform tasks once reserved only for the med-school set. Rowell, a pediatric nurse practitioner, remembers hearing of her colleagues in infant intensive-care units being allowed to insert breathing tubes down the throats of delicate babies. "It was an incredible thing," she says. These trends are continuing today as residents' exhausting schedules are further cut back—for reasons of patient safety.

As nurses are asked to do more, they are also trained to do more. In 1980, 60 percent of nurses received their basic education through on-the-job training courses in hospitals. That number was cut in half by 2000. During that same time, the share of nurses earning associate's degrees more than doubled, to 40 percent. The number of nurses pursuing master's degrees and doctorates has tripled over the past two decades—by 2000, one in 10 registered nurses had made the leap. And the number of doctoral programs nationwide has grown from 52 in 1990 to 93 today. By 2015, the American Association of Colleges of Nursing wants all nurses doing advanced practice work that now requires at least a master's degree—this includes nurse practitioners, clinical nurse specialists, nurse midwives, and nurse anesthetists—to hold a doctorate of nursing practice.

Nurses already have rigorous training. Most undergraduate nursing schools require students to take a variety of courses, from statistics to biology, before they can even enter the nursing program. Once in a program, students' classes include anatomy and ethics, and they must complete several practicums. After earning an undergrad degree, every student must pass a nationally standardized test before officially becoming a nurse.

Rules governing what nurses are allowed to do with this training vary by state. "You sometimes push for 15 years, so hard, just to take one baby step," says Marie-Annette Brown, a nursing professor at the University of Washington who helped lobby to give nurses in her state more power to prescribe medication. Before such efforts, nurses could not prescribe controlled substances like morphine without doctors' supervision. That's still the case, though, in 37 states that demand a doctor's sign-off. But 27 states allow nurses to open their own private practices without a doctor in the house. Many nurses still have trouble, however, persuading insurance companies to reimburse them for their work.

Whole-patient care. But more than the ability to prescribe drugs, nurses are pushing to practice a breed of care that bears their unique imprint. "More so than doctors, we focus on health promotion, the strategy of teaching our patients how to live healthier lives," Brown explains. This means helping patients manage their symptoms and chronic conditions and avoid health pitfalls like poor diets. "If I have a child with diabetes, I try to teach him to self-regulate his condition," says Rowell. "But I also tell him to . . . engage in after-school activities; don't be afraid to live a normal life."

Looking at the whole patient is critical for oncology nurse Ann Welsh. As a senior nurse in the chemotherapy division at the University of Pittsburgh Medical Center's Hillman Cancer Center, she is doing more than she could have imagined when she graduated with an associate's degree in nursing from Northern Virginia's Marymount University in 1973. At that point, she was given a little white nurse's cap and went to work on the National Cancer Institute's pediatric ward. "Early on, I was the liaison between the doctor and the patient," she explains. But through the 1980s, as more cancer treatment moved to an outpatient basis, her responsibilities grew to coordinating most aspects of her patients' therapy regimen. With her understated manner, Welsh, 51, gently administers chemo to hundreds of patients. Though she closely collaborates with Hillman's oncologists, there's nary a doctor in sight while she sets up complicated webs of intravenous drips for folks like Mario Urlini, a hearty 83-year-old who's getting his last treatment for non-Hodgkin's lymphoma. While Welsh connects the clear tubes, Urlini's wife, Grace, looks on and says: "I went through two major operations, and I never found a nurse like Ann."

Welsh also trouble-shoots problems with patients, such as sorting out whether an older man with worrisomely low blood counts can undergo needed cataract surgery. Patients call Welsh, not their physicians, with day-to-day concerns. "I know them inside and out, so I can assess if there is a big change," she says. "We can take better care of our patients if we use our own judgment, provided we know them and fully understand the course of their disease." As for that nurse's cap? "I wore it for a few months," says Welsh. Then she threw it out.

Balancing act. The ability of nurses to balance complex elements of long-term care while acting as counselors has made them a good fit as case managers. They work as the primary contact for patients, like the elderly, who have complex treatment plans. They can help eliminate frustrating redundancies among various doctors' offices and keep a constant record of detailed medical regimens. In San Francisco, gerontology expert and nurse Monika Pettross started her own case-manager business four years ago. One year into the venture, she began working with an elderly woman whose family found a suicide note in her home. Pettross, 32, discovered that some of the patient's medications could badly interact and lead to depression. After suggesting changes to the woman's prescriptions, she helped place her in a therapy program and counseled the family on how best to offer support. After two months, the woman was looking for a part-time job. Pettross still works with her today. "This work is so important I can't take off a badge in my free time and walk away from it," says Pettross.

Pettross is on to something. As people live longer and the nation's 78 million baby boomers approach retirement, more nurses are taking on elder care. The federal National Institute of Nursing Research is devoting millions of research dollars to the topic. One project explored the effect of delivering education and follow-up care at home to older patients hospitalized for heart trouble, says institute director Patricia Grady. Home delivery decreased the number of visits to the hospital, saving $4,845 in Medicare expenses per patient.

Nurses are also walking the cutting edge of health technology. They are at the vanguard in shifting hospitals away from mountains of problematic paper patient charts to automated, computer-based systems that cut down on delays and errors, says Scott Young, director of health information technology at the federal Agency for Health Research and Quality. About an hour south of Pittsburgh at Uniontown Hospital, Chief Nursing Officer Rebecca Ambrosini is one of those pioneers. Tired of 40-plus-page medical charts that "were never where we needed them when we needed them," she helped make a monumental change. After two years of work, nearly 85 percent of Uniontown's patient information is now available in a user-friendly desktop program called PowerChart. Only about 13 percent of U.S. hospitals have done anything like this.

The Uniontown system not only records patients' vital stats, like allergies and blood pressure, but also automatically sends doctors' orders to nurses, organizes reminders on necessary lab work, and sets up schedules free of hassles like double booking. The program also fires off automatic requests for help from social workers and even interpreters if one is needed. "Before, you had to pick up a phone and call for this assistance—if you remembered," says Darlene Ferguson, a critical-care nurse who is now Uniontown's director of clinical informatics and runs the system on a day-to-day basis. "We would grab a paper towel, write down what we needed, and stuff it in a pocket."

New ventures. At first, this high-tech onslaught made some Uniontown nurses a little nervous. "Some people talked about quitting" because they were worried about using computers, says registered nurse Donna Martin. But Ambrosini and Ferguson ran hours of workshops to get everyone up to speed. Now, wireless PowerChart workstations on wheels line the beige hallways at the 230-bed facility. Nurses roll the computers into patients' rooms to take information and update charts with a keyboard and mouse rather than paper and pencil. There also are computer stations that are built into the walls and fold up like Murphy beds.

Nurses are also gaining ground in one of the fastest-growing areas of medical practice: genetics. According to Grady, of the nursing research institute, it makes sense for nurses to be involved in this fast-moving field because while genes may put patients at risk for health problems, changes in lifestyle and habits—a topic close to nurses' hearts—can help mitigate them. One recipient of the institute's funding is Lorraine Frazier, who holds a doctorate in nursing, a postdoc in genetics, and a teaching position at the University of Texas at Houston College of Nursing. Several years ago, when she first told a molecular medicine researcher that she wanted to work with him, "he was surprised," she says. "No nurse had ever asked to do that. Then, he asked me what nurses do." Now he knows. Frazier runs studies looking at how genetics shape the risks faced by patients with unstable heart disease; the results will eventually help tailor treatments for these patients.

For all its advances, nursing is still bedeviled by one old problem: lack of respect. When it comes to funding for nursing research, for instance, the money is a trickle. The nursing institute's budget makes up just 0.5 percent of the overall research pot at the National Institutes of Health. And according to a recent survey from VHA, a Texas-based healthcare cooperative, disrespect often takes a more direct, virulent form: abuse by doctors. Hospital nurses report vicious arguments. This toxic environment causes a breakdown in communication that can adversely affect care.

Down in Grantsville, nurse practitioner Ritchie hopes that the research that is going on will not just win more respect one day but will also help her patients with chronic illnesses. In her green cargo pants, black turtleneck (from which hangs her ever present stethoscope), and lug-soled boots, Ritchie ducks from crowded office to examining room to the "extra medicine" supply closet as country music plays on a boombox. She tells Harry Curry she'll go to Wal-Mart and buy him more of the salve he likes to rub on his dry, scaly shins. Then she tells another patient, Delberta Hickman, that it's OK to trust in Jesus to heal her diabetes but that medicine can help, too. She gives a 15-year-old girl free antibiotics and an enormous stack of condoms. "I plan to stay here," says Ritchie. "I couldn't leave my patients."

It's a simple thing really, says Uniontown's Ferguson: "A nurse always wants to make things better."

Critical Thinking

1. How can nurses combat the age-old problem of lack of respect?

2. What are the four generations of nursing today?

Mitigating the Impact of Hospital Restructuring on Nurses

The Responsibility of Emotionally Intelligent Leadership

GRETA CUMMINGS, LESLIE HAYDUK AND CAROLE ESTABROOKS

A decade of North American hospital restructuring in the 1990s resulted in the layoff of thousands of nurses, leading to documented negative physical and emotional health effects. Nurses who remained employed experienced significant reductions in job satisfaction and quality of care (Cummings & Estabrooks, 2003; Sochalski, 2001). Research results suggested that the negative impact on nurses carried over to patients in the form of reduced quality of care and increased patient mortality (Aiken, Clarke, Sloane, Sochalski, & Silber, 2002; Blegen, Goode, & Reed, 1998; Kovner & Gergen, 1998; Needleman, Buerhaus, Mattke, Stewart, & Zelevinsky, 2002). What remained unclear was whether all nurses experienced the effects of hospital restructuring to the same degree, or whether nurses working in environments that reflected emotional competence by the nursing leadership experienced reduced effects.

Daniel Goleman and his colleagues have written extensively on emotional intelligence (EI), recently asserting that while leadership attributes include analytic intelligence, task completion, and organizational skills, the primary role must extend to effectively responding to their own and other' emotions (Goleman, Boyatzis, & McKee, 2002). They claimed that the most effective leaders were those with high EI, who portrayed resonant leadership. This leadership reflects the art of hearing their workers' negative feelings yet responding empathically. In times of change and even chaos, an effective leader needs to be empathic and supportive, and demonstrate a wide range of EI competencies (Goleman et al., 2002). Empathy, or the ability to comprehend another person's feelings and to reexperience them oneself, has been reported as a central component of EI (Salovey & Mayer; 1990) and the key to successful resonant leadership (Goleman et al., 2002). Empathic leaders are attuned to a wide range of emotional signals, allowing them to sense the felt, but unspoken, emotions in another person or group (Goleman et al., 2002; Wolff, Pescosolido, & Druskat, 2002).

- *Background:* A decade of North American hospital restructuring in the 1990s resulted in the layoff of thousands of nurses, leading to documented negative consequences for both nurses and patients. Nurses who remained employed experienced significant negative physical and emotional health, decreased job satisfaction, and decreased opportunity to provide quality care.
- *Objective:* To develop a theoretical model of the impact of hospital restructuring on nurses and determine the extent to which emotionally intelligent nursing leadership mitigated any of these impacts.
- *Methods:* The sample was drawn from all registered nurses in acute care hospitals in Alberta, Canada, accessed through their professional licensing body ($N = 6,526$ nurses; 53% response rate). Thirteen leadership competencies (founded on emotional intelligence) were used to create 7 data sets reflecting different leadership styles: 4 resonant, 2 dissonant, and 1 mixed. The theoretical model was then estimated 7 times using structural equation modeling and the seven data sets.
- *Results:* Nurses working for resonant leaders reported significantly less emotional exhaustion and psychosomatic symptoms, better emotional health, greater workgroup collaboration and teamwork with physicians, more satisfaction with supervision and their jobs, and fewer unmet patient care needs than did nurses working for dissonant leaders.
- *Discussion:* Resonant leadership styles mitigated the impact of hospital restructuring on nurses, while dissonant leadership intensified this impact. These findings have implications for future hospital restructuring, accountabilities of hospital leaders, the achievement of positive patient outcomes, the development of practice environments, the emotional health and well-being of nurses, and ultimately patient care outcomes.
- *Key Words:* emotional intelligence • hospital restructuring • leadership • structural equation modeling

The current study sought to determine the extent to which EI (resonant) leadership mitigated the detrimental consequences of hospital restructuring to nurses.

Relevant Literature and Research
Leadership and Emotional Intelligence

Emotionally intelligent leaders inspire by engaging emotions, passions, and motivations that reveal the possibility of achieving goals that might not otherwise be seen. They work through emotion to mobilize teams, coach performance, inspire motivation, or create a vision for the future. Goleman and colleagues' (2002) view of EI has been based on four domains, self-awareness, self-management, social awareness, and relationship management, each consisting of several competencies. The self-awareness and self-management domains reflect *personal competence* in understanding and managing personal emotions. *Social competence* (the ability to develop and manage relationships with other) is composed of social awareness and relationship management domains. Goleman and colleagues reframed these EI competencies to reflect *leadership* competencies, using them to describe and distinguish six leadership styles. Four leadership styles (visionary, coaching, affiliative, and democratic) were termed *resonant* because they demonstrated high levels of EI, and two styles (pace setting and commanding) were *dissonant* because they failed to demonstrate EI. Resonant leaders' messages are tuned to their own and others' feelings as they build harmony and positive working climates. Dissonance, an unpleasant, harsh experience in both musical and human terms, references a lack of harmony and being emotionally "out of touch" with employees. Dissonant leadership undermines the emotional foundations that support and promote staff success (Goleman et al., 2002).

In this study, Goleman and colleagues' EI competencies were matched to theoretical descriptions of their six leadership styles.[1] For example, Goleman and colleagues (2002) described visionary leadership as being able to move people toward shared dreams, having empathy for and developing relationships with others, sharing knowledge to empower other to innovate, having integrity (transparency), and continually reminding people of the greater purpose of their work. These characteristics were required to be present before a nurse in this study was classified as working in a visionary leadership environment. Likewise, as pace-setting and commanding leadership is not known for empathy or developing others, these latter characteristics were required to be absent in environments defined by dissonant leadership. Goleman and colleagues' descriptions of coaching, affiliative, and democratic styles also informed the determination of environments reflecting their requisite characteristics. Coaching leadership focuses primarily on developing others and achieving high levels of individual performance, affiliative leadership builds strong relationships, and democratic leadership builds consensus and promotes innovation, teamwork, and collaboration.

Whether the most appropriate means of assessing leadership competence is through direct ability testing, self-assessment, or by the perceptions of those actually working for the leader has been discussed broadly. Most researchers have concluded that having actual workers rate their leaders provides the best construct validity (Bass & Avolio, 1991; Dasborough & Ashkanasy, 2002; Dunham, 2000; Kellett, Humphrey, & Sleeth, 2002; Xin & Pelled, 2002). This tradition was followed when nurses' responses to specific survey questions were analyzed in this study.

Moderating or Mitigating Action of Leadership

While considerable literature has supported the notion that EI contributes to effective leadership (Freshman & Rubino, 2002; McColl-Kennedy & Anderson, 2002; Robbins, Bradley, Spicer, & Mecklenburg, 2001; Snow, 2001), the literature on leadership as a moderator of effects has reported mixed results (de Vries, Roe, & Taillieu, 2002; Gavin & Hofmann, 2002; Pirola-Merlo, Haertel, Mann, & Hirst, 2002; Villa, Howell, Dorfman, & Daniel, 2003). Research on the role and responsibility of leadership in mitigating (lessening in force or intensity) the consequences of massive organizational change to employees was not found. Recent doctoral studies investigating EI and leadership among nurses have not considered leadership as partially mitigating the effects of hospital restructuring on nurses (Graves, 2000; Molter, 2002; Tjiong, 2002; Vitello-Cicciu, 2002).

A decade of restructuring and downsizing placed prolonged pressures on the nurses of Alberta and provided an opportunity to examine whether various styles of leadership differentially protected nurses from the effects of budget cutting.

Methods
Survey and Sample

The Alberta Nurse Survey of Hospital Characteristics (Giovannetti, Estabrooks, & Hesketh, 2002), one provincial component of the Canadian portion of the International Survey of Hospital Staffing and Organization of Patient Outcomes (Aiken et al., 2001), was used for this analysis. The survey, completed in 1998, reported on various organizational attributes and the state of Alberta nurses' physical and emotional well-being (Giovannetti et al., 2002). The survey had 139 questions in seven categories (employment characteristics, nursing work index, burnout inventory, staffing, details of the last worked shift, quality of care, demographic characteristics), and questions that addressed restructuring, violence in the workplace, and the use of information resources (Giovannetti et al., 2002). Alberta was the second Canadian province to undergo regionalization of its health authority structures in the mid-1990s, collapsing 283 hospital and health care boards into 17 regional health authorities. This, along with other initiatives in health care reform, led to the layoff of thousands of nurses in Alberta throughout the 1990s (Maurier and Northcott, 2000). The Alberta portion of the Canadian Nurse Survey was used because it was the only province in which specific questions were asked about local hospital restructuring, including how many times each nurse was laid off or was required to change nursing units as a result of hospital restructuring.

The study sample was drawn from all registered nurses working in acute care hospitals in Alberta, Canada. Nurses were invited to participate through their professional licensure association. The final sample included 6,526 nurses (53% response rate); demographic comparisons showed no significant difference from the acute care nursing population (Giovannetti et al., 2002).

Data Sets

Seven data sets were created, each consisting of data from nurses who practiced in environments reflecting different leadership styles. There were 13 questions chosen from the nurse survey to reflect EI leadership competencies and sort each case into one or more data sets.[2] Thirteen leadership competencies were used because few survey questions captured how leaders managed their own emotions (*personal competence*). Goleman and colleagues (2002) argued that only six to eight EI competencies were common even for successful leaders. Each of our resonant and dissonant leadership styles was defined by six to eight competencies. The required *presence* or *absence* of each competency was determined to fit each leadership style. Nurses had indicated the degree to which each particular statement described their current work environment using a 4-point Likert-type scale (from *strongly agree* to *strongly disagree*). By reporting on the presence or absence of a variety of work environment features, nurses had provided information that identified the styles of their nursing leaders. Therefore, *presence* of a leadership competency in the nurse's work environment was identified by the nurse's response of "agree" or "strongly agree" that the specific statement described their work environment. A nurse's response of "disagree" or "strongly disagree" was taken as *absence* of evidence of that leadership competency in their work environment. A nurse's survey data were included in the data set describing a specific style of leader (eg, visionary) if that nurse reported both the presence of all the leadership characteristics required by that style and the absence of all the characteristics contraindicating that style. The cases that fit the four resonant styles were selected out of the main data set and transferred into the four resonant data sets. Then the cases that fit the two dissonant styles were selected out and all remaining cases were placed into the Mixed Leadership Styles data set. Leadership Empathy, as indicated by "administration that listens and responds to employee concerns," was the competency that differentiated the dissonant and resonant groups.

Information on the supervisory environments was not complete enough to unambiguously classify the leadership environment of some nurses as reflecting only one leadership style. However, consistent with Goleman and colleagues' concepts, leaders portray different styles depending on the situation at hand—visionary when inspiration is called for and democratic when consensus team building is needed. Therefore, if a nurse's response pattern was compatible with the characteristics of two different leadership styles, that nurse's data was included for analysis in both data sets. Such multiple classifications appeared only within the resonant leadership styles or within the dissonant styles, not between them. Fortunately, the major differences in outcomes appear between the three distinct groups of resonant, dissonant, and mixed leadership styles; therefore cross-classifications between the groups did not confound the important results. Differences in the means and standard deviations for each variable across all seven styles

confirmed that each data set reflected a different population of nurses.[3] The sample sizes for each of the six leadership styles ranged from 699 to 1,065. Nurses who worked in environments that reflected leadership styles other than the four resonant or two dissonant styles were placed together in a seventh group of "mixed" leadership styles. Initially, over half ($N = 3,868$) of the entire Alberta survey data set were classified into this group. In order to ensure that the statistics sensitive to sample size were at least relatively comparable, 1,065 cases were randomly selected from the mixed leadership data set for analysis.

Model Development

A theoretical model was developed that portrayed causal relationships between hospital restructuring (background causal variables) and effects on nurses (outcome variables), and that used the results of a systematic review of the research literature to determine these relationships (Cummings & Estabrooks, 2003). The causal relationships among the outcome variables were derived from the literature and the primary author's leadership experience during hospital restructuring in a large tertiary care hospital. The key elements of the model are described below, and the theoretical model underlying the research is posted at the Editor's Web page (http://nursing-research-editor).

The key elements of the model are described below, and the theoretical model underlying the research is posted at the Editor's Web page (http://nursing-research-editor).

Causal Variables (Hospital Restructuring)

The causal variables included the number of restructuring events occurring in the hospital, being laid off in the past 5 years, changing units in the past 5 years, along with several demographic variables (years worked in a hospital, parttime/full-time status, and age). Gender was included as a control variable, that is, as having no effect. Summing positive responses to seven questions that asked whether specific hospital restructuring events, such as loss of the senior nursing position without replacement or an increase in the number of patients assigned to each nurse, had occurred in their hospital was used to derive the number of restructuring events experienced by each nurse. The number of times that each nurse reported being laid off or changing units, the number of years worked in that hospital, and the nurse's age were entered as reported. Work status was coded 1 for part-time and 2 for full-time.

Nursing Outcome Variables

The nursing outcome variables included nurse reports of freedom to make important patient care decisions, emotional health, satisfaction with time to spend with patients, teamwork between physicians and nurses, nursing workgroup collaboration, satisfaction with supervision, satisfaction with financial rewards, job mobility options, job security, and job satisfaction. The degree to which each feature was perceived to be present in the respondent's workplace had been answered on a 4-point Likert-type scale. Intent to quit was measured on a 3-point scale. Nurses had recorded their degree of emotional exhaustion, psychosomatic symptoms, and professional efficacy on 6-point scales from *never* (0) to *every day* (6). An important variable, unmet patient care needs (a proxy measure for quality of care used by Sochalski [2001]), was derived by summing the number of patient care tasks (maximum of 8) that were deemed necessary by the nurse but were left unattended by the end of the last worked shift.

Each concept in the theoretical model was indexed to a single indicator from the nurse survey. On the basis of the authors' judgment of how accurately the specific indicator reflected the corresponding underlying latent concept, an adjustment was made for the quality of each indicator by assigning 2–30% of its variance as error. The percentages of measurement error were determined by carefully examining how closely each latent variable in our theoretical model was being measured by its indicator in the data sets.[4] Thus, a compensation for problematic wordings, lack of clarity in some questions, and other measurement concerns was made. Pairwise covariance matrices were created because listwise deletion would have resulted in the loss of too many cases.[5]

Model Estimation and Testing Results

The same model was estimated (using Lisrel 8.20 maximum likelihood estimation; Jöreskog & Sörbom, 1996) for each of the seven leadership style data sets. The chi-square (X^2) for the seven models ranged from 205.21 ($p < .001$) to 340.26 ($p < .001$) and the adjusted goodness of fit index (AGFI) ranged from .928 to .945, indicating substantial inconsistencies between the models and the data sets (Hayduk, 1987). The seven models were examined carefully to locate model modifications that were theoretically tenable and that could be made uniform across the seven leadership models. Three criteria were used: the change had to be reasonable theoretically; the modification indices for the relevant coefficients had to be greater than 7 in three or more models, or greater than 10 in two or more models; and reciprocal effects that would have resulted in underidentified models were avoided.[6] Modifications were applied to all models, not merely those with the substantial modification indices. This consistency reduced the likelihood of capitalizing on chance sampling fluctuations that might have existed across the seven data sets. If an effect truly was not required, it would merely lead to a null coefficient estimate and hence would not harm the model, but would cost a degree of freedom. Seventeen additional coefficients were added to the model for estimation using these decision rules.

The large decrease in chi square after these modifications (Table 1) indicated a substantially improved, but not completely acceptable, model fit (Jöreskog & Sörbom, 1996; Hayduk, 1987). The risk of biased estimates due to model misspecification had to be balanced against the risk of bias resulting from inserting "effects" corresponding to chance sampling fluctuations in the covariances. The requirement of inserting changes consistent across the seven models had provided substantial protection against bias due to sampling-fluctuation–induced controls during the numerous changes that had already been made, but the risk of improper-control biases would have increased markedly had unique changes to each model been applied. Most of the models were only one or two modifications away from chi-square fit, but the likely small size of the estimates and the markedly increased risk of improper control bias led the authors to believe that the estimates in the slightly ill-fitting models were the best estimates attainable.

Analysis

The analysis occurred in two stages. Initially, the estimated coefficients for the effects of hospital restructuring on nursing outcomes for each leadership style (effects within each leadership group) were analyzed. Then the impact of leadership styles on the nursing outcome variables (i.e., differences among leadership groups) was analyzed.

The Impact of Hospital Restructuring on Nursing Outcomes

The direction and significance of 50 effects of hospital restructuring on nursing outcomes are reported in Table 2. The results of model estimation included 16 effects that were significant, of which 6 were significant in all seven leadership styles. The two largest and most significant effects were found in seven leadership styles: (a) the direct relationship between the number of hospital restructuring events and the reported number of patient care needs left unattended, and (b) the direct relationship between full-time status and increased emotional exhaustion in nurses. Greater numbers of hospital restructuring events also led to nursing reports of greater emotional exhaustion, deterioration in emotional health, and disruption to workgroup collaboration. The only differential effects between full-time and part-time nurses were that full-time status not only led to more emotional exhaustion but also more satisfaction with their job and time to spend with patients, than did part-time status. The number of times that nurses changed units had no significant effect, and the number of times that nurses were laid off resulted only in the perception of decreased job security.

The direction and significance of 63 causal relationships among the outcome variables are reported in Table 3. Thirty-nine effects were significant, and, of these, nine were significant in all leadership styles. Emotional exhaustion led to the greatest number of significant outcomes, including more psychosomatic symptoms and unmet patient care needs, and deterioration in emotional health, satisfaction with financial rewards, and job satisfaction. Job security and satisfaction with supervision improved nurses' emotional health. Professional efficacy and freedom to make important patient care decisions led to fewer patient care needs being left unattended. The more the patient care needs that were not met, the lower the nurses' satisfaction with the time to spend with patients, which further

Table 1   Fit of the Initial and Final Models

	Visionary	Coaching	Affiliative	Democratic	Mixed	Pace Setting	Commanding
Theoretical model							
Chi square	232.954	259.840	205.208	295.731	340.258	253.020	273.417
Significance	$p < .001$	$p < .001$	$p < .001$	$p < .001$	$p < .001$	$p < .001$	$p < .001$
Degrees of freedom	113	113	113	113	113	113	113
AGFI	0.934	0.939	0.945	0.945	0.937	0.928	0.933
Following 17 modifications							
Chi square	129.717	138.804	122.625	155.885	205.642	147.318	175.487
Significance	$p = .012$	$p = .002$	$p = .034$	$p < .001$	$p < .001$	$p < .001$	$p < .001$
Degrees of freedom	96	96	96	96	96	96	96
AGFI	0.957	0.962	0.966	0.966	0.955	0.949	0.949
N	699	851	716	1,065	1,065	674	799

Note. AGFI = adjusted goodness-of-fit index.

Table 2   Significant and Nonsignificant Effects of Hospital Restructuring Variables on Nursing Outcomes

Nursing Outcome Variable	Causal Variable						
	Hospital Restructuring Events	Times Nurse Changed Units	Part-Time/ Full-Time Status	Years of Experience in Current Hospital	Times Nurse Laid Off	Age	Gender
1. Unmet patient care needs	+*	ns	ns			−	
2. Freedom to make important patient care decisions	ns		ns*		ns*	+/−	
3. Professional efficacy	ns*	ns*	ns*				
4. Satisfaction with time to spend with patients		ns*	+/−				
5. Emotional exhaustion	+	ns	+*	ns			
6. Psychosomatic symptoms	ns	ns*		ns*	ns*	−*	
7. Emotional health	−						
8. Teamwork between physicians and nurses	ns*	ns			ns*		
9. Nursing workgroup collaboration	−	ns*		ns*	ns	ns*	
10. Job security	ns		ns*	+*	−*	−	
11. Satisfaction with supervision					ns*		
12. Satisfaction with financial rewards				ns*		ns	
13. Job mobility options				−		−	
14. Job satisfaction	ns*	ns*	+/−	ns*	ns*		
15. Intent to quit	ns*	ns	ns*	−*			

Note. + = A significant positive effect was estimated in at least two of three leadership-style groups (resonant, dissonant, and mixed). − = A significant negative effect was estimated in at least two of three leadership-style groups (resonant, dissonant, and mixed). ns = Estimated effect was not significant in all, or at least most, of the leadership styles. * = Effect was consistent across all seven leadership styles. Blank = Effect was not estimated.

Table 3 Significant and Nonsignificant Relationships Among Nursing Outcome Variables

	Nursing Outcome Variable														
Nursing Outcome Variable	1	2	3	4	5	6	7	8	9	10	11	12	13	14	15
1. Unmet patient care needs		−	−		+	+						ns*			
2. Freedom to make important patient care decisions						ns*	+								
3. Professional efficacy		ns*		ns*	ns*	ns*	ns								
4. Satisfaction with time to spend with patients	−*	+													
5. Emotional exhaustion		ns		ns					−		−				
6. Psychosomatic symptoms					+*		+							−	
7. Emotional health				+	−*					+	+				
8. Teamwork between physicians and nurses		+*		+			ns								
9. Nursing workgroup collaboration	−	+		+*		ns*									
10. Job Security				−											
11. Satisfaction with supervision		+		+					+						
12. Satisfaction with financial rewards		ns		ns	−*			ns	ns						
13. Job mobility options					−					+*				+	
14. Job satisfaction	ns*	ns	+	+	−*		+*	+	ns		ns	+			
15. Intent to quit	ns*	ns	ns			+				−	ns*	ns*	+	−	

Note. + = A significant positive effect was estimated in at least two of three leadership-style groups (resonant, dissonant, and mixed). − = A significant negative effect was estimated in at least two of three leadership-style groups (resonant, dissonant, and mixed). ns = Estimated effect was not significant in all, or at least most, of the leadership styles. * = Effect was consistent across all seven leadership styles. Blank = Effect was not estimated.

reduced nursing workgroup collaboration. Freedom to make important patient care decisions enhanced teamwork between physicians and nurses, leading to increased job satisfaction.

Impact of Leadership Styles

Determining the impact of specific leadership styles involved examining the degree to which nurses experienced the consequences of hospital restructuring depending on the leadership style characterizing their work environment. This analysis was enabled by graphing each leadership style's effect coefficient with the means of the two variables contributing to that particular effect (Figure 1). Figure 1a is discussed in some detail and the remainder is summarized.

The slope of each of the seven lines depicted in Figure 1a is the estimated effect coefficient of hospital restructuring events on unmet patient care needs for each leadership style. These show that hospital restructuring led to reported increases in unmet patient care needs for all nurses surveyed. However, the placement of each line is determined by the means of both variables (hospital restructuring events and unmet patient care needs) for each specific leadership style and illustrates the degree to which the outcome variable (unmet patient care needs) differed by leadership style. Therefore, the differences between the seven lines are the impact of leadership styles as reported by nurses on the relationship between hospital restructuring and unmet patient care needs. As the number of

hospital restructuring events increased, so did the number of unmet patient care needs at the end of the shift for all nurses. Yet, despite experiencing relatively similar numbers of hospital restructuring events, the nurses who worked in dissonant leadership environments (pace setting and commanding) reported 3 times the number of unmet patient care needs than those who worked in resonant leadership environments (visionary, coaching, affiliative, and democratic). This graph depicts the mitigation, or lessening in force or intensity, of the impact of hospital restructuring on unmet care needs by resonant leadership, and the intensification of this same effect by dissonant leadership, when compared with mixed leadership (the sample norm).

Although the impact of nurse having to change nursing units did not significantly alter the number of unmet patient care needs, the impact of resonant leadership on these patient care needs was still evident (Figure 1b). The effect of hospital restructuring events on nurses' reported emotional exhaustion (Figure 1c), emotional health (Figure 1d), and workgroup collaboration (Figure 1e) were all mitigated by resonant leadership. While all nurses reported an increase in emotional exhaustion following hospital restructuring events, those who worked in dissonant leadership environments experienced these effects weekly, whereas nurses working in resonant leadership environments consistently reported experiencing emotional exhaustion on a monthly basis. The emotional health of nurses was markedly different between the dissonant and resonant leadership groups. Nurses working under resonant leadership

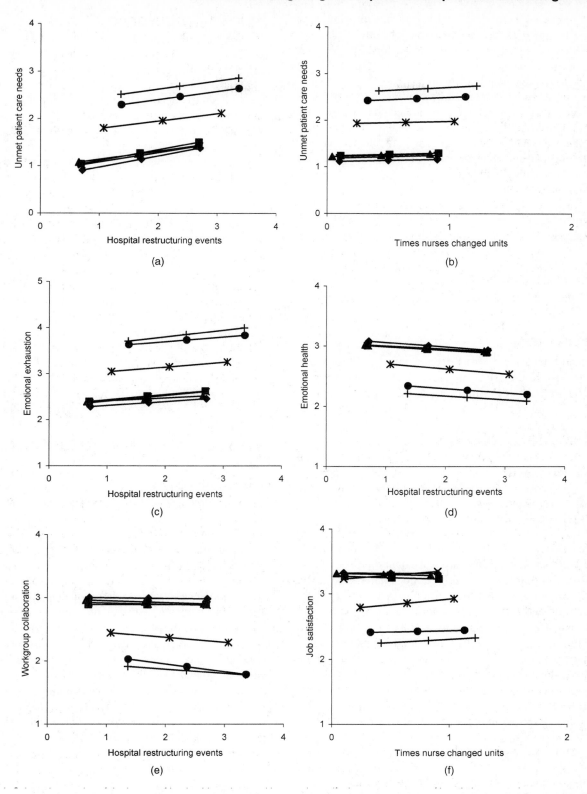

Figure 1 Selected examples of the impact of leadership styles to mitigate or intensify the consequences of hospital restructuring on nurses. ◆, Visionary; ■, Coaching; ▲, Affiliative; ×, Democratic; *, Unknown; ●, Pacesetting; I, Commanding.

reported improved emotional health over the previous year compared to the deteriorating emotional health reported by nurses under the influence of dissonant leadership. The effect of changing units on nurses' job satisfaction (Figure 1f) was mitigated also by resonant leadership.

Six causal relationships are illustrated among several nursing outcome variables that were mitigated also by resonant leadership. Figure 2a illustrates that not being able to attend to patient care needs had a negative impact on the nurses' satisfaction with time to spend with patients. This effect was greater for nurses working in dissonant leadership environments. Figure 2b shows that the effect of nurses' emotional exhaustion on their psychosomatic symptoms was very strong, irrespective of the leadership styles. The effects that were mitigated more clearly by resonant leadership were nurses' job satisfaction on their reported psychosomatic symptoms (Figure 2c), satisfaction with time to spend with patients on nursing workgroup collaboration (Figure 2d), and nurses' emotional exhaustion and emotional health on their job satisfaction (Figure 2e and f, respectively).

Discussion

In this study, leadership was reflected through EI. Although these findings are diverse, three areas are highlighted further: leader–staff relationships, implications for organizational policy, and study limitations that should guide additional research.

Leader–Staff Relationships

The results of this theoretical model estimation showed that all nurses felt the effects of hospital restructuring; however, nurses who worked in resonant leadership environments reported fewer negative effects. This would lead to greater satisfaction and more "emotional resilience" with which to provide quality care. This was evidenced by fewer necessary patient care needs being left unattended. The current findings suggest that resonant leaders used their emotional skills to understand what individual employees or teams were feeling during difficult times, thereby building trust through listening, empathy, and responding to staff concerns. The results suggest that after layoffs, resonant leaders be expected to work with remaining staff to understand their issues, their increased workload and emotional turmoil resulting from the layoff of colleagues and changes in practice patterns. After hospital restructuring, the results suggest these resonant leaders also would continue to invest in staff development and to consider nurses' freedom to make important patient care decisions a high priority.

The results suggest that dissonant leadership would not be tuned to staff members' emotional needs or focus on developing or maintaining relationships with them. Leaders exhibiting mixed leadership styles may be perceived as somewhere between the resonant and dissonant leadership styles. Also these leaders may demonstrate a resonant style in one situation and a dissonant style in another. The large standard deviations in the means of the mixed leadership style group[7] suggest that the latter was the case, and at minimum that the EI behaviors of this undifferentiated leadership group were diverse.

Study Limitations

We examined the translation validity (Trochim, 2003) of our process to use Goleman and colleagues' description of six leadership styles into our research. Many of the 13 questions used referred to specific behaviors of the nurse's supervisor or manager, adding to construct validity. Other questions were chosen on the basis of the primary author's judgment in collaboration with an external expert in clinical outcome and health system research and the third author. To ensure that each question appropriately reflected a specific EI competency, only questions that described a specific characteristic that is a frontline leader's responsibility and that could be perceived by a staff nurse as an EI competence were selected. Concurrent validity (Trochim, 2003) of the three theoretical groupings of leadership styles (resonant, mixed, and dissonant) was supported by using the required presence or absence of 13 EI leadership competencies to sort cases into one of the three mutually exclusive groups. The significant difference in study results for each of the three groupings of leadership styles suggests that we had achieved discrimination between these three groupings.

Future research into the mitigation of the consequences of adverse events should be done prospectively, and by using instruments designed to measure the EI of leaders as perceived by followers. It is recommended also that future research examine the influence of both the nursing manager's and the senior nursing leader's styles on nurses and on the nursing work environment.

Implications for Organizational Policy

The findings suggest that resonant leadership would result in better quality of care by frontline providers. Hiring resonant leaders or providing training for existing leaders hence becomes a priority consideration for chief executives and nursing administrators, albeit recognized that screening for and assessing EI competencies in the workplace is still problematic (Matthews, Zeidner, & Roberts, 2002). The findings also suggest that nurses who reported characteristics of resonant leadership also reported enhanced teamwork between physicians and nurses, nursing workgroup collaboration, and the freedom to make important patient care decisions—all of which are important aspects of nursing practice environments. Health Canada (2000) has advocated that healthcare organizations reduce their occupational health and financial risk by establishing an organizational climate of fairness, purpose, and trust, in which staff wellness is a priority, leading to greater staff satisfaction. The findings suggest that resonant nursing leadership is a key—but missing—ingredient in this model for achieving these goals in hospitals.

Researchers and developers of EI training programs have identified that it is possible to learn how to increase EI. Earlier studies have shown that by wanting to learn and choosing to make a sustained, intentional behavioral change, people can change their performance on a complex set of competencies that distinguish outstanding managers (Boyatzis, 2001). The Consortium for Research on EI has summarized empirical findings on the best mode of learning EI competencies and has guidelines for developing training programs (Goleman et al., 2002).

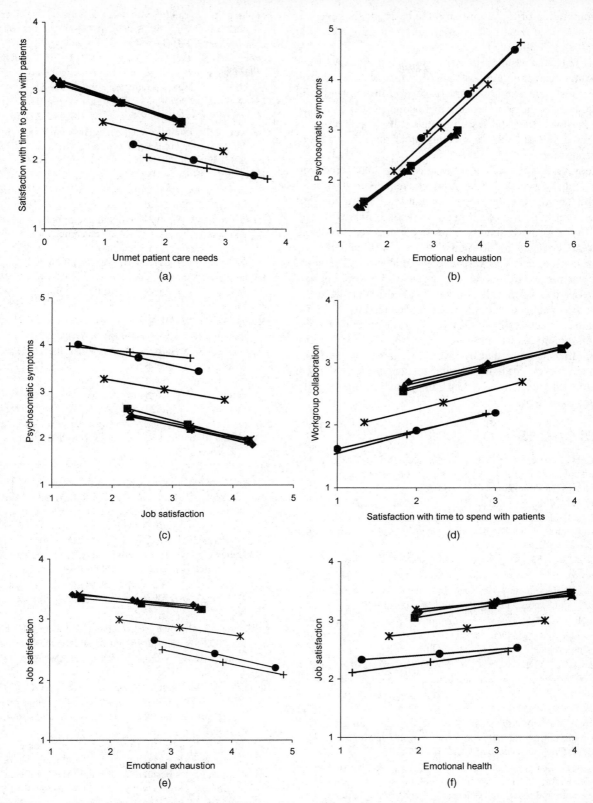

Figure 2 Selected examples of the impact of leadership styles to mitigate relationships among the nursing outcome variables. ◆, Visionary; ■, Coaching; ▲, Affiliative; ×, Democratic; *, Unknown; ●, Pacesetting; |, Commanding.

The incorporation of EI training into the basic nursing curriculum has also been identified as essential for nursing education (Evans & Allen, 2002).

It has not been implied that by employing resonant leaders, hospitals can mitigate—and thereby justify—the adverse effects of restructuring. Resonant leadership in this study did not eliminate the negative effects of hospital restructuring on nurses; however, they did lessen some of the negative effects that resulted and did so to a greater extent than dissonant or mixed leadership.

A theoretical model of causal relationships between hospital restructuring events and negative consequences to nurses' work and health was developed and tested. These findings indicated that numerous detrimental effects to nurses' health and ability to provide quality care to patients resulted from widespread changes to hospitals. Nurses who experienced negative consequences of restructuring and worked in resonant leadership environments experienced these effects to a much lesser degree than those who worked under the influence of dissonant leadership. Resonant leadership mitigated most of the effects of hospital restructuring on their nurses, while dissonant leadership intensified these same effects. These findings suggest that by investing energy into relationships with nurses, resonant nursing leaders positively affect the health and well-being of their nurses, and, ultimately, the outcomes for patients.

References

Aiken, L. H., Clarke, S. P., Sloane, D. M., Sochalski, J., Busse, R., Clarke, H., et al. (2001). Nurses' reports on hospital care in five countries. *Health Affairs, 20*(3), 43–53.

Aiken, L. H., Clarke, S. P., Sloane, D. M., Sochalski, J., & Silber, J. H. (2002). Hospital nurse staffing and patient mortality, nurse burnout and job dissatisfaction. *Journal of the American Medical Association, 288*(16), 1987–1993.

Bass, B. M., & Avolio, B. J. (1991). *Multifactor Leadership Questionnaire.* Palo Alto, CA: Consulting Psychologist Press.

Blegen, M. A., Goode, C. J., & Reed, L. (1998). Nurse staffing and patient outcomes. *Nursing Research, 47*(1), 43–50.

Boyatzis, R. E. (2001). *Unleashing the power of self-directed learning.* Retrieved October 25, 2002, from www.eiconsortium.org/research/self-directed_learning.htm.

Cummings, G., & Estabrooks, C. A. (2003). The effects of hospital restructuring including layoffs on nurses who remained employed: A systematic review of impact. *International Journal of Sociology and Social Policy, 8–9,* 8–53.

Dasborough, M. T., & Ashkanasy, N. M. (2002). Emotion and attribution of intentionality in leader-member relationships. *Leadership Quarterly, 13*(5), 615–634.

de Vries, R. E., Roe, R. A., & Taillieu, T. C. B. (2002). Need for leadership as a moderator of the relationships between leadership and individual outcomes. *Leadership Quarterly, 13*(2), 121–137.

Dunham, T. J. (2000). Nurse executive transformational leadership found in participative organizations. *Journal of Nursing Administration, 30*(5), 241–250.

Evans, D., & Allen, H. (2002). Emotional intelligence: Its role in training. *Nursing Times, 98*(27), 41–42.

Freshman, B., & Rubino, L. (2002). Emotional intelligence: A core competency for health care administrators. *Health Care Manager, 20*(4), 1–9.

Gavin, M. B., & Hofmann, D. A. (2002). Using hierarchical linear modeling to investigate the moderating influence of leadership climate. *Leadership Quarterly, 13*(1), 15–33.

George, J. M. (2000). Emotions and leadership: The role of emotional intelligence. *Human Relations, 53*(8), 1027–1055.

Giovannetti, P., Estabrooks, C. A., & Hesketh, K. L. (2002). *Alberta Nurse Survey Final Report (Report No. 01-02-TR).* Edmonton, AB: University of Alberta, Faculty of Nursing.

Goleman, D., Boyatzis, R., & McKee, A. (2002). *The New Leaders: Transforming the Art of Leadership Into the Science of Results.* London, England: Little, Brown.

Graves, M. L. M. (2000). Emotional intelligence, general intelligence, and personality: Assessing the construct validity of an emotional intelligence test using structural equation modeling. *Dissertation Abstracts International, 61*(4B). (UMI No. 2255).

Hayduk, L. (1987). *Structural equation modeling with LISREL: Essentials and advances.* Baltimore: Johns Hopkins University Press.

Health Canada. (2000). *Best advice on stress risk management in the workplace.* Ottawa: Government Services Canada.

Jöreskog, K. G., & Sörbom, D. (1996). LISREL 8: User's reference guide. Chicago, IL: SPSS Inc.

Kellett, J. B., Humphrey, R. H., & Sleeth, R. G. (2002). Empathy and complex task performance: Two routes to leadership. *Leadership Quarterly, 13*(5), 523–544.

Kovner, C., & Gergen, P. J. (1998). Nurse staffing levels and adverse effects following surgery in U.S. hospitals. *Journal of Nursing Scholarship, 30*(4), 315–321.

Matthews, G., Zeidner, M., & Roberts, R. D. (2002). *Emotional Intelligence: Science & Myth.* Cambridge, MA: The MIT Press.

Maurier, W. L., & Northcott, H. C. (2000). Job uncertainty and health status for nurses during restructuring of health care in Alberta. *Western Journal of Nursing Research, 22*(5), 623–641.

McColl-Kennedy, J. R., & Anderson, R. D. (2002). Impact of leadership style and emotions on subordinate performance. *Leadership Quarterly, 13*(5), 545–559.

Molter, N. C. (2002). Emotion and emotional intelligence in nursing leadership. *Dissertation Abstracts International, 62*(10B). (UMI No. 4470).

Needleman, J., Buerhaus, P., Mattke, S., Stewart, M., & Zelevinsky, K. (2002). Nurse-staffing levels and the quality of care in hospitals. *New England Journal of Medicine, 346*(22), 1715–1722.

Pirola-Merlo, A., Haertel, C., Mann, L., & Hirst, G. (2002). How leaders influence the impact of affective events on team climate and performance in R&D teams. *Leadership Quarterly, 13*(5), 561–581.

Robbins, C. J., Bradley, E. H., Spicer, M., & Mecklenburg, G. A. (2001). Developing leadership in healthcare administration: A competency assessment tool/Practitioner application. *Journal of Healthcare Management, 46*(3), 188–202.

Salovey, P., & Mayer, J. D. (1990). Emotional intelligence. *Imagination, Cognition and Personality, 9*(3), 185–211.

Snow, J. L. (2001). Looking beyond nursing for clues to effective leadership. *Journal of Nursing Administration, 31*(9), 440–443.

Sochalski, J. (2001). Quality of care, nurse staffing, and patient outcomes. *Policy, Politics, & Nursing Practice, 2*(1), 9–18.

Tjiong, L. A. (2002). The relationship between emotional intelligence, hardiness and job stress among registered nurses. *Dissertation Abstracts International, 62*(11B). (UMI No. 5039).

Trochim, W. (2002). *Research methods knowledge base.* Retrieved July 6, 2003, from http://trochim.human.cornell.edu/kb/-measval.htm

Villa, J. R., Howell, J. P., Dorfman, P. W., & Daniel, D. L. (2003). Problems with detecting moderators in leadership research using moderated multiple regression. *Leadership Quarterly, 14,* 3–23.

Vitello-Cicciu, J. M. (2002). Leadership practices and emotional intelligence of nursing leaders. *Dissertation Abstracts International, 62*(11B). (UMI No. 5039).

Wolff, S. B., Pescosolido, A. T., & Druskat, V. U. (2002). Emotional intelligence as the basis of leadership emergence in self managing teams. *Leadership Quarterly, 13*(5), 505–522.

Xin, K. R., & Pelled, L. H. (2002). Supervisor-subordinate conflict and perceptions of leadership behaviour: A field study. *Leadership Quarterly, 14,* 25–40.

Notes

1. Additional information provided by the authors expanding this article can be found at the editor's Web site (http://nursingresearch-editor).

2. See Footnote 1.

3. See Footnote 1.

4. See Footnote 1.

5. The covariance matrices for seven leadership styles may be obtained from the primary author at gretac@ualberta.ca.

6. Four coefficients in the original model that satisfied the first decision rules but that would have created loops were not acted upon.

7. See Footnote 1.

Critical Thinking

1. How has hospital restructuring in the 1990s resulted in the lay-offs of thousands of nurses? Relate this to the nursing shortage facing the profession today.

GRETA CUMMINGS, PhD, RN, is Assistant Professor, Faculty of Nursing; **LESLIE HAYDUK,** PhD, is Professor, Department of Sociology; and **CAROLE ESTABROOKS,** PhD, RN, is Associate Professor, Faculty of Nursing, University of Alberta, Edmonton, Alberta, Canada.

From *Nursing Research,* Vol. 54, No. 1, January/February 2005, pp. 2–12. Copyright © 2005 by Lippincott, Williams & Wilkins. Reprinted by permission via Rightslink.

Predictors of Professional Nursing Practice Behaviors in Hospital Settings

Milisa Manojlovich, PhD, RN, CCRN

Many hospital nurses perform isolated, routine tasks instead of higher level skills that require independent judgment and thinking (Shorr, 2000). Shorr commented on the loss of a professional focus in nursing, stating, "the behavior of many nurses can only be described as 'mechanical'" (p. 90). Understanding the nature of nursing practice behaviors is important for many reasons, but two reasons in particular are relevant. First, the current nursing shortage suggests that not only have nurses been highly dissatisfied with their positions (Cumbey & Alexander, 1998) but also the profession of nursing has not been as attractive a career choice as it once was (Buerhaus, Staiger, & Auerbach, 2000). The task-centered focus of much nursing work may be in part to blame, and one answer to the nursing shortage may lie in providing opportunities for a more variable and autonomous job. Second, nursing practice, as exemplified by a focus on task-centered behaviors, may also adversely affect patient outcomes. Research has shown that patient outcomes improved when the hospital organization supported professional practice characteristics, rather than task-centered behaviors (Aiken, Clarke, & Sloane, 2000).

Three essential attributes of professional nursing practice were identified by Scott, Sochalski, and Aiken (1999) in their review of the original magnet hospital research as well as other descriptive studies, which used the magnet hospital framework. The attributes included (a) the ability of the nurse to establish and maintain therapeutic relationships with patients (usually through a primary nursing delivery model); (b) nursing autonomy and control over the practice environment; and (c) collaborative relationships with physicians at the nursing-unit level (Scott et al., p. 10). Aiken and colleagues later conceptualized autonomy, control over the practice environment, and nurse-physician collaboration as hospital characteristics (Aiken, Havens, & Sloane, 2000).

It was likely that nurses frequently were unable to use their professional training, training that focused on autonomous practice and independent decision making, because they were controlled by both organizational and medical divisions of labor (Alexander, 1982). Without the authority to make decisions, nurses may not be able to work effectively, "in a manner consistent with professional standards" (Alexander, p. 21). Providing decision support to nurses at the point of care has more recently been identified as promoting patient safety (Page, 2004).

- **Background:** Many hospital nurses perform isolated, routine tasks, rather than use their professional training, because they are subject to control by organizational and medical divisions of labor. The environment may interfere with a nurse's ability to practice autonomously and according to professional standards.

- **Objectives:** The purpose of the study was to explore how certain factors in the environment and personal characteristics interact to affect hospital nursing practice behaviors.

- **Methods:** The study used a nonexperimental, comparative design. Surveys were sent to a random sample of 500 nurses throughout the state of Michigan. Three instruments, measuring structural empowerment, self-efficacy for nursing practice, and professional practice behaviors, were included. Path analysis was used for statistical analysis.

- **Results:** Three hundred sixty-four nurses responded (73%), of whom 251 provided usable protocols for the final analysis. Environmental factors (structural empowerment) contributed both directly to professional practice behaviors as well as indirectly through self-efficacy. Self-efficacy mainly exerted its effect as a mediator in the relationship between environmental factors and practice behaviors. Support for the proposed theoretical model was mixed, although the proposed model fit the data well ($X^2 = 11.02$ [(5, $N = 251$), $p < .05$, CFI $= .999$, NNFI $= .991$, RMSEA $= .069$]). An alternative model emerged from the data analysis.

- *Discussion:* Nurses may practice more professionally when the environment provides opportunities and power through resources, support, and information. Self-efficacy may contribute to professional practice behaviors, especially in an environment that has the requisite factors that provide empowerment.
- *Key Words:* professional nursing practice behaviors • self-efficacy • structural empowerment

Organizational systems that encourage nurses to use their professional skills may provide a rewarding career choice for current students (Bednash, 2000). Theoretical (Kanter, 1993) and empirical studies suggest that when the organization provides opportunity and access to authority, employees are more satisfied and more effective on the job (Laschinger & Havens, 1997; Laschinger, Shamian, & Thomson, 2001). Providing bedside nurses with control over both the content and the context of their practice may increase their involvement in decision making (Laschinger, Sabiston, & Kutszcher, 1997), which has been deemed essential to professional practice (Alexander, 1982). However, a continued lack of control over the content and context of their work would suggest that power remains an elusive attribute to many nurses. The complex issue of task-centered (instead of profession-centered) behavior should be investigated. Therefore, the purpose of this study was to explore how certain factors in the environment and personal characteristics interact to affect nurse practice behaviors.

Review of the Literature

Structural empowerment is a concept that refers to four social structure factors in the environment that, when available, promote employee effectiveness and satisfaction. These factors are opportunity, information, resources, and support (Kanter, 1993). Structural empowerment has been studied in the nursing literature and linked to many nurse outcomes, some of which may be associated with professional practice behaviors. Sabiston and Laschinger (1995) demonstrated that a significant, positive relationship existed between job-related empowerment and nurse autonomy. In a secondary analysis of two earlier studies, it was demonstrated that nurses perceived significantly greater control over both the content

and the context of their practice when they had access to work empowerment structures (Laschinger et al., 1997). Earlier autonomy and control over the practice environment had been identified already as hallmarks of professional nursing practice (Schutzenhofer & Musser, 1994).

Very little published research has examined the concept of self-efficacy for nursing practice, although nursing has studied self-efficacy as it relates to nurse managers, nursing students, and patient care (Haas, 2000). *Self-efficacy* has been defined as the belief in one's abilities to mobilize the motivation, cognitive resources, and courses of action needed to exercise control over one's work (Bandura, 1997). Caring has been identified as being pivotal to the nursing role as well as being the essence of nursing (Wolf, Giardino, Osborne, & Ambrose, 1994). *Caring self-efficacy* has been defined as "nurses' beliefs in their abilities to express caring orientations, attitudes, and behaviors and to establish caring relationships with clients or patients" (Coates, 1997, p. 54). Therefore, caring self-efficacy may be appropriate to use as an indicator of self-efficacy for nursing practice.

Self-efficacy has been linked to empowerment as a crucial stage of the empowerment process (Kramer & Schmalenberg, 1993), suggesting that a relationship may exist between empowerment and self-efficacy. Access to both information and resources—two components of structural empowerment—has been found to enhance self-efficacy as well (Spreitzer, 1996). Laschinger and Shamian (1994) found a positive relationship between managerial self-efficacy and job-related empowerment, suggesting a possible link between organizational structures and personal levels of efficacy necessary to carry out managerial roles.

Although the relationship between professional nursing practice behaviors and self-efficacy has not previously been investigated, a few studies have suggested that there may be a relationship between these two concepts. Self-efficacy beliefs have been demonstrated to aid in efficient analytic thinking in complex decision-making situations (Bandura, 1989). In addition, perceived self-efficacy has been found to contribute to higher levels of productivity and improved performance (Gist & Mitchell, 1992).

Research has shown also that personal self-efficacy determines the levels of motivation by determining the amount of effort to exert on a task or how long to persist in the face of obstacles (Bandura, 1989). Considering the complex nature of modern health care environments, nurses may benefit from increased levels of self-efficacy. To identify the appropriate course of action and function effectively, professionals must have an understanding of and control over the activities associated with their job (Alexander, 1982) as well as a belief in their ability

(Bandura, 1997). Wide variability in reaction to the same environment occurs because individuals are influenced by their perceptions of the environment, and not objectivity (Spreitzer, 1996). Thus, a nurse's perception of the work environment and a sense of self-efficacy are important to understand variation in practice behaviors.

Research Aims and Study Questions

The following specific aims and related research questions were proposed for study:

Aim 1: To examine the effects of social structural factors in the work environment, known collectively as structural empowerment, on professional nursing practice behaviors. What is the relationship between structural empowerment and professional nursing practice behaviors?

Aim 2: To examine the effects of a personality variable, self-efficacy, on both environmental factors (structural empowerment) and professional nursing practice behaviors. What is the relationship between self-efficacy and structural empowerment? What is the relationship between self-efficacy and professional nursing practice behaviors?

Theoretical Framework

The theoretical framework developed for this study blended two well-established theories to explain variation in professional nursing practice. The theory of structural empowerment (Kanter, 1993) recognized the environment for its ability to influence behavior whereas Social Cognitive Theory (SCT) maintained that interpretation of the environment was dependent, in part, upon self-efficacy (Bandura, 1997). These theories provided two distinct perspectives, because they have shown variation in human work behavior arising from different causes.

Both the work environment and the manner in which it is interpreted affect nursing practice behaviors. Some environments are empowering because they allow workers flexibility in getting the job done. Other work environments may not be empowering, yet there are individuals who manage to be effective on the job.

It may be that these individuals believe in their ability to mobilize the necessary psychological resources to complete the job. Many individuals believe themselves to be effective in multiple spheres of influence (e.g., parents, wood carvers, cooks, car drivers). Yet, in the workplace, these same people no longer believe they can be effective. On one hand, the hospital environment, or any work

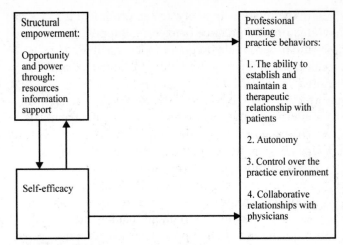

Figure 1   Initial theoretical model.

environment, without empowering structures, is likely to diminish perceptions of self-efficacy over time. On the other hand, a certain amount of self-efficacy may be necessary to be able to use empowering structures that are present. Therefore, the relationship between empowerment and self-efficacy may be reciprocal. The interaction between environmental structures, as expressed by structural empowerment, and the psychological attribute of self-efficacy may explain variation in nursing practice, from a focus on tasks at one end of the spectrum to practice behaviors more consistent with professional standards at the other. On the basis of the theoretical framework described earlier, a conceptual model was tested in this study (Figure 1).

Methods
Design

This descriptive study used a comparative survey design. Approval was obtained from the Institutional Review Board (IRB) before the study was initiated.

Sample and Procedures

The sample consisted of 500 nurses randomly selected from a list of 1,509 names provided by the Michigan Nurses Association (MNA). A list of medical-surgical nurses was specified from the MNA to achieve a more homogeneous sample. Of the 365 individuals who responded, 308 provided usable surveys and demographic information and signed informed consent forms. Ten of the 308 surveys were completed by nurses who did not work in hospitals. These 10 cases were dropped from the sample, because the study purpose was to investigate the effect of environmental and personal influences on hospital nurses only. To achieve further sample homogeneity, managers, supervisors, and nursing faculty were not included in the final sample; thus, an additional 32 cases were dropped.

Dillman's (2000) method of implementing surveys was adapted for this study. The survey responses were self-reported in paper-and-pencil instruments. Three contacts were made to obtain responses.

Variables and Instruments
Structural Empowerment

Structural empowerment was measured by three empowerment scales: the Conditions for Work Effectiveness Questionnaire-II (CWEQ-II), the Job Activities Scale II (JAS-II), and the Organizational Relationships Scale II (ORS-II). All scales measured various aspects of Kanter's concept of empowerment (Kanter, 1993). The CWEQ-II is a 12-item, four-subscale measure. The four subscales are Opportunity, Information, Support, and Resources. The JAS-II is a three-item measure of Kanter's concept of formal power whereas the ORS-II measures Kanter's concept of informal power. All three scales, 19 items in total, have demonstrated high internal consistency in multiple studies, ranging from .78 to .93 (Laschinger, Almost, & Tuer-Hodes, 2003). Content and construct validity of the CWEQ-II have been established (Laschinger, Finegan, Shamian, & Wilk, 2001) and the factor structure of both the JAS-II and ORS-II scales have been validated. A total empowerment score can be created by summing all six subscales (range = 6 – 30). Only the total score was used in this analysis.

Self-Efficacy

The Caring Efficacy Scale (CES) was used to measure caring self-efficacy, and is a 30-item self-report tool, arranged in a 6-point Likert-type format (Coates, 1997). Items are balanced between positive and negative content. Scores are summed and averaged, with higher numbers associated with higher efficacy beliefs. The Cronbachs α reliability coefficients have ranged from .85 to .92 (Meretoja & Leino-Kilpi, 2001). The scale has been tested for content (via expert groups) and concurrent validity (Meretoja & Leino-Kilpi), although it has yet to be subjected to factor analysis (Coates). Sample scale items of the CES include, "I can usually create some way to relate to most any client/patient," and "Clients/patients can tell me most anything and I won't be shocked."

Professional Nursing Practice

Professional nursing practice was measured by the Nursing Activity Scale (NAS). With this 30-item self-report 4-point Likert-type scale, scores can range from 60 to 240 and are based on the sum of weighted item scores. Continuing work with the NAS has yielded α reliability coefficients of .81 to .92 (Schutzenhofer & Musser, 1994) as well as test-retest reliability ($r = .79$). Content validity was established during the initial item generation process and after revision by consulting with nursing experts. Construct validity was established through factor analysis in which three strong factors emerged: individual development of the nurse; development of the professional role; and development of the role of patient advocate (K. Kelly, personal communication, October 13, 2002). A summary of all instruments and α coefficient reliabilities for this study is provided in Table 1.

Control variables: Because educational level (Johnson, 1988) and years of work experience (McCloskey & McCain, 1988) have been associated with more professional nursing practice, these variables were added to the model as control variables. No study has examined specialty certification's effect on practice behaviors, although a recent study has suggested that "certification may give nurses the means or opportunity to practice in a manner likely to improve outcomes" (Cary, 2001, p. 49). Therefore, specialty certification was also added as a control variable. The addition of control variables was necessary to correctly specify the model used in path analysis (Kline, 1998). The demographic questionnaire was used to collect information on these variables.

Instrumental variable: An instrumental variable is related to one variable, but not the other, in a reciprocal relationship (Heise, 1975). Global empowerment was chosen as the instrumental variable because of its strong relationship to structural empowerment in previous research and because, theoretically, it was expected to be less related to self-efficacy. Global empowerment was created from two additional items of the CWEQ, separate from and not included in the total scale. Because its only purpose was to be able to correctly identify the nonrecursive model for path analysis, global empowerment was not otherwise included in any analysis.

Table 1 Instrument Reliability Coefficients

Scale	Number of Items	Cronbach α
NAS	30	.90
CES	30	.89
CWEQ	19	.90

Note. NAS = Nursing Activity Scale; CES = Caring Efficacy Scale; CWEQ = Conditions for Work Effectiveness Questionnaire.

Data Analysis

Data analysis was performed using the Statistical Package for the Social Sciences (SPSS, 2001), version 11.0, and Analysis of Moment Structures (AMOS) statistical software programs (Information Technology Services [ITS], 2001). Descriptive analyses of the study sample and variables were conducted. Inferential statistics included correlations, reliability assessments of study instruments, Sobel's tests, and path analysis. Path analysis, a causal modeling technique, was used to test the theoretical model presented in Figure 1. To determine whether self-efficacy acted as a mediator in the relationship between empowerment and professional practice behaviors, both path analysis and Sobel's tests were performed. Baron and Kenny (1986) recommend use of Sobel's test to determine significant mediating effects. The level of significance chosen for this study was .05.

Results

The nurses in the sample ranged in age from 21 to 75 years (M = 45.42), had an average of 20 years' experience in nursing, and had spent more than 10 years in both their positions (M = 10.27) and institutions (M = 15.02). The sample consisted of individuals who were mainly female (96.2%), and White (92.1%). Most participants were educationally prepared at either the associate (36.8%) or baccalaureate (37.6%) level, with the remainder being diploma (14.3%), master's (10.5%), or doctorally prepared (0.8%) nurses. Most nurses worked full-time (59.8%).

A correlation matrix (Table 2) was generated to begin an understanding of the relationships among study variables. The NAS scale used to measure professional

Table 2 Pearson Correlation Coefficients for Study Variables

	1	2	3	4	5	6
1 NAS						
2 CES	.45*					
3 CWEQ	.31*	.17*				
4 Education	.26*	.07	.16*			
5 Specialty certification	.17*	.04	.05	_.01		
6 Years' experience	_.01	_.05	.07	.01	.10	

Note. NAS (Nurse Activity Scale) measures professional practice behaviors. CES (Caring Efficacy Scale) measures self-efficacy. CWEQ (Conditions for Work Effectiveness Questionnaire) measures structural empowerment.

*p + .01 level (two-tailed test).

practice behaviors was moderately related to both the empowerment (r = .32, p < .01) and self-efficacy (r = .45, p + .01) scales, suggesting that both structural empowerment and self-efficacy were positively related to professional practice behaviors. Of the control variables, education (r = .26, p + .01) and specialty certification (r = .17, p + .01) were also significantly related to professional practice behaviors.

All cases with any missing data were deleted, with the result that 251 cases were used to test the model. The theoretical model was nonrecursive, because of the possible reciprocal relationship between structural empowerment and self-efficacy. Therefore, to identify the model (Heise, 1975), an instrumental variable (global empowerment) was used. As predicted, global empowerment was highly correlated to the total structural empowerment score (r = .77, p = .01), but not significantly related to self-efficacy (r = .11).

Testing the theoretical model involved several stages. In the first stage of model testing, a model was generated that consisted of the main variables of interest (structural empowerment, self-efficacy, and professional practice behaviors) as well as control variables (type of nursing education, specialty certification, and years of work experience).

To determine model adequacy, the model was evaluated for its goodness of fit to the data (Kline, 1998). The Bentler Comparative Fit Index (CFI) indicated the relative proportion in improvement of the overall fit of the proposed model to a null model (Kline). Values for the CFI range from 0 (*poor fit*) to 1 (*perfect fit*); higher values are desirable. The Bentler-Bonnett Non-Normed Fit Index (NNFI), also known as the Tucker-Lewis Index (TLI), is similar to the CFI, but penalizes for model complexity. Values close to 1 indicate good fit. Finally, Steiger's Root-Mean-Square Error of Approximation (RMSEA) is an index that penalizes for lack of model parsimony; lower values are desirable. In addition, values + .06 are considered indicative of good model fit (Kline). This model fit the data well (Figure 2), as demonstrated by a X^2 of 11.02 ([5, N = 251], p = .05, CFI = .999, NNFI = .991, RMSEA = .069).

In model testing, path significance is based on values of the critical ratio (cr), which is the ratio of the unstandardized parameter estimate to the standard error of that estimate. Critical ratios >1.64 were considered to be significant, using a one-tailed test because a direction was proposed for each relationship. Both structural empowerment (β = .20, cr = 3.76) and self-efficacy (β = .40, cr = 7.48) were significant predictors of professional practice behaviors. In fact, self-efficacy was a stronger predictor of practice behaviors than

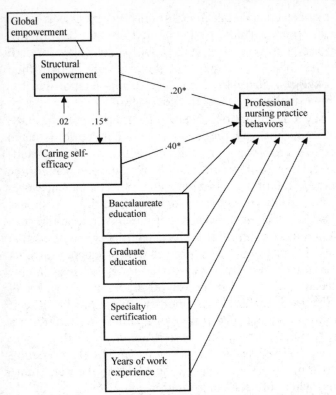

Figure 2   Predictors of professional practice behaviors. Path coefficients are standardized. An asterisk indicates a significant path coefficient.

structural empowerment. The path from structural empowerment to self-efficacy was significant as well ($\beta = .15$, cr $= 1.74$). However, the path from self-efficacy to structural empowerment was not significant ($\beta = .02$, cr $= .32$).

To determine whether self-efficacy mediates the relationship between structural empowerment and professional practice behaviors, additional analyses were conducted. A new model was generated wherein the path from self-efficacy to structural empowerment was removed, as was the instrumental variable and its effect on structural empowerment.

An examination of critical ratios in all three paths demonstrated that the direct path from structural empowerment to professional practice behaviors was significant (cr $= 3.73$), as were both indirect paths (the path from structural empowerment to self-efficacy [cr $= 2.65$] and the path from self-efficacy to professional practice behaviors [cr $= 7.48$]). All component paths have to be significant in order for indirect effects and direct effects to be considered significant (Kline, 1998). However, the product of indirect paths was not greater than the direct path. These results indicate that the direct impact of structural empowerment on practice behaviors was greater than its

indirect impact. To demonstrate complete mediation, the previously significant direct path would have to become insignificant (Baron & Kenny, 1986).

A series of equality-constrained models were generated as an additional test to determine if self-efficacy mediated the relationship between structural empowerment and professional practice behaviors. The basic mediator model without any path constraints was just identified, and therefore, had no degrees of freedom and a X^2 equal to zero. In the model created to test mediation, the paths from structural empowerment to self-efficacy and from self-efficacy to professional practice behaviors were both constrained to zero at the same time. When this model was compared to the model without any path constraints imposed, the X^2 difference was significant ($X^2 = 57.45$, $df = 2$), indicating that there were indirect paths through self-efficacy and that self-efficacy did mediate the relationship between structural empowerment and professional practice behaviors.

The Sobel test, performed as an additional validation of the mediation effect (Preacher & Leonardelli, 2001), was significant ($p = .013$). The results of all models tested to this point therefore failed to support the theoretical model. However, an alternative theoretical model (Figure 3) emerged from the data analysis.

Discussion

The findings indicate that structural empowerment contributed to professional practice behaviors directly as well as indirectly through self-efficacy. The results failed to support the notion of a reciprocal relationship between structural empowerment and self-efficacy. A stronger contribution from self-efficacy to professional behaviors than from structural empowerment to

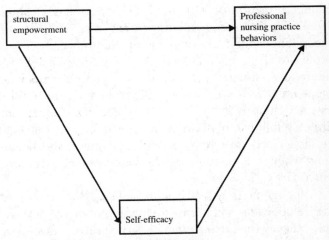

Figure 3   Revised theoretical model.

professional practice behaviors are also suggested. The contribution of self-efficacy to professional practice behaviors indicated a moderate effect, whereas structural empowerment's contribution to practice behaviors indicated a small to moderate effect. Path coefficient values + .10 may indicate a small effect, values around .30 a medium effect, and values >.50 a large effect (Kline, 1998).

The model predicted 30% of the variance in professional practice behaviors and provided mixed support for the theoretical model. Both structural empowerment and self-efficacy directly and significantly contributed to practice behaviors, and, further, structural empowerment contributed directly and significantly to self-efficacy. The finding through path analysis that structural empowerment is not only related to, but may cause, professional behaviors, adds more information about the outcomes of structural empowerment. Laschinger and Havens (1996) were able to demonstrate relationships between structural empowerment and work effectiveness as well as between structural empowerment and perceived control over nursing practice. Work effectiveness may be viewed as a positive contributor to professional behaviors, whereas control over nursing practice is one of the characteristics of professional practice behaviors used in this study.

The significant relationship extending from self-efficacy to professional behaviors is important because it provides information about the determinants of professional practice behaviors and about the sources of those determinants. The current research has concluded that environmental and personal factors influenced the development of professional practice behaviors, and is consistent with SCT (Bandura, 1997).

The findings show that self-efficacy partially mediated the relationship between structural empowerment and professional behaviors, and are consistent with Kanter's theory of structural empowerment (Kanter, 1993). The indirect, mediating effect of self-efficacy on the relationship between structural empowerment and professional practice behaviors may be responsible for the relative smaller direct effect of structural empowerment on practice behaviors. Baron and Kenny (1986) assert that as a mediator effect becomes greater, the direct relationship between independent and outcome variables becomes smaller, with full mediation occurring when the relationship between independent and outcomes variables is zero.

Although no previously published study has examined the relationship between structural empowerment and self-efficacy for nursing practice, structural empowerment

may precede and contribute to self-efficacy, as was found in the present study. Hospitals that provide opportunity, information, resources, and support may engender staff nurses' beliefs that they can exercise control over their work lives. Theoretical evidence of this relationship has been established by Laschinger (1996).

Several limitations were identified in this study. Concepts that are processes rather than static, isolated occurrences were measured with a single self-assessment. Longitudinal study would better determine if these relationships exist over time. Several measures were taken to reduce social desirability response bias that can occur when participants rate their own behaviors. For example, participants were assured that their identity would be protected. Consistency artifact was minimized by varying the order of instruments placed in each packet (Podsakoff & Organ, 1986). Using a list from the MNA is another limitation, because it could be argued that nurses who choose to belong to a professional organization are different from those who do not. However, many hospitals in Michigan require MNA membership because the MNA is the bargaining unit, which offsets this limitation somewhat.

On one hand, organizational changes to improve the hospital work environment for nursing practice may be simplified by using structural empowerment as a blueprint for improvement. On the other hand, it may be that the impetus for a satisfactory work environment has to be initiated by staff nurses, rather than by hospital administrators. Such an approach has been recommended by the American Academy of Colleges of Nursing (Miller et al., 2002), highlighting the importance of the practice environment to professional nursing practice. Nurses may be able to actively seek work environments that foster more professional practice behaviors by asking about structural empowerment factors during the job interview process. Alternatively, nurses who are established in a particular hospital setting may conduct an environmental assessment of their work environment, appraising structural environment factors that may already be present, but previously ignored.

References

Aiken, L. H., Clarke, S. P., & Sloane, D. M. (2000). Hospital restructuring: Does it adversely affect care and outcomes? *Journal of Nursing Administration, 30*(10), 457–465.

Aiken, L. H., Havens, D. S., & Sloane, D. M. (2000). The magnet nursing services recognition program. *American Journal of Nursing, 100*(3), 26–35.

Alexander, J. A. (1982). *Nursing unit organization: Its effects on staff professionalism* (Vol. 4). Ann Arbor, MI: UMI Research Press.

Bandura, A. (1989). Human agency in social cognitive theory. *American Psychologist, 44*(9), 1175–1184.

Bandura, A. (1997). *Self-efficacy: The exercise of control.* New York, NY: W. H. Freeman.

Baron, R. M., & Kenny, D. A. (1986). The moderator-mediator variable distinction in social psychological research: Conceptual, strategic, and statistical considerations. *Journal of Personality and Social Psychology, 51,* 1173–1182.

Bednash, G. (2000). The decreasing supply of registered nurses: Inevitable future or call to action? *JAMA, 283*(22), 2985–2987.

Buerhaus, P. I., Staiger, D. O., & Auerbach, D. I. (2000). Implications of an aging registered nurse workforce. *JAMA, 283*(22), 2948–2954.

Cary, A. H. (2001). Certified registered nurses: Results of the Study of the Certified Workforce. *American Journal of Nursing, 101*(1), 44–52.

Coates, C. J. (1997). The Caring Efficacy Scale: Nurses' self-reports of caring in practice settings. *Advanced Practice Nursing Quarterly, 3*(1), 53–59.

Cumbey, D. A., & Alexander, J. W. (1998). The relationship of job satisfaction with organizational variables in public health nursing. *Journal of Nursing Administration, 28*(5), 39–46.

Dillman, D. A. (2000). *Mail and Internet surveys: The tailored design method* (2nd ed.). New York: John Wiley.

Gist, M. E., & Mitchell, T. R. (1992). Self-efficacy: A theoretical analysis of its determinants and malleability. *The Academy of Management Review, 17*(2), 183–211.

Haas, B. K. (2000). Focus on health promotion: Self-efficacy in oncology nursing research and practice. *Oncology Nursing Forum, 27*(1), 89–97.

Heise, D. R. (1975). *Causal analysis.* New York: John Wiley. Information Technology Services (ITS). (2001). *Introduction to structural equation modeling using AMOS.* Retrieved September 15, 2004, from University of Texas at Austin website: www.utexas.edu/cc/stat/tutorials/amos

Johnson, J. H. (1988). Differences in the performances of baccalaureate, associate degree, and diploma nurses: A meta-analysis. *Research in Nursing and Health, 11,* 183–197.

Kanter, R. M. (1993). *Men and women of the corporation* (2nd ed.). New York: Basic Books.

Kline, R. B. (1998). *Principles and practice of structural equation modeling.* New York: The Guilford Press.

Kramer, M., & Schmalenberg, C. (1993). Learning from success: Autonomy and empowerment. *Nursing Management, 24*(5), 58–64.

Laschinger, H. K. S. (1996). A theoretical approach to studying work empowerment in nursing: A review of studies testing Kanter's theory of structural power in organizations. *Nursing Administration Quarterly, 20*(2), 25–41.

Laschinger, H. K., Almost, J., & Tuer-Hodes, D. (2003). Workplace empowerment and magnet hospital characteristics: Making the link. *Journal of Nursing Administration, 33*(7/8), 410–422.

Laschinger, H. K., Finegan, J., Shamian, J., & Wilk, P. (2001). Impact of structural and psychological empowerment on job strain in nursing work settings: Expanding Kanter's model. *Journal of Nursing Administration, 31*(5), 260–272.

Laschinger, H. K., & Havens, D. S. (1996). Staff nurse work empowerment and perceived control over nursing practice: Conditions for work effectiveness. *Journal of Nursing Administration, 26*(9), 27–35.

Laschinger, H. K., & Havens, D. S. (1997). The effect of workplace empowerment on staff nurses' occupational mental health and work effectiveness. *Journal of Nursing Administration, 27*(6), 42–50.

Laschinger, H. K., Sabiston, J. A., & Kutszcher, L. (1997). Empowerment and staff nurse decision involvement in nursing work environments: Testing Kanter's theory of structural power in organizations. *Research in Nursing and; Health, 20,* 341–352.

Laschinger, H. K., & Shamian, J. (1994). Staff nurses' and nurse managers' perceptions of job-related empowerment and managerial self-efficacy. *Journal of Nursing Administration, 24*(10), 38–47.

Laschinger, H. K., Shamian, J., & Thomson, D. (2001). Impact of magnet hospital characteristics on nurses' perceptions of trust, burnout, quality of care, and work satisfaction. *Nursing Economics, 19*(5), 209–219.

McCloskey, J. C., & McCain, B. E. (1988). Variables related to nurse performance. IMAGE: Journal of Nursing Scholarship, 20(4), 203–207.

Meretoja, R., & Leino-Kilpi, H. (2001). Instruments for evaluating nurse competence. *Journal of Nursing Administration, 31*(7/8), 346–352.

Miller, K. L., Bradley, C., Jones, R., McCausland, M. P., Potempa, K., Rendon, D., et al. (2002). *Hallmarks of the professional nursing practice environment.* Washington, DC: American Association of Colleges of Nursing.

Page, A. (Ed.). (2004). *Keeping patients safe: Transforming the work environment of nurses.* Washington, DC: National Academies Press.

Podsakoff, P. M., & Organ, D. W. (1986). Self-reports in organizational research: Problems and prospects. *Journal of Management, 12*(4), 531–544.

Preacher, K. J., & Leonardelli, G. J. (2001). *Calculation for the Sobel Test.* Retrieved September 15, 2004, from the Ohio State University website: www.unc.edu/~preacher/

Sabiston, J. A., & Laschinger, H. K. S. (1995). Staff nurse work empowerment and perceived autonomy: Testing Kanter's theory of structural power in organizations. *Journal of Nursing Administration, 25*(9), 42–50.

Schutzenhofer, K. K., & Musser, D. B. (1994). Nurse characteristics and professional autonomy. *IMAGE: Journal of Nursing Scholarship, 26*(3), 201–205.

Scott, J. G., Sochalski, J., & Aiken, L. (1999). Review of magnet hospital research: Findings and implications for professional nursing practice. *Journal of Nursing Administration, 29*(1), 9–19.

Shorr, A. S. (2000). Has nursing lost its professional focus? *Nursing Administration Quarterly, 25*(1), 89–94.

Spreitzer, G. (1996). Social structural characteristics of psychological empowerment. *Academy of Management Journal, 39*(2), 483–504.

SPSS. (2001). *SPSS base 11.0 user's guide.* Chicago: Author.

Wolf, Z. R., Giardino, E. R., Osborne, P. A., & Ambrose, M. S. (1994). Dimensions of nurse caring. *IMAGE: Journal of Nursing Scholarship, 26*(2), 107–111.

Critical Thinking

1. What are the three essential attributes of professional nursing practice?

MILISA MANOJLOVICH, PHD, RN, CCRN, is Assistant Professor, School of Nursing, University of Michigan, Ann Arbor.

This study was supported by a grant from the Midwest Nursing Research Society (2003 Dissertation Grant).

The author acknowledges members of the dissertation committee: Shaké Ketefian, EdD, RN, FAAN, Chair and Professor, Director of the Office of International Affairs, and Violet Barkauskas, PhD, RN, FAAN, Associate Professor, both from the School of Nursing at the University of Michigan; Heather Laschinger, PhD, RN, Professor, Associate Director of Nursing Research and Chair of Graduate Nursing Programs, from the School of Nursing at the University of Western Ontario; and Lance Sandelands, PhD, Professor and Chair of the Department of Organizational Psychology, from the Department of Psychology at the University of Michigan. The guidance and support provided by all committee members are greatly appreciated.

The Winning Job Interview
Do Your Homework

BELINDA E. PUETZ, PhD, RN

You've found the ideal position, one apparently custom-made for you. Whether online, in a journal, magazine, or newspaper, or by word of mouth, you've discovered what you believe to be the job perfectly suited to your knowledge, experience, and skills in nursing—in teaching or management or on a nursing staff. Now you must proceed to the next and crucial step in obtaining the position: the interview.

The prospect of a job interview can intimidate even the most seasoned of nurses. Of course, few people relish being put on the spot and made to feel insecure, both of which commonly occur during job interviews. But preparedness can help a great deal in making the experience less threatening and even enjoyable. The process, however, entails some preparation.

Doing Your Homework

Before the job interview, learn as much as you can about the possible employer by contacting former or current employees, consulting the employer's Web site, and reading newspapers and business journals published in the employer's city or state. Take plenty of notes, and then identify two or three points to inquire about during the interview, such as the following:

- Is the facility applying for Magnet status?
- Has there been recent expansion at the facility?
- Is the employer trying a new product line?

These are examples of points you can raise to impress the interviewer favorably, but be sure to have accurate information and review your notes directly before the meeting. Once the interview is scheduled, find out who will conduct it and request materials such as the employer's annual report and strategic plan, and of course, a full description of the position.

Learn about the person who will interview you: is it the director of human resources or the chief nursing officer? Practice pronouncing the interviewer's name correctly—upon introduction, you'll have only one chance to get it right.

If you learn from the recruiter, or whoever arranged the interview, that a group of people will conduct the interview, you might benefit by acting out, in advance, a likely scenario, particularly if it's your first interview. Ask a group of colleagues to act as your interviewers, asking questions such as "What do you know about this position?" and "Why would you like to work here?" Afterward, ask for comments about your performance. The confidence you display and your ready ability to respond to questions can be of great value to you. Even if yours is to be a one-to-one interview, such planning can help you to be at ease during it.

A word about dressing for the interview: you won't have a second chance to make a good first impression, so dress in a businesslike manner. Traditional dress is the most appropriate at an interview for employment; a dark two-piece, matched suit often is recommended as a safe and conservative approach for both men and women.

Presenting Yourself Well

Be on time, or even a few minutes early, for the appointment. Bring additional copies of your résumé, a notebook and pen, and a list of references. Be sure that you've prepared the individuals on this list: have permission to use their names and know that they're willing to comment on your professional performance.

Greet the interviewer with a firm handshake, look her directly in the eye, and hand her a copy of your résumé, so that she won't have to search for it if it's not at hand. Maintain eye contact throughout the interview; don't let your eyes wander, even though you may be tempted to take in the setting. If the interviewer notices that you are looking around while she is reading your résumé, for example, you may appear to be either disinterested or intrusive. Making and maintaining eye contact conveys confidence and honesty—two qualities that will be viewed as assets.

If you're offered a drink, accept it. While you might think that handling a cup or glass is the last thing you need to do during an interview, you might use the time spent sipping on a beverage to consider your response to a question, or simply to pause while you regroup.

As the interviewer speaks, smile and nod occasionally, but avoid bobbing the head. Sit upright and when asked a question, respond forthrightly, but be mindful of not interrupting. You're at the interview to sell yourself and your skills, so appear to be enthusiastic. Laughter is acceptable; generally, employers seek employees with whom they can get along and who act in a manner thought to be collegial. Your responses to questions can demonstrate that quality. Remember to breathe normally. Anxiety can cause you to hold your breath, exacerbating anxiety, but inhaling and exhaling normally will help you to be calm and confident.

Handling Difficult Questions

Interviews generally begin with "lead-in" questions, such as "Will you tell me a little about yourself?" When each is proposed, listen carefully and be sure that you understand what is being asked. Ask for clarification, if necessary. Pause to develop your response, and go directly to the point, without extraneous details unless they are asked for. Be candid, but don't offer unsolicited information. Offer a brief description of the position you most recently held and of your professional skills and training. Don't offer personal information, concerning your hobbies or interests, for example, unless asked for it.

Be prepared to respond to typical interview questions, such as "Where would you hope to find yourself in five, 10, or 15 years?" "What is your ultimate career goal?" or "Why do you want to leave your current job?" Rehearse answers to those questions, as well as to others such as "In which areas do you most need to improve?" "What is your greatest weakness?" or "If you could change one thing about yourself, what would it be?" When responding, always accentuate your skills and strengths, without acknowledging personal weakness—"I often find that the high expectations I have of myself compel me to work extra hard," for example. Further, consistently represent yourself as one who solves problems, not creates them.

You may be presented with a scenario to which you are asked to respond. In such an instance, answer with a focus on the positive approaches that you would take. Be prepared to describe situations in which you have encountered success—or failure—and both what you learned from the experiences and how you might do things differently in the future.

You might be asked difficult questions, such as "Why should we hire you?" "Are you looking for other positions?" and "Because you don't have some of the experience, skills, or educational background we're looking for, shouldn't I wonder what it is that makes you a good candidate for the position?" Answer with a focus on your particular and pertinent skills and experience. If those don't emerge from your

employment history, be prepared to describe a problem that you encountered in college or in a social setting, and how you analyzed and solved it.

Remember, too, that communication in an interview is reciprocal. You may propose questions of your own—"What are the greatest problems faced by this unit?" "What would you expect me to accomplish at this job?" or "What are some of the accomplishments of others holding the same position?" But ask them after pausing to take a breath or a sip of your drink, so that you don't seem defensive.

Consider, too, asking the interviewer about himself—for example "What do you like best about working here?" and "Why did you accept a position here?" Generally, people like to talk about themselves. But don't ask such questions if the interviewer inquires, "Do you have any questions?" The ones to propose at that point should focus on the position, the employer, or the industry, such as, "What are the three challenges the person taking this position will be faced with in the first 100 days?"

If you provide an inappropriate answer at any time, remain calm and state that you're nervous (that is expected and will be forgiven). Then begin again. Don't dwell on the mistake or apologize more than once for it. Appear confident, even if you're not.

The discussion of salary at a job interview presents a difficulty, and it's best not to be the one to raise the issue of remuneration. But if you're asked what salary you expect, be prepared to respond in asking the range allotted to the position, then state that you expect the salary to be within it. Avoid naming a specific monetary figure until you are offered the position.

In a group interview, look directly at the person asking the question and direct your response to him. Avoid looking around the table at others who may be preparing their own questions or who may for another reason not be attentive, which may distract you. Address each person by name. Make a drawing of the meeting table and mark the attendees' names on it so that you can more easily address each one. Remember that there are no standard responses to interview questions; choose the one that's most clear and pertinent, and that casts the best possible light on you.

If you are asked to bring samples of your work to the interview, ask which types would be most helpful and then select those that reflect your best efforts. If applying for a managerial position, consider bringing a sample of a budget or published article that you've written. If applying for a staff position, consider bringing a sample of a patient care plan. And if applying for a position as an educator, consider bringing a continuing education course or a lecture outline. In each instance, be sure that the confidentiality of patients involved has been withheld.

Conclude the interview by stating that you're very interested in the position, and ask about the next steps in the application process.

Following Up

Later on the day of the interview, send a letter thanking the interviewer and stating that you enjoyed the meeting and that you'd be very pleased to be considered for the position. Include a note that involves the interviewer personally; for example: "The information that you provided me about the hospital makes me even more eager to work there." In the case of a group interview, send a different note to each participant. Although e-mail communication will suffice, a personal note on a card or on business stationery with letterhead is preferable.

Unless the interviewer suggested otherwise, if you don't receive a response within a week, telephone to inquire whether there's any additional information that you can provide to assist in the consideration of your application. If the hiring decision hasn't been made, ask when it is expected to be, then pursue the matter again on that proposed date.

Telephone several times, if necessary; persistence shows interest and commitment. Sometimes such decisions become delayed, and the continued demonstration of your interest could secure the position.

Critical Thinking

1. What are several things an applicant can do to prepare for a job interview for a position in nursing?

BELINDA E. PUETZ is president and CEO of Puetz & Associates, Inc., Pensacola, FL.

Meeting the Challenges of Stress in Healthcare

Dan Johnston, PhD

Healthcare is a stressful profession, and takes its toll at physical, emotional, and mental levels. Healthcare professionals not only work long hours, but also face the continual challenge of doing more with less. Recently this situation has been made even more difficult because of staffing shortages, on-going budget cuts, and increasing in levels of patient acuity.

Sicker patients necessitate a heightened vigilance for signs of anything that might go wrong, but this need competes with requirements for more and more documentation. Workers are finding that as abilities for sustained concentration and memory are strained, the risk for errors increases.

In addition, the healthcare environment itself is one of constant change and adaptation, which often leads to communication problems and an increased potential for staff conflict. Mix in the many demands of daily home life and the result can be an overwhelming risk for on the job problems of stress, burnout, and depression.

What can be done to ease the risks of this crisis-prone situation? The solution lies in healthcare professionals understanding that to effectively care for other people, you must take good care of yourself. "Good care" of yourself requires the utilization of basic resiliency skills such as developing and maintaining an attitude of optimism and hope, overcoming the daily effects of stress, and bringing appropriate levels of fun and enjoyment in life.

What is resiliency? In physics, it is the ability of a material to quickly return to its original form after being bent, stretched, or twisted. Psychological resiliency is a similar concept. It is the ability of people to return to normal by bouncing back from the ups and downs of life. Levels of psychological resiliency depend upon three key features.

The first is attitude. Resilient people consistently choose an optimistic outlook on life. An optimistic attitude is one of hopeful expectation for positive results. This is a flexible optimism that does not discount the negative events of life, but intentionally and realistically looks for the best outcome in any situation. It is this looking for the best that pulls resilient people through hard times and puts them back into shape. A positive attitude reduces the potential for stress and depression.

Our attitude is always determined by our inner Voice of Conscience. This is the Voice that continually evaluates and interprets the events around us with the result that it is not primarily the events alone that influence us. Rather, it is our interpretation of them. It is what we are busy telling ourselves about the daily happenings of life. We all have such an inner Voice, and it is more or less optimistic.

Our Optimistic Voice tells us the best, while the Pessimistic Voice points out the worst. If this inner Voice is already positive, we can strengthen it. If the Voice is negative, we can change it. We can become "intentional" optimists, however, hard work is required. Being negative is easier, but not as healthy.

The key skill in having a positive attitude is talking sense to yourself. You must learn to catch the Pessimistic Voice in action and challenge it. Don't let it drone on and on about how terrible everything is, and how nothing will ever improve. Start paying attention to what you say to yourself. You may discover automatic negative responses of which you are unaware.

Listen for the voice of doom and despair and confront it. When you hear it saying, "This situation is impossible to cope with," ask, "Is this true? How do I know? What is the evidence?" When you hear, "Nothing ever goes my way," call time-out. Is it really true that nothing has ever gone your way? No! This is negative exaggeration, and you can challenge this self-talk. If you do so, your attitude will improve, and a positive attitude is not only good for you but also for the people (family, coworkers, and patients) around you.

The second element of resiliency lies in knowing how to manage stress. Life is naturally stressful and resilient people know how to take purposeful action to control it. They avoid whatever stress they can by saying "No" and setting limits, but they also practice unwinding from stress. Such unwinding may be through physical exercise, as with a daily workout at the gym, or it might involve the practice of meditation, tai chi, or yoga. Unwinding from everyday stress can be as simple as taking a slow, mindful walk through the neighborhood.

The key requirement in managing stress is that you must take responsibility for doing something relaxing every day; no matter how busy that day may be. Do so and you will not only be less stressed, but you will become more efficient at what you do both on the job and at home.

The third characteristic of resilient people is that they enjoy life by making the intentional choice to participate in it. Each day should provide us a sense of joy and accomplishment. It is important to both have fun and to get something done.

Resilient individuals accept the fact that after a difficult day at work, motivation may be down, and they may not "feel" like going to a movie, exercising, or attending a party. However, they also know that it is important to do these things whether or not they feel like it. They understand that action can precede motivation, and that doing the things that they usually enjoy is the best way to recapture their energy.

Hopefully, your work gives you a sense of achievement as well as enjoyment, but this may not always be true. Luckily, our daily accomplishments and joys do not have to be sensational and may be as basic as making the bed and petting your dog. It is primarily the appreciation of such simple joys and accomplishments that keeps life in balance. Resilient people know this so they intentionally engage in the daily practice of enjoyment as a way of keeping their energy up.

Resilient healthcare professionals know the importance of taking care of themselves so that they can better take care of others. They understand the value of maintaining a positive outlook, controlling stress, and enjoying what they do, and they practice these skills every day.

Make sure that you know and practice your resiliency skills. Take better care of yourself, and you will take better care of others.

Critical Thinking

1. What are some of the emotional and psychological issues experienced by nurses today?

DAN JOHNSTON, PhD is affiliated with Mercer Health Systems in Macon, Georgia and is Assistant Professor of Psychiatry and Behavioral Science for Mercer University School of Medicine. He is creator of Awakenings website (www.lessonsforliving.com) and author of Lessons for Living: Simple Solutions for Life's Problems from Dagali Press.

UNIT 10

Culture and Cultural Care

Unit Selections

54. **Ethics and Advance Care Planning in a Culturally Diverse Society,** Megan-Jane Johnstone and Olga Kanitsaki
55. **Jewish Laws, Customs, and Practice in Labor, Delivery, and Postpartum Care,** Anita Nobel, Miriam Rom, Mona Newsome-Wicks, Kay Engelhardt, and Ann Woloski-Wruble
56. **Understanding Transcultural Nursing,** Nursing Career Directory

Learning Outcomes

After reading this unit, you should be able to:

- Discuss how ethnicity may impact advance care planning in the healthcare environment.

- List the Jewish customs, laws, and practices that are taken into account during labor.

- List some common behaviors and how they are perceived by people of various cultural groups.

Student Website

www.mhhe.com/cls

Internet References

Cultural Diversity in Healthcare
www.ggalanti.com/cultural_profiles
Cultural Diversity in Nursing
www.culturediversity.org
Diversity Rx
www.diversityrx.org
Transcultural Nursing Society
www.tcns.org
University of Washington Harborview Medical Center
http://ethnomed.org

Transcultural nursing was introduced by Dr. Madeline Leininger approximately 30 years ago. Her theories are more important today to the delivery of healthcare than ever before. In other units, the diversity of nurses and patients has been discussed. It is essential that the face of nursing be as diverse as the population that we serve. This will assist in decreasing the stigmatism that is often associated with a culturally diverse patient population. Many groups are not represented in the same proportions as their numbers in the general population. The American Association of Colleges of Nursing urges recruitment of people from a wide array of ethnic and racial backgrounds in order to find more people to enter the profession.

Students and licensed nurses must be introduced to the concepts of transcultural nursing. The nurse-patient relationship can be severely hampered due to lack of understanding of a patient's cultural beliefs and personal needs. Many cultures have customs and beliefs that will impact patient decisions in medical treatment and frequently hinder progressive patient outcome. Treatment modalities will differ from the birthing process to death and illness because of cultural beliefs. It is also important to identify where one's cultural belief begins and when acculturation occurs. Transcultural nursing does not require a nurse to reject his or her own beliefs, but it does require sensitivity to cultural differences as the individual patient and family's needs are identified and focused upon. It is essential that nursing curricula incorporate cultural diversity, proper assessment, and appropriate intervention and communication. This will ensure that nursing students will become nurses who can identify, relate, communicate, and break the cultural barriers that most nurses encounter.

A culturally sensitive nurse who has the ability to deliver care to all patients regardless of their cultural background will be a valued asset to any healthcare agency in which he or she is employed.

Ethics and Advance Care Planning in a Culturally Diverse Society

Emerging international research suggests that in multicultural countries, such as Australia and the United States, there are significant disparities in end-of-life care planning and decision making by people of minority ethnic backgrounds compared with members of mainstream English-speaking background populations. Despite a growing interest in the profound influence of culture and ethnicity on patient choices in end-of-life care, and the limited uptake of advance care plans and advance directives by ethnic minority groups in mainstream health care contexts, there has been curiously little attention given to cross-cultural considerations in advance care planning and end-of-life care. Also overlooked are the possible implications of cross-cultural considerations for nurses, policy makers, and others at the forefront of planning and providing end-of-life care to people of diverse cultural and language backgrounds. An important aim of this article is to redress this oversight.

MEGAN-JANE JOHNSTONE, PhD, RN AND OLGA KANITSAKI, PhD, RN

In recent years, there has been increasing recognition in the international health professional literature of the profound influence of culture and ethnicity on patient choices in end-of-life care. This has been especially so in Australia, Canada, the United States, and the United Kingdom, which are renowned for their culturally and linguistically diverse populations and where there has been only a limited uptake of advance care plans and advance directives (also called advance health planning) by ethnic minority groups. Despite a growing interest in cross-cultural considerations in end-of-life care in Australia and other multicultural countries, cross-cultural perspectives have rarely been considered in public policy and legislative frameworks pertinent to advance care planning. A comprehensive search of multiple electronic databases (e.g., Medline, PubMed, CINAHL, EBSCO, Google Scholar) for all years has similarly found that only limited attention has been given to cross-cultural considerations in end-of-life care planning in mainstream health professional and related literature as well. In this article, attention is given to challenging this oversight by providing a critical cross-cultural examination of the mainstream bioethical underpinnings of advance care planning and advance directives and their possible implications when applied without modification in a culturally diverse society.

The Right to Refuse Medical Treatment

Under common law and in most civil jurisdictions in common law countries, all adults have the right to refuse medical treatment, including lifesaving treatment (Ashby & Mendelson, 2003; Lewis, 2001; Mendelson & Jost, 2003; Stewart, 2006). Underpinning this common law right is the presumption that every adult has the mental capacity to consent or refuse to consent to any medical intervention "unless and until that presumption is rebutted" (Ashby & Mendelson, 2003, p. 261). Sometimes, people who are suffering from serious illnesses and who are dying lose their capacity to make prudent and responsible life choices and, because of this, will not be able to exercise their right to decide. On such occasions, treatment decisions invariably fall to someone else, for example, the patient's proxy (family/friends) or the health care team. Proxies and health care professionals who find themselves in a surrogate decision-making role may not always know what to decide, however. This is especially so in cases where the patient's wishes are not known or, if known, are open to a variety of interpretations and hence dispute. Problems can also arise where proxies express certainty regarding the preferences of the patient when, in fact, there is no clear basis for their opinions or where preferences collide with the values and beliefs of professional caregivers (Johns, 1996).

During the decades of the 1980s through to the 1990s, a strong consensus emerged in the health professional literature that "advance planning" has a constructive role to play in—and can be an effective guide to—end-of-life decision making and that the most appropriate instrument for communicating such planning is the *advance directive* (Jordens, Little, Kerridge, & McPhee, 2005). There were high expectations that advance directives would become prevalent and would effectively guide end-of-life decisions (Hammes & Rooney, 1998). Significantly, these expectations have not been realized. Research and practical experience to date strongly suggest that, for the most part, advance directives are underused in clinical contexts and appear not to play a major role in guiding treatment decisions to withhold or withdraw life-sustaining treatment (Ashby & Mendelson, 2003; Cook et al., 2001; Dembner, 2003; Johns, 1996; Lynn & Teno, 1995; Martin, Emanuel, & Singer, 2000; Perkins, 2007). In response to this situation some have called for a new approach—notably one that places less emphasis on signing documents and more emphasis on improving processes of communication and "preparing patients and families for the uncertainties and difficult decisions of medical crises" (Perkins, 2007, p. 51). This call has prompted a change in thinking about processes for planning end-of-life care and a shift away from the notion and nomenclature of *advance directives* to that of *advance care planning*. To this end, the issue of advance directives has been reframed as a matter of *planning* future care and treatment (as opposed to merely *consenting* to it) and, more recently, as a matter of formally "respecting patient choices" notably to *refuse* future care and treatment that has been deemed medically futile by professional caregivers (Jordens et al., 2005; Perkins, 2007; Seal, 2007; Shanley & Wall, 2004).

Defined as a "process of communication between a person and the person's family members, health care providers and important others about the kind of care the person would consider appropriate if the person cannot make their own wishes known in the future" (Shanley & Wall, 2004, p. 32), *advance care planning* and now its counterpart *respecting patient choices* are being increasingly devised and implemented as formal "programs" in health care contexts. The primary purpose of such programs, however, is still strongly linked to an advance directive framework insofar as their key aim is to improve advance directive *use* through providing a supportive framework for advance care planning—primarily by equipping nurses as "Respecting Patient Choices (RPC) consultants" to systematize the expression of patient choices (Seal, 2007, p. 29).

The Cultural Imperative in Advance Care Planning

Emerging international research is increasingly suggesting that there are significant ethnic disparities in factors influencing end-of-life decision making and care, including the uptake of advance care planning and directives (Crawley, 2005; Crawley, Marshall, Lo, & Koenig, 2002; Degenholtz, Arnold, Meisel, & Lave, 2002; Krakauer, Crenner, & Fox, 2002; Krakauer & Truog, 1997; Kwak & Haley, 2005; Perkins, Geppert, Gonzales, Cortez, & Hazuda, 2002). This has led some researchers to suggest, contrary to mainstream thought, that choices around end-of-life decision making "may be more related to ethnicity and culture than to age, education, socio-economic status, or other variables" (Baker, 2002, p. 34).

Despite a growing interest in improving knowledge and understanding of the influence that culture and ethnicity has on people's choices on whether to complete advance directives (Baker, 2002), substantive consideration of cross-cultural perspectives and "cultural imperatives" in public policy and related discourse concerning end-of-life decision making and care is, on the whole, conspicuously missing. One reason for this relates to what Renteln (2004, p. 6) calls the "presumption of assimilation," that is, an attitude on the part of majoritarians that individuals from "Othered" cultures should conform to a single monolithic (national) standard of behavior, with considerations of a person's cultural background being deemed irrelevant (see also Werth, Blevins, Toussaint, & Durham, 2002). Given the profound influence that culture has on

how people perceive, experience, and practice health, and on how health care is planned, delivered, and evaluated, a critical examination of the ethical underpinnings of advance care planning and delivery in multicultural health care contexts is warranted.

Ethical Underpinnings of Advance Care Planning and Advance Directives

The conceptualization and operationalization of advance care planning and advance directives have their origins in years of bitter ethical debates, controversies, and legal battles surrounding end-of-life medical treatment and emergent tensions in interpretations of the professional ethos of "First, do no harm" (Baker, 2002). Historically, the "First do no harm" principle was uniformly taken to mean that "life should be prolonged at all costs." However, during the late 1970s and early 1980s, in the aftermath of the now famous North American cases, Karen Quinlan (circa 1976) and Nancy Cruzan (circa 1983), as well as a number of other less publicized rulings at the time, an unprecedented consensus emerged between the medical and legal professions and the general public that prolonging life at all costs was *doing harm* (Baker, 2002; Wolpe, 1998). As court decisions, first in the United States and later in other overseas jurisdictions (e.g., Australia, the United Kingdom), started to consistently favor "patient's rights to refuse unwanted treatment," the issue of patient autonomy gained center stage in public debate on end-of-life treatment, which in turn paved the way for major legislative reform enabling patients to refuse "futile" medical treatment and/or to have such treatment withheld or withdrawn at their request (Baker, 2002; Johnstone, 2009; Wolpe, 1998).

The common law right of adults to refuse medical treatment, and the interpretation of this right in contemporary advance care planning and advance directive policies and programs, have as their cornerstones the following principles and related core values:

- Patient autonomy
- Informed decision making
- Truth telling
- Control over the dying process

These principles are an expression of mainstream Western bioethics and its theoretical stance of "principlism" (the view that ethical decision making and problem solving is best undertaken by appealing to "universal" moral principles such as autonomy, beneficence, nonmaleficence, and justice; Beauchamp & Childress, 2009; Wolpe, 1998). Despite being the subject of considerable debate and controversy since first gaining preeminence in the 1970s, ethical principlism remains the fundamental framework of bioethics, with autonomy being privileged as *the* preeminent principle in Western bioethical thought and as *the* principal guide to ethical decision making in health care—particularly in countries such as Australia, Canada, the United States, and the United Kingdom, which have been profoundly influenced by the field of contemporary bioethics (Johnstone, 2009; Wolpe, 1998). Indeed, so preeminent is autonomy as a principle that, as Bowman (2004, p. 664) points out, "many bioethicists equate autonomy with personhood, as if autonomy exists independently of specific beliefs and commitments."

Not all agree, however, that autonomy is and even ought to be considered as the preeminent principle in bioethics. Its value even in the doctrine of informed consent is contested, with some philosophers noting that although autonomy is *a* value underlying the doctrine of informed consent it is not *the* value, nor an *absolute* value. As Faden and Beauchamp (1986) point out, at best autonomy is only a prima facie value, and to regard it as having overriding value would be both historically and culturally "odd."

In recent years, there has been an increasing interest in the mediating influence that culture has on bioethical thought, discourse, and decision-making processes in health care. Even so, there remains little understanding that the dominant and highly individualistic theoretical stance taken in regard to the ethics of advance care planning, advance directives, and respecting patient choices may be foreign and even antithetical to patients and families from cultural groups who do not share or accept the principles and value assumptions that underpin much of mainstream Western bioethics discourse on end-of-life decision making (Candib, 2002). This is most evident in the way in which proponents of advance care planning and related policies and programs continue to minimize and even neglect altogether the significant differences in the way in which people of different cultural backgrounds perceive, experience, and explain illness and the influence these processes may and do have on the interpretations and meanings they give to the content and processes of end-of-life decision making (Bowman, 2004).

Cultural Differences in End-of-Life Decision Making and Advance Care Planning

Empirical studies exploring cross-cultural differences in end-of-life decision making are limited in scope, number, and geographic location. Nonetheless, of those that have been reported, it has been consistently shown that people of minority cultural and linguistic backgrounds tend to:

- Complete advance directives and advance care plans less frequently than does the majoritarian population (and may even reject these outright)
- Prefer family/group decision making
- Regard advance directives as an "intrusive" legal mechanism that interferes with their responsibilities as family members to care for their loved ones (most regard intimate health care decisions at the end stage of life "as the purview of the family and physician, not the state or "the system""—Bito et al., 2007, p. 260)
- Indicate preferences for "aggressive" (proactive), life-prolonging treatment to be administered "regardless of the state of their illness and even when there is no medical hope of recovery" (Baker, 2002, p. 33)
- Be less trusting of health care policies and the health care system (especially if previously treated in a manner they perceived and experienced as being prejudicial, discriminatory, and dismissive of their cultural lifeways; Baker, 2002; Bito et al., 2007; Blackhall, Murphy, Frank, Michel, & Azen, 1995; Doorenbos & Nies, 2003; Dupree, 2000; Fan, 1997; Giger, Davidhizar, & Fordham, 2006; Karim, 2003).

Research has also shown that patients of minority cultural and language backgrounds are fearful that if they complete advance care plans and advance directives in a mainstream health care context, they may be left to die in instances where further medical intervention could improve their health outcomes. Thus, they hold fears that not only might they not benefit from policies on advance care planning and advance directives but that these "might actually work to their detriment" (Baker, 2002, p. 33).

Despite these findings and other emerging research implicating cultural racism in ethnic disparities in health care (including end-of-life decision making), providers are remarkably unaware of—and are surprised by—the extent to which patients and involved families of ethnic minority backgrounds may distrust the system and feel unsafe when in it (Candib, 2002; Johnstone & Kanitsaki, 2007, 2008; Krakauer et al., 2002; Smedley, Stith, & Nelson, 2003). Providers are also often bewildered and frustrated by what they see as a negative attitude and at times hostile resistance toward advance care planning by patients and involved families of diverse cultural backgrounds and their apparent inability to understand and accept the intended benefits (and benevolence) of advance care planning and related respecting patient choices programs. Patients and involved families, meanwhile, are often left bewildered and frustrated by the inability of *providers* to understand and accept the benevolence and justice of *their* actions, which have as their sole objective protecting the life and well-being of their loved ones—whatever the medical prognosis.

The Limits of Bioethics and Ethical Principlism in a Culturally Diverse Society

An important, although largely unrecognized, dimension of end-of-life decision making in cross-cultural health care contexts relates to the conspicuous lack of awareness among health service providers, legal professionals, and policy makers of the limits of Western bioethics and ethical principlism and the paradoxical risks these pose to human rights when applied without modification in a culturally diverse society (Bowman, 2004; Elliott, 2006; Hanssen, 2004; Johnstone, 2009; Neves, 2004; Werth et al., 2002). Equally problematic is the conspicuous lack of recognition given to the principles and standards expressed in the human right to cultural liberty and its application to and in health care domains.

The right to cultural liberty encompasses the right that all people have to maintain their ethnic, linguistic, and religious identities—otherwise referred to as *cultural rights* (Fukuda-Parr, 2004). Respecting people's cultural rights does not mean that it is imperative for health care providers to blindly accept *all* cultural traditions—especially those that are widely recognized as being harmful (e.g., female genital mutilation; Johnstone, 2009). This is because culture is not a *static process* encompassing a frozen set of values, beliefs, and practices but rather a process that is "constantly recreated as people question, adapt and redefine their values and practices to changing realities and exchanges of ideas" (Fukuda-Parr, 2004, p. 4). Accordingly, as Fukuda-Parr (2004, p. 4) explains, recognizing people's claims to the right to cultural liberty is *not* about "preserving values and practices as an end in itself with blind allegiance to tradition" but about expanding individual choice and the "capability of people to live and be what they choose, with adequate opportunity to consider other options."

Patient Autonomy

The principle of autonomy is widely regarded in Western bioethics "as the intellectual and moral foundation of the discipline" (Bowman, 2004, p. 666; see also Wolpe, 1998). An important related concept of autonomy is the Western notion of *individualism,* which as Bowman (2004, p. 666) explains is underpinned by "a belief in the importance, uniqueness, dignity, and sovereignty of each person and the sanctity of each individual life." This highly individualistic and culturally relative concept characterizes the person (patient) as a solitary competent individual who possesses a sphere of protected activity or privacy free from unwanted interference; by this view, although influence is acceptable, coercion in any form is not (Kuczewski, 1996). On this point, Kuczewski explains that

> Within this zone of privacy, one is able to exercise his or her liberty and discretion. Within this protected sphere take place disclosure, comprehension, and choice, which express the patient's right of self-determination . . . The person is opaque to others and therefore the best judge and guardian of his or her own interests. Although the physician may be the expert on the medical "facts," the patient is the only individual with genuine insight into his [sic] private sphere of "values." Because treatment plans should reflect personal values as well as medical realities, the patient must be the ultimate decision-maker. (p. 30)

In Australian, Canadian, the United States and the United Kingdom health care contexts, autonomy is generally seen as *empowering.* It is for this reason that autonomy (both as a concept and as a principle) has been accorded a central place in advance care planning—the primary objective of which is to preserve patient autonomy in end-of-life decision making, especially in regard to refusing unwanted "futile" medical treatment. Its universal application as the moral basis of end-of-life decision making is not, however, appropriate for some cultural groups and may in fact cause more harm than good (Bito et al., 2007; Elliott, 2006; Johnstone, 2009; Werth et al., 2002). International research has shown, for instance, that some ethnic groups (e.g., Greek, Italian, Chinese, Ethiopian) do not regard autonomy as empowering at all but as isolating and burdensome to patients who are often too sick and too uninformed about their condition to be able to make meaningful choices (Blackhall et al., 1995). Some have gone even further to suggest that, in mainstream health care contexts where an ethos of individualism dominates, insisting on patient autonomy without taking into account cultural imperatives in its expression is paternalistic, hostile, morally imperialistic, and may even be racist (Candib, 2002).

Although research in this area is limited, emerging evidence is consistently showing that in the ethnic groups studied (e.g., East Asian, Bosnian immigrant, Japanese, Korean Americans, Mexican Americans, African Americans, Middle Eastern Lebanese, Greek Australians), group consensus and decision making by family are more valued than individual decision making (Bito et al., 2007; Blackhall et al., 1995; Candib, 2002; Fan, 1997; Gebara & Tashjian, 2006; Kanitsaki, 1993, 1994; Searight & Gafford, 2005). A key conclusion drawn from the examples given in these reports is that, in many minority ethnic groups, *family-sovereignty* and *family-determination* supersede *individual-sovereignty* and *self-determination* as the preeminent principles for guiding decision making at the end of life (see also Fan, 1997). Moreover, what guides familial decision making in end-of-life contexts are *family love and care* not *abstract principles and detached reasoning.*

The values underpinning the different culturally mediated stances toward end-of-life decision making are equally instructive. For example, East Asian and Japanese people hold deep respect for *harmony* and *interdependence* ("We should make everything right with harmony") rather than individual autonomy (Bito et al., 2007, p. 259; see also Fan, 1997). In keeping with this value, to cause disharmony is a thing of shame; accordingly, people who subscribe to the value of harmony and interdependence tend to avoid confrontation with others (Bito et al., 2007; Fan 1997). Thus, unlike Western bioethical thought, which privileges individual autonomy, as Bito et al. (2007, p. 259) explain, this model "subjugates autonomy to the more compelling value of collective well-being."

Other researchers have drawn similar conclusions. For example, in their study on ethnicity and attitudes toward autonomy involving a survey of 200 participants from four different ethnic groups (e.g., European American, African American, Korean American, and Mexican American), Blackhall et al. (1995, p. 825) similarly found that "for those who hold the family-centred model, a higher value may be placed on the harmonious functioning of the family than on the autonomy of its individual members."

In Australia, a small grounded theory study of Greek-born Australians likewise found *interdependence* not independence to be the norm. It further found that the value of interdependence encompassed the expectation that care and assistance would be provided to family members *as needed and without being requested* (Kanitsaki, 1993). In contrast to the views held by nurses and other health providers, informants in this study regarded autonomy as "the most dangerous and disruptive forces within the finely webbed, interdependent, and balanced structure of their Greek-born families" (Kanitsaki 1993, p. 29). An important outcome of this study was the robust depiction of the Greek family lay-therapeutic care model (summarized in Table 1), which continues to provide an important framework for guiding ethical end-of-life decision making in this ethnic group today.

In light of these considerations, there is much to suggest that health service providers and policy makers need to broaden their view of autonomy as conventionally defined and to accept that *respect for persons* (and *respecting patient choices*) involves much more than merely giving arbitrary deference to an ideological notion and value of individualism and the sovereignty (supreme power and authority) of the individual. At the very least, a just view of autonomy must minimally include a discretionary (not an arbitrary) observance of the principle that includes giving due respect to the cultural values that people bring with them to the decision-making process and also respecting people's right to cultural liberty, namely, to maintain their ethnic, linguistic, and religious identities.

Decision Making

In Western bioethics, the principle of autonomy prescribes that people ought to be respected as self-determining choosers and that others are obliged to accept their choices—even if they do not agree with them and even if they think that they are foolish—provided they do not interfere prejudicially with the significant moral interests of others. In contradistinction to the independent and self-contained decision-making model embraced by mainstream Western bioethics, for some minority ethnic groups it is the *family* that has

Table 1 Expression of Greek Family Lay-Therapeutic Care Model

- Intense interest, sensitivity, and action aimed at promoting and achieving the health both of individual family members and the family as a whole
- Constant intense involvement, with each person being finely tuned cognitively and emotionally to anticipate the needs and hopes of the others
- Sharing, sensing, anticipating, and identifying threats—individual and/or family—including fears and weaknesses that endanger life goals and health (the aim here is the early detection of threats to initiate specific caring actions that will protect the individual and family)
- Expecting that each individual will perceive dangers, will warn family members of potential problems, and will take protective action
- Standing close and hovering—observing intensely, collecting information, creating meaning, and discussing the significance of initial indications of a change in the family member's behavior or health problem; giving advice, support, and encouragement when vulnerable
- Taking direct action as appropriate to protect against potential loss of important family or individual goals and thus promote health
- Family advocacy, including coaxing, "giving presence" (being there), sacrifice, praying, requesting, bargaining, fostering hope and courage, and focusing on future possibilities (adapted from Kanitsaki, 1993, pp. 33–35)

ultimate authority (sovereignty) to decide not the *individual*. By this view, it is the family that has—and is expected to have—the *primary* role in decision making in end-of-life care (Bito et al., 2007). Despite the principle of autonomy commanding respect for patient choices (decision making), providers are often extremely resistant and even hostile to the idea that patient disengagement from decision making (and the abdication of his/her autonomy to family) may not only be an autonomous act but, paradoxically, the *ultimate* expression of autonomous decision making (Orona, Koenig, & Davis, 1994). Here, a patient's autonomous choice to abdicate his/her decision-making authority to family may stand as an *ultimate* act of autonomous decision making because it tests the very tenets of autonomy, notably, that respect is due to the self-determining choices of an individual—including *the choice to not have to choose*, that is, to abdicate his/her autonomy to another (Candib, 2002).

Cultural norms prescribe and permit the protective abdication of autonomous decision making to family by sick individuals in end-of-life care contexts. Although providers might struggle with this abdication, they are nonetheless morally bound to accept it—even if they do not agree with it and even if they think it foolish and unwise (Johnstone, 2009). To not accept this would be to permit autonomy to fail at its own rhetoric and the system of ethical thought that is otherwise dependent on it to justly guide decision making in end-of-life care contexts.

Disclosures/Truth Telling

It is a commonly accepted maxim in mainstream Western bioethical discourse that people have the "right to know" and ipso facto should (and must) be told *the truth*—no matter how awful or how adverse the consequences of doing so might be. To this end, there is a firm requirement that people will be given "the facts" of their situation in clear and unequivocal terms. What is not always appreciated, however, is that beliefs about and attitude toward truth telling in clinical contexts are largely determined by culture (Bowman, 2004; Johnstone, 2009). Moreover, what information is

disclosed, how, when, where, and by whom are also all matters that are strongly mediated by cultural considerations (Johnstone, 2009; Kanitsaki, 1989, 1993, 1994).

All cultures have strict taboos surrounding what is and what is not appropriate to talk about in given contexts. In some cultures, illness is viewed as a punishment and accordingly discussions about the diagnosis and prognosis of certain illnesses (e.g., cancer) may be forbidden, a situation that can pose significant challenges for health care providers (Karim, 2003). Furthermore, in some cultures, talking about end-stage illnesses and plans for managing care at the end stage of life can be interpreted as a portent of death and, hence, something to be avoided, which again can pose significant challenges for health care providers.

In the case of "bad news" being given in a mainstream health care context, there are high expectations among some cultural groups (e.g., Greek, Italian, Chinese, Asian Indian Hindus, Middle Eastern Lebanese, Bosnian migrants/refugees, Indigenous peoples) that the family will assume a protector-advocate and gatekeeping role to ensure that if and when disclosures are made they are made in a manner that does not deny hope and/or plunge their ill loved one into a state of soul destroying and hopeless despair (Backhall et al., 1995; Carrese & Rhodes, 1995, 2000; Doorenbos & Nies, 2003; Gebara & Tashjian, 2006; Johnstone, 2009; Kagawa-Singer & Blackhall, 2001; Kanitsaki, 1994; Karim, 2003; McGrath, Ogilvie, Rayner, Holewa, & Patton, 2005; Searight & Gafford, 2005; Tse, Chong, & Fok, 2003).

In East Asian cultures, for example, it is widely regarded as inappropriate and wrong to talk about diagnoses of end-stage illnesses, such as cancer. In an important essay examining the principle of *self-determination* versus *family determination* as viewed from the ethical traditions of China, Japan, South Korea, Taiwan, and Hong Kong, Fan (1997) explains that

> Because of the ethics of family-sovereignty, it is considered extremely rude and inappropriate if a physician directly informs the patient about a diagnosis of a terminal disease rather than a family representative. If a student doctor does this carelessly, his/her in-charge resident would doubt

his/her qualification to become a *real* physician. [italics added] (p. 319)

In the context of seeking informed consent, Fan (1997, p. 319) further explains that, in keeping with the family-sovereignty model, it is a *family representative* (not the patient) who is obligated to "sustain the burden of listening to and discussing with the physician, communicating with the patient, consulting with her family members and finally signing the formal consent form" (see also Tse et al., 2003).

In the case of Indigenous peoples, although research in this area is scant, findings have likewise emphasized the centrality of family and, more particularly, the importance of "telling the right story" to the "right people" (Carrese & Rhodes, 1995, 2000; McGrath et al., 2005). For example, in a study exploring Navajo perspectives on information disclosures by biomedical providers and the possible limitations of dominant Western bioethical perspectives in this population, Carrese and Rhodes (1995, 2000) found that Navajo people feel great discomfort when talking about "bad things" and "negative information" concerning things such as the diagnoses of a serious and life-threatening disease or impending death. This discomfort relates to a strong cultural value that group members need to think and speak in a positive way ("Think in the beauty way" and "Talk in the beauty way") and avoid thinking and speaking in a negative way ("Don't talk like that"). The researchers further found that to impose negative information directly onto an ill individual was not only inappropriate but a "dangerous violation" of traditional Navajo values that could have far reaching consequences (Carrese & Rhodes, 2000). This was so even in the case of routine disclosures of the possible risks associated with surgery. For example, in one case involving a Navajo man who was to undergo triple bypass surgery, as his daughter explained to the researchers:

> The surgeon told him that he may not wake up, that this is a risk of every surgery. For the surgeon this was very routine, but the way my Dad received it, it was almost like a death sentence, and he never consented to the surgery. (Carrese & Rhodes, 1995, p. 828)

In a small descriptive study of Aboriginal people living in the Northern Territory of Australia, researchers similarly found that if information was not communicated in the "right way" through the "right story" to the "right people," the sick person could be left frightened and anxious (McGrath et al., 2005). The researchers also found that if providers did not take care to communicate end-of-life matters in a culturally meaningful way, their discussions could be misinterpreted with serious consequences. For example, in one case involving a man in need of palliative care, the end-of-life discussion that providers had with a family were interpreted by them to mean that he was already dying. The researchers report that, based on what the family understood they had been told, and in keeping with their traditional practices, the family took the man home "and painted him up and put him under a tree for 3 days because they thought he was going to die right there and then" (McGrath et al., 2005, p. 310).

It is acknowledged that truth telling in mainstream Western biomedical and bioethical thought is widely regarded as being an ethically imperative, clinically right, and legally responsible and accountable thing to do. What is not always appreciated, however, is that truth telling can be seriously problematic on account of its

propensity to stimulate what has been described in the medical anthropological literature as the *nocebo phenomenon* (from the Latinate *noceo,* "I hurt," and the Greek *nosis,* "disease"), defined by Helman (1990, p. 257) as "the negative effect on health of beliefs and expectations—and therefore the exact reverse of the 'placebo' phenomenon."

From a cross-cultural perspective, truth telling and the impost of unconsented disclosures of "bad things" is problematic primarily because of the injurious *nocebo* effect it can trigger. This is why culturally inappropriate disclosures of certain types of information may be viewed as being tantamount to being a portent of death and even "wishing death" on a person. Culturally inappropriate disclosures are also problematic on account of their being perceived and experienced by recipients as authoritarian, paternalistic, disrespectful, and demeaning (Candib, 2002). It is for this reason that patients of minority ethnic backgrounds may also view disclosures by attending health professionals with shock and dismay—even though the providers making the disclosures may be well-intended, compassionate, and highly ethical human beings (Candib, 2002).

To help mitigate the impact of disclosures that might inadvertently trigger the nocebo phenomenon, in contradistinction to conventional models of truth telling, family members and acculturated health care providers will use ambiguous and equivocal terms (e.g., terms such as *inflammation, cyst,* or *growth* rather than the word *cancer*) or speak in the third person or refer to a "generalized other" (e.g., "Some people have these troubles" rather than "This is what is going to happen to you") when disclosing a bad diagnosis, possible risks, or a poor prognosis to a patient (Candib, 2002, p. 222; Carrese & Rhodes, 2000, p. 94).

Control Over Dying

As with truth telling and the other dimensions of advance care planning, beliefs about the locus of control in regard to death and dying and end-of-life care are also profoundly influenced by a person's culture (Bowman, 2004). In many cultural groups, the power to end life is viewed as inappropriate for human decision making (Candib, 2004). Keeping loved ones alive, on the other hand, is seen as being not only appropriate for human decision making but also morally imperative (Bito et al., 2007). In East Asian cultures, for example, family sovereignty works *in accepting treatment;* it is, however, another matter entirely for the family to withhold or withdraw treatment on behalf of the patient (Fan, 1997).

In contrast, in mainstream Western biomedical and bioethical thought, control over dying is regarded as an ultimate expression of individual autonomy and something that ought to be respected. To this end, hospital-based *respecting patient choices* policies and programs have been or are in the process of being implemented. It is important to clarify, however, that these respecting patient choices programs are fundamentally concerned with supporting patients and families to *refuse* the provision of life-prolonging treatment not *request* such treatment. This aim is underscored by the fact that the courts in most common law jurisdictions have to date been extremely reluctant to affirm a positive right to health care (Flood, Gable, & Gostin, 2005). This lack of support for a positive right to health has significant implications for family-centered models of care, because it stands to interfere with and also seriously violate the cultural lifeways and norms that require sons and daughters to fulfill their filial duties to parents and individual family members to fulfill their family advocacy obligations to each other (see Kanitsaki, 1993, 1994).

Conclusion

Many patients of minority cultural and language backgrounds do not have advance directives or advance care plans and, indeed, may consider the idea as being quite strange and alien (Candib, 2002). Moreover, even if a patient has an advance care plan, research has suggested that if it has not been discussed with family members beforehand, the family will not uphold it (Fan, 1997; Matsui, 2007). If this situation is to change, and patients of minority cultural and language backgrounds are to be meaningfully engaged in end-of-life decision making, there are a number of considerations that need to be taken into account.

First, policy makers, health service providers, and cultural care theorists need to make a stringent effort to ensure that policies and programs pertinent to end-of-life planning and care are properly and appropriately informed by cross-cultural and culture care considerations and do not reflect only the values and beliefs of the dominant culture (Baker, 2002). To this end, advance care planning and respecting patient choices policies and programs need to reflect pertinent knowledge, understanding, and sensitivity (right attitudes) toward culturally diverse worldviews and meanings associated with health, illness, life, and death. In light of the views advanced in this article (and summarized in Table 2), this must include broadening mainstream conceptualization and understanding of the principles of autonomy, informed consent, truth telling, cultural liberty, and the role of families and family love in end-of-life decision making and care. As Baker (2002) correctly points out

> What is perceived and promoted as advantageous to one group may not hold the same beneficial value to another group with a different frame of reference, a different value system, and different life experiences. A broadened view of autonomy that includes respect for cultural values, and diverse beliefs around life and death issues needs to be recognised by policy makers, if policies are to be embraced and utilized by a more diverse range of [people]. (p. 36)

An important first step toward advancing a cultural inclusive approach to end-of-life decision making is for policy makers and health service providers to recognize that (as summarized in Table 2)

- autonomy can be conceptualized and expressed variously as *autonomy-in-relation* as well as *autonomy-as-separateness;*
- assumptions about truth telling are not universally appropriate or acceptable;
- decision making can occur along a continuum, which ranges from an individual-centered *self-determination* approach to a family-centered *family-determination* approach;
- a family-determination/orientated approach takes as its focus the initiation of *conversations that ensures that their loved ones feel cared for, loved, and supported* rather than the formalized writing of directives and plans driven by a "system" mediated interpretation of patient autonomy; and

- the principles of privacy and confidentiality ("secrecy"), commonly regarded as "normal" and expected in majority populations, may be inappropriate and wrong in ethnic minority groups who otherwise require access to confidential patient information to inform family-centered decision making and determinations of care.

Given the intimate role of family in decision making and end-of-life care, it also requires recognition that in contexts where there is strong family involvement, advance care planning may be perceived by both the patient and his or her family as being unnecessary and even as highly intrusive, impractical, unhelpful and, hence, as undermining of the family-caring relationship (Candib, 2002).

Second, policy makers and health service providers must stop regarding involved families as being an intrusion or interference and instead respect them as an essential and integral part of the therapeutic relationship. It needs to also be recognized that families can play a major role in helping prevent the inadvertent stimulation of the nocebo phenomenon in patients and its potentially disastrous consequences to the patient's health. To this end, the patient's family needs to be respected as having the surrogate authority to decide—in the moral interests of and for the well-being of their sick loved one—*whether, when, where, how, and by whom* information about the diagnosis of a serious illness and poor prognosis will be given (Johnstone, 2009). Taking such an approach will help avoid undermining the sick person's hope (about getting better and being able to go on living a meaningful life) and thereby help also to maximize the person's ability to continue making important life-interested choices. By maintaining hope the patient's autonomy will also be promoted. This is because, unless hope is maintained, there will be *nothing left to choose for* and, hence, nothing left for which autonomy could find meaningful expression (Johnstone, 2009).

Third, policy makers and health service providers need to be very careful in how they speak about cross-cultural considerations in advance care planning and end-of-life decision making. Of particular concern is the importance of avoiding terms such as *cultural barriers*. As Bowman (2004) correctly cautions, the term *cultural barriers* (which tends to mean the perceived cultural obstacles posed by "Othered" cultures, not the dominant culture) "implies that culture blocks access to the resolution of the ethical issues, implying that there is something *universal* on the other side of the barrier," raising the further question of "is there?" (p. 666). The term *language barriers* should also be avoided, because this presupposes that all that is required to "overcome the barriers" is the use of an interpreter. This, however, fails to take into account the complexities of meanings that may be operating in the situation at hand.

Fourth, policy makers and health service providers need to understand that there are multiple layers of complexity that have to be navigated when advancing a cross-cultural approach to end-of-life decision making, including the matter of the "multiple jeopardies" that may be associated with being a member of a given ethnic or Indigenous minority group. For instance, it is known that being "ethnic" can place a person at risk of disparate care and treatment in a mainstream health care setting, being "ethnic and gay" can place that person at double risk, and being "ethnic, gay, and HIV-positive" at triple risk, and so forth. Adding to this complexity is the modification of attitudes, beliefs, and practices that can occur

Table 2 Comparison of Core Value Assumptions and Meanings Underpinning Conflicting Worldviews in Advance Care Planning/Advance Directives

Core Values	Dominant (Western Bioethical–Biomedical) Cultural Worldview	Minority ("Othered") Cultural Worldviews
Autonomy	• *Individualism,* encompassing right to self-determination	• *Collectivism,* encompassing right to family determination
Decision making	• *Independence* and independent decision making *free of the unwanted interference and/or influence of others*	• *Interdependence* and interdependent decision making with *full participation by family/partner/chosen others in consultation with professional care givers*
	• Autonomy *inalienable*	• Autonomy *may be abdicated to loved ones*
Truth telling	• *Full disclosure* and *clarity*	• *Partial disclosure* and *ambiguity* (e.g., avoid use of term *cancer* and use other terms such as *inflammation and cysts*)
Control over dying	• *Living will* encompassing power ("empowerment") to plan for and direct future treatment at the end of life; *letting die*	• *Willing ill* (seen as "portent of death") and disempowering, with power to end life viewed as inappropriate for human decision making; *letting live*
	• Emphasizes quality time and not having to worry about "the cost of futile treatment"	• Emphasizes meaning of life as a *shared life* + quantity of time to share this life and not having to worry about others "pulling the plug"
Advance care planning/directives (ethos and purpose)	• Liberal, antipaternalistic, and *respectful of patient choices* (fosters hope by fostering choice)	• Authoritarian, strongly paternalistic, and *disrespectful of patient choices* (denies choice by determining what autonomy is + denying hope)
	• Nonintrusive, practical, helpful, and constructive tool for *reducing patient–family conflict*	• Highly intrusive, impractical, unhelpful, and destructive tool *used by "the system" to reduce patient–family control when their wishes conflict with those of health professional staff*
	• *Reduces family stress* by removing responsibility for decision making	• *Increases family stress* by rejecting caring participation and removing responsibility for decision making
	• Emphasis on writing and operationalizing advance care *directives that ensures autonomy upheld*	• Emphasis on initiating and sustaining advance care *conversations that ensures patient feels cared for, loved, and supported*
Decisional authority	• Biomedical/bioethical reasoning and explanation	• Familial relationship, love, and understanding
Life stage	• Enacted only *after* individual loses capacity to decide (literally at the "end" stage of life)	• Enacted *before* individual loses capacity to decide—that is, the moment a loved one becomes ill, irrespective of capacity to decide (literally at "all" stages of life)

through acculturation and the tensions these may give rise to in contexts where other family members continue to subscribe to traditional values (Bito et al., 2007; Blackhall et al., 1995; Dupree, 2000; Hyun, 2002; Konishi & Davis, 1999; Matsui, 2007; Matsumura et al., 2002; Tarn et al., 2005; Tse et al., 2003). Adding yet another layer of complexity to this issue is the problem of stereotyping people from different cultural backgrounds and wrongly assuming heterogeneity of lifeways and viewpoints concerning end-of-life care.

Finally, policy makers and health service providers need to be very aware of the risk of unintended consequences that can flow from the development and operationalization of given public policies and related programs (advance care planning being a case in point). As Bridgman and Davis (2004, p. 7) caution, often the side effects of a policy are discovered only after the policy has been implemented, when they "undermine the policy's effect or create new, complex problems." It would be extremely problematic, for example, if a situation emerged whereby patients and families were not permitted to decide to contribute to end-of-life decision making *unless* they have agreed to a formalized advance care plan or advance directive.

Nurses, allied health professionals, policy makers, and culture care theorists have a fundamental responsibility to ensure the effective planning, delivery, and evaluation of end-of-life care delivered to patients of minority cultural and language backgrounds. How best to fulfill these responsibilities, however, is not something that can be decided in isolation of the social-cultural contexts, cultural lifeways, life experiences, and ethical values of the people who will be at the receiving end of the decisions that are ultimately made. Nor is this something that can be decided reliably in the absence of evidence-based policy. To remedy this situation, a robust program of cross-cultural and culture-care research investigating end-of-life care practices and policies needs to be progressed and the findings used to inform future policies and practices. By taking such an evidence-based approach, importantly end-of-life practices may be improved for *all* patients and families not only those of ethnic minority backgrounds.

References

Ashby, M., & Mendelson, D. (2003). Natural death in 2003: Are we slipping backwards? *Journal of Law and Medicine, 10*, 260–264.

Baker, M. E. (2002). Economic, political and ethnic influences on end-of-life decision-making: A decade in review. *Journal of Health & Social Policy, 14*, 27–39.

Beauchamp, T., & Childress, J. (2009). *Principles of biomedical ethics* (6th ed.). New York: Oxford University Press.

Bito, S., Matsumura, S., Singer, M. K., Meredith, L. S., Fukuhara, S., & Wenger, N. S. (2007). Acculturation and end-of-life decision making: Comparison of Japanese and Japanese American focus groups. *Bioethics, 21*, 251–262.

Blackhall, L. J., Murphy, S. T., Frank, G., Michel, V., & Azen, S. (1995). Ethnicity and attitudes toward patient autonomy. *Journal of the American Medical Association, 274*, 820–825.

Bowman, K. (2004). What are the limits of bioethics in a culturally pluralistic society? *Journal of Law, Medicine & Ethics, 32*, 664–669.

Bridgman, P., & Davis, G. (2004). *The Australian policy handbook* (3rd ed.). Sydney, Australia: Allen & Unwin.

Candib, L. (2002). Truth telling and advance planning at the end of life: Problems with autonomy in a multicultural world. *Family Systems & Health, 20*, 213–228.

Carrese, J. A., & Rhodes, L. A. (1995). Western bioethics on the Navajo reservation: Benefit or harm? *Journal of the American Medical Association, 274*, 826–829.

Carrese, J. A., & Rhodes, L. A. (2000). Bridging cultural differences in medical practice. *Journal of General Internal Medicine, 15*, 92–96.

Cook, D., Guyatt, G., Rocker, G., Sjokvist, P., Weaver, B., Dodek, P., et al. (2001). Cardiopulmonary resuscitation directives on admission to intensive-care unit: An international observational study. *Lancet, 358*, 1941–1945.

Crawley, L. (2005). Racial, cultural, and ethnic factors influencing end-of-life care. *Journal of Palliative Medicine, 8*(Suppl. 1), S58–S69.

Crawley, L., Marshall, P. A., Lo, B., & Koenig, B.A. (2002). Strategies for culturally effective end-of-life care. *Annals of Internal Medicine, 136*, 673–679.

Degenholtz, H. B., Arnold, R. A., Meisel, A., & Lave, J. R. (2002). Persistence of racial disparities in advance care plan documents among nursing home residents. *Journal of the American Geriatrics Society, 50*, 378–381.

Dembner, A. (2003, September 11). "Do not resuscitate" instructions often ignored, overlooked. *Boston Globe*, p. A.1.

Doorenbos, A. Z., & Nies, M.A. (2003). The use of advance directives in a population of Asian Indian Hindus. *Journal of Transcultural Nursing, 14*, 17–24.

Dupree, C. Y. (2000). The attitudes of black Americans toward advance directives. *Journal of Transcultural Nursing, 11*, 12–18.

Elliott, A. C. (2006). Health care ethics: Cultural relativity of autonomy. *Journal of Transcultural Nursing, 12*, 326–330.

Faden, R. R., & Beauchamp, T. L. (1986). *A history and theory of informed consent.* New York: Oxford University Press.

Fan, R. (1997). A report from East Asia: Self-determination vs. family-determination: Two incommensurable principles of autonomy. *Bioethics, 11*, 309–322.

Flood, C., Gable, L., & Gostin, L. (2005). Introduction: Legislating and litigating health care rights around the world. *Journal of Law, Medicine and Ethics, 33*, 636–640.

Fukuda-Parr, S. (Ed.). (2004). *Human Development Report 2004: Cultural liberty in today's diverse world.* New York: United Nations Development Programme.

Gebara, J., & Tashjian, H. (2006). End-of-life practices at a Lebanese hospital: Courage or knowledge? *Journal of Transcultural Nursing, 17*, 381–388.

Giger, J. N., Davidhizar, R. E., & Fordham, P. (2006). Multi-cultural and multi-ethnic considerations and advanced directives: Developing cultural competency. *Journal of Cultural Diversity, 13*, 3–9.

Hammes, B., & Rooney, B. (1998). Death and end-of-life planning in one midwestern community. *Archives of Internal Medicine, 158*, 383–390.

Hanssen, I. (2004). From human ability to ethical principle: An intercultural perspective on autonomy. *Medicine, Health Care and Philosophy, 7*, 269–279.

Helman, C. (1990). *Culture, health and illness.* London: Wright.

Hyun, I. (2002). Waiver of informed consent, cultural sensitivity, and the problem of unjust families and traditions. *The Hastings Center Report, 32*, 14–24.

Johns, J. (1996). Advance directives and opportunities for nurses. *Image: Journal of Nursing Scholarship, 28*, 149–153.

Johnstone, M. (2009). *Bioethics: A nursing perspective* (5th ed.). Sydney, Australia: Churchill Livingstone/Elsevier.

Johnstone, M., & Kanitsaki, O. (2007). An exploration of the notion and nature of "cultural safety" and its applicability to the Australian health care context. *Journal of Transcultural Nursing, 18*, 247–256.

Johnstone, M., & Kanitsaki, O. (2008). Cultural racism, language prejudice and discrimination in hospital contexts: An Australian study. *Diversity in Health and Social Care, 5*, 19–30.

Jordens, C., Little, M., Kerridge, I., & McPhee, J. (2005). Ethics in medicine, from advance directives to advance care planning: Current legal status, ethical rationales and a new research agenda. *Internal Medicine Journal, 35*, 563–566.

Kagawa-Singer, M., & Blackhall, L. J. (2001). Negotiating cross-cultural issues at the end of life: You got to go where he lives. *Journal of the American Medical Association, 286,* 2993–3001.

Kanitsaki, O. (1989). Cross-cultural sensitivity in palliative care. In P. Hodder & A. Turley (Eds.), *The creative option of palliative care: A manual for health professionals* (pp. 68–71). Melbourne, Australia: Melbourne City Mission.

Kanitsaki, O. (1993). Transcultural human care—Its challenge to and critique of professional nursing care. In D. A. Gaut (Ed.), *A global agenda for caring* (pp. 19–45). New York: National League for Nursing Press.

Kanitsaki, O. (1994). Cultural and linguistic diversity. In J. Romanini & J. Daly (Eds.), *Critical care nursing: Australian perspectives* (pp. 94–125). Sydney, Australia: W.B. Saunders/Bailière Tindall.

Karim, K. (2003). Informing cancer patients: Truth telling and culture. *Cancer Nursing Practice, 2,* 23–31.

Konishi, E., & Davis, A. J. (1999). Japanese nurses' perceptions about disclosure of information at the patients' end of life. *Nursing and Health Sciences, 1,* 179–187.

Krakauer, E. L., Crenner, C., & Fox, K. (2002). Barriers to optimum end-of-life care for minority patients. *Journal of the American Geriatrics Society, 50,* 182–190.

Krakauer, E. L., & Truog, R. D. (1997). Case study: Mistrust, racism, and end-of-life treatment. *The Hastings Center Report, 27,* 23–25.

Kuczewski, M. (1996). Reconceiving the family: The process of consent in medical decision making. *The Hastings Center Report, 26,* 30–37.

Kwak, J., & Haley, W. E. (2005). Current research findings on end-of-life decision making among racially or ethnically diverse groups. *Gerontologist, 45,* 634–641.

Lewis, P. (2001). Rights discourse and assisted suicide. *American Journal of Law, Medicine & Ethics, 27,* 45–99.

Lynn, J., & Teno, J. (1995). Advance directives. In W. T. Reich (Ed.), *Encyclopedia of bioethics* (rev. ed., pp. 572–577). New York: Simon & Schuster Macmillan.

Martin, D., Emanuel, L., & Singer, P. (2000). Planning for the end of life. *Lancet, 356,* 1672–1676.

Matsui, M. (2007). Perspectives of elderly people on advance directives in Japan. *Journal of Nursing Scholarship, 39,* 172–176.

Matsumura, S., Bito, S., Liu, H., Kahn, K., Fukuhara, S., Kagawa-Singer, M., et al. (2002). Acculturation of attitudes toward end-of-life care. *Journal of General Internal Medicine, 17,* 531–539.

McGrath, P., Ogilvie, K. F., Rayner, R. D., Holewa, H. F., & Patton, M. A. (2005). The "right story" to the "right person": Communication issues in end-of-life care for Indigenous people. *Australian Health Review, 29,* 306–316.

Mendelson, D., & Jost, T. S. (2003). A comparative study of the law of palliative care and end-of-life treatment. *Journal of Law, Medicine & Ethics, 31,* 130–145.

Neves, M. P. (2004). Cultural context and consent: An anthropological view. *Medicine, Health Care and Philosophy, 7,* 93–98.

Orona, C., Koenig, B., & Davis, A. (1994). Cultural aspects of nondisclosure. *Cambridge Quarterly of Healthcare Ethics, 3,* 338–346.

Perkins, H. S. (2007). Controlling death: The false promise of advance directives. *Annals of Internal Medicine, 147,* 51–57.

Perkins, H. S., Geppert, C. M. A., Gonzales, A., Cortez, J. D., & Hazuda, H. P. (2002). Cross-cultural similarities and differences in attitudes about advance care planning. *Journal of General Internal Medicine, 17,* 48–57.

Renteln, A. (2004). *The cultural defense.* New York: Oxford University Press.

Seal, M. (2007). Patient advocacy and advance care planning in the acute hospital setting. *Australian Journal of Advanced Nursing, 24,* 29–36.

Searight, H. R., & Gafford, J. (2005). "It's like paying with your destiny": Bosnian immigrants' views of advance directives and end-of-life decision-making. *Journal of Immigrant Health, 7,* 195–203.

Shanley, C., & Wall, S. (2004). Promoting patient autonomy and communication through advance care planning: A challenge for nurses in Australia. *Australian Journal of Advanced Nursing, 21,* 32–38.

Smedley, B., Stith, A., & Nelson, A. (Eds.). (2003). *Unequal treatment: Confronting racial and ethnic disparities in health care.* Washington, DC: National Academies Press.

Stewart, C. (2006). Advance directives; disputes and dilemmas. In I. Freckelton & K. Peterson (Eds.), *Disputes and dilemmas in health law* (pp. 38–53). Sydney, Australia: Federation Press.

Tarn, D. M., Meredith, L. S., Kagawa-Singer, M., Matsumura, S., Bito, S., Oye, R. K., et al. (2005). Trust in one's physician: The role of ethnic match, autonomy, acculturation, and religiosity among Japanese and Japanese Americans. *Annals of Family Medicine, 3,* 339–347.

Tse, C. Y., Chong, A., & Fok, S. Y. (2003). Breaking bad news: A Chinese perspective. *Palliative Medicine, 17,* 339–343.

Werth, J., Blevins, D., Toussaint, K., & Durham, M. (2002). The influence of cultural diversity on end-of-life care and decisions. *American Behavioral Scientist, 46,* 204–219.

Wolpe, P. R. (1998). The triumph of autonomy in American bioethics: A sociological view. In R. DeVries & J. Subedi (Eds.), *Bioethics and society: Constructing the ethical enterprise* (pp. 38–59). Upper Saddle River, NJ: Prentice Hall.

Critical Thinking

1. How can ethnicity impact advance care planning in the healthcare environment?

From *Journal of Transcultural Nursing,* October 2009, pp. 405–416. Copyright © 2009 by Sage Publications. Reprinted by permission via Rightslink.

Jewish Laws, Customs, and Practice in Labor, Delivery, and Postpartum Care

Many communities throughout the world, especially in the United States and Israel, contain large populations of religiously observant Jews. The purpose of this article is to provide a comprehensive, descriptive guide to specific laws, customs, and practices of traditionally, religious observant Jews for the culturally sensitive management of labor, delivery, and postpartum. Discussion includes intimacy issues between husband and wife, dietary laws, Sabbath observance, as well as practices concerning prayer, communication trends, modesty issues, and labor and birth customs. Health care professionals can tailor their practice by integrating their knowledge of specific cultures into their management plan.

ANITA NOBLE DNSc, CNM ET AL.

Childbearing practices are highly influenced by cultural values and beliefs (Andrews & Boyle, 2003; Callister, Seminic, & Foster, 1999; Campinha-Bacote, 2003; Galanti, 1997; Kater, 2000; Leininger & McFarland, 2002; Mattson, 2000; Purnell & Paulanka, 2003; Schuiling & Sampselle, 1999; Weber, 1996). Families of different cultures incorporate their beliefs, values, and practices into the childbirth experience. Culture is defined by Spector (2009) as, "a non physical trait which include beliefs, attitudes, values and customs and are shared by a group of people and passed from one generation to the next" (p. 348). Within this meta-communication system of nonphysical traits, a belief can also be a person's religion, which is "the belief in a divine or superhuman power who is worshipped and obeyed and seen as creator or ruler of the universe" (p. 352). The labor, delivery, and postpartum units are prime examples of the interface between culture/religion, childbearing practices, and health care management. This may create a challenge for a Westernized health care system.

Judaism is a monotheistic religion whose history dates back to the biblical forefather, Abraham. Through a series of historical events that began at creation dated 3760 BCE (Year 1 on the Jewish calendar), the Jewish people received a divinely ordained code that included the Ten Commandments and a Written and Oral Law in1280 BCE (American-Israeli Cooperative Enterprise, 2008). Judaism is both a religion and culture (Selekman, 2003). Today, in 2009, (Year 5769 on the Jewish calendar), there are approximately 12 to 14 million Jews in the world today, with the greatest concentrations being in the United States and Israel. According to the Jewish People Policy Planning Institute (Tal, 2007), the four top centers of the Jewish population are North America (5,649,000), Israel (5,393,000), Europe/non–Former Soviet Union (non-FSU; 1,155,000), and the FSU (357,000).

The largest ethnic groups of Jews descend from either European origin (*Askenazim*) or Mediterranean/Middle Eastern origin (*Sephardim/Edot Mizrach;* Dobrinsky, 1986; Selekman, 2003). There are other smaller Jewish ethnic groups throughout the world. In the United States, there are three main levels of Jewish religious practice: Orthodox, Conservative, and Reform (Lewis, 2003; Selekman, 2003;

Shuzman, 2004). Smaller affiliated groups present a varied picture of degrees of religious observance. Within the Orthodox Jewish group itself, there are additional categories, for example, the Ultra-Orthodox, which can be divided into the non-Chassidic and Chassidic sects, where the men are most recognizable by their year-round black attire, uncut side locks (*peyote* or *peyis*), and beard (Schwartz, 2004). The Orthodox group, on the whole, adheres to the strictest interpretation of the teachings of the Torah (Five Books of Moses, Prophets, and Writings) and Jewish law (*halacha;* Lutwak, Ney, & White, 1988; Schwartz, 2004; Selekman, 2003; Shuzman, 2004).

In Israel, the division of religious observance can be delineated as Ultra-Orthodox (*Haredi*), Religious, Traditional, and Secular. The Ultra-Orthodox (*Haredi*) person can be described as one who conducts his lifestyle in accordance to Ultra-Orthodox customs, beliefs, and practices.

> Ultra-Orthodox religious practice applies stringencies on the basic Jewish law to ensure adherence to the religious law and similarly applies such stringencies to dealing with the outside world to ensure protection from negative (anti-Ultra-Orthodox) influences. Knowledge is primarily sought through religious study and religious life-style alone. (Zarembski, 2002, p. 12)

A Religious person can be defined as "a person who conducts his customs, beliefs and practices in a manner that combines integration into the general society with religious life. Knowledge is attained through non-religious and religious study" (Zarembski, 2002, p. 11).

A Traditional person refers to one "who observes religious Jewish ritual practices out of respect to religious command. Traditional practice differs from religious practice in that it does not necessarily place Jewish law in the forefront of everyday decision making" (Zarembski, 2002, p. 12).

A Secular person refers to a person who may distance himself or herself from "the connection between Jewish ritual and divine commandment. Secular Jews may be sub-divided into those that do not perform any Jewish ritual or perform Jewish ritual as a cultural rather than religious requirement" (Zarembski, 2002, p. 11).

The descriptions above offer a two-dimensional compartmentalization of the different strains of the Jewish religion. In reality, there can be overlap, nuanced differences, and individual interpretations, which make the groups similar or very different. Therefore, it can be summarized that Judaism can be viewed as both a religion and a culture (Selekman, 2003).

There are some transcultural nursing books and articles that describe certain Jewish practices (Andrews & Boyle, 2003; Galanti, 1997; Leininger & McFarland, 2002; Lewis, 2003; Lutwak et al., 1988; Schwartz, 2004; Selekman, 2003; Shuzman, 2004; Spector, 2009). It is important to note that there is much fluidity in observance among the various Jewish groups. Jewish women and their families may observe different degrees of observance. It is recommended that the health care professional perform a cultural assessment and dialogue with the client and her family to ascertain their needs in terms of their religious observance. This article will describe selected Jewish laws, customs, and practices pertaining to the labor, delivery, and postpartum periods. A cultural assessment questionnaire and culturally appropriate provider responses are offered. It is important to note that though the laws, customs, and practices presented in this article are formally Orthodox/Ultra-Orthodox traditions, there are many Jews who practice some or all of what is described below, though they do not identify themselves as Orthodox/Ultra-Orthodox.

Selected Jewish Laws Pertaining to Intrapartum and Postpartum Period

Health care professionals who provide family-centered care to the childbearing family recognize the integral role of cultural values, beliefs, and practices. Providing culturally competent care includes cultural knowledge. There ligiously observant Jewish woman, and family, in labor, delivery, and postpartum needs to have care that allows her to abide by Jewish laws, customs, and practices that guide everyday life as well as those that pertain to childbearing. These cultural issues include adherence to the laws of intimacy issues between husband and wife or *niddah* (Lewis, 2003; Lutwak et al., 1988), dietary laws or *kashrut,* and observance of the Sabbath. The following is a comprehensive descriptive guide to practices concerning prayer, communication trends, modesty issues, and labor and birth customs. This guide focuses on selected common laws, customs, and practices and how they are defined within traditionally religious observance.

Laws of Niddah

"Niddah" means "removed or separated" generally in the context of Jewish Marital Laws (*Taharat HaMishPacha*). This word has been mistranslated as "unpure," which is incorrect and does not represent the essence of the term. The laws of *niddah* require the abstinence of physical contact between husband and wife in such cases that the wife has bleeding originating from the uterus (Tendler, 1977). Observant Jewish women who are menstruating or experiencing physiologic uterine bleeding adhere to the laws of *niddah*. The time frame is set from the point of menstruation until seven days after the bleeding has stopped. In labor, a woman may be considered to be in a state of *niddah* if the woman has any of the following: appearance of "bloody show," rupture of membranes, or active bleeding from the cervix. There are some religious authorities that deem *niddah* status for women who have contractions that cause difficulty walking without assistance, or are fully dilated even if no bleeding is noted (Nishmat Women's Online Information Center, 2006; Webster, 1997).

In general, the religiously observant couple will refrain from physical contact (see Table 1).

The husband will not customarily view the actual birth, but he may be present in the room where the delivery occurs and stand in a place where he cannot view the birth. The couple may require that the genital area be draped for the delivery. The nurse and midwife caring for laboring patients who observe the laws of *niddah* can assist by providing all comfort measures that require physical contact. The health care professional with a clear understanding of this religious practice should not misinterpret the nonphysical contact between husband and wife as a relationship issue. On the other hand, knowing this religious practice, the health care professional needs to take cues from the couple as to their degree of adherence and personal choices.

Laws of Dietary Specifications (Kashrut)

The Jewish dietary laws are complex, requiring avoidance of nonkosher foods, proper preparation of meat, not eating milk and meat together, maintaining separate utensils and dishes for meat and milk meals, and a specified waiting time after consumption of meat until dairy may be eaten (Orthodox Union, 2004). In addition, the observant consumer searches for kashrut symbols on packaged food labels to indicate that those foods were prepared according to dietary laws. Presently, kosher meals may be requested in health care institutions in the United States through outside kosher food services (Selekman, 2003). The meals are double sealed to allow for heating in nonkosher heating units. This procedure assures kashrut law maintenance from time of preparation through distribution. The health care professional should allow the patient or family to remove any wrapping. Though the laws of kashrut are set, there are differences in customs and practice. Therefore, the nurse should ask the client for any kashrut preferences during the initial interview. In Israel, the hospital kitchens have only kosher food available with a kashrut supervisor maintaining kosher dietary laws. The hospitals also offer patients meals prepared by outside caterers that are supervised by Ultra-Orthodox kashrut supervisors acceptable to most Ultra-Orthodox sects. Nevertheless, the nurse and midwife should be aware that some members of the Ultra-Orthodox community will not eat these foods and prefer to have food brought to the hospital by family members (Table 1).

Sabbath and Holiday Observance

The concept of a "day" according to traditional Jewish religion begins the night before, at sundown at the time of the evening prayers. The Jewish Sabbath (*Shabat, Shabbos*) begins at sundown on Friday evening and concludes approximately an hour after sunset on Saturday night (Lutwak et al., 1988). The Jewish Sabbath has its origins in the Bible when God created the world in 6 days and rested on the 7th (Genesis, 2:2). Jewish women begin the Sabbath by lighting the Sabbath candles. Sabbath observance includes special prayer services in the synagogue, festive meals, time devoted to family, and religious studies (Schwartz, 2004). The Sabbath-observant Jew will refrain from specific actions that are not permitted on the Sabbath, such as writing, going to work, riding in a car, turning on and off or moving electrical appliances, using the telephone, and cooking food (Lewis, 2003; Shema Yisrael Torah Network, n.d.). These prohibitions are based on actions that were considered"creative work". Just as, according to biblical tradition, God rested from creation on the 7th day, observant Jews rest from creative activities. These activities are determined by what was considered creative activities in relation to the Jewish Tabernacle and subsequently part of the service in the Temple of biblical times. For example, fire was used in the service and, therefore, igniting

Table 1 Observant Jewish Customs, Laws, and Practices During Labor, Delivery, and Postpartum

Practice	Husband	Wife	Provider's Appropriate Response
Prayer	Prayer mandatory three times a day (morning, evening, and night prayers)	May choose to pray but not mandatory for a woman in labor	Promote a supportive, nonjudgmental environment
	Morning prayer requires a talit (prayer shawl) and *tefilin* (phylacteries); may prefer to say psalms while wife is in labor	May prefer to say psalms	Allow couple to observe religious practices
			Allow husband to pray within the designated prayer time frame; husband may leave to attend prayer service
	Will not speak to others during most forms of prayer unless there is an emergency situation		Differentiate between religious practices and relational issues
			If not urgent, do not converse with person during prayer
	Will pray facing towards Jerusalem		
	Will ritually wash hands with a cup of water before bread is eaten	Will ritually wash hands with a cup of water before bread is eaten	Provide cup near sink
	Will say Grace after meals after eating bread	Will say Grace after meals after eating bread	
	Will ritually wash hands with a cup after using the bathroom	Will ritually wash hands with a cup after using the bathroom	Provide cup and basin near woman's bedside after the delivery
		Will do ritual hand-washing after delivery	
Communication	Ultra-Orthodox custom: May prefer not to look or speak directly to a woman who is not his wife; may look at wife when speaking to midwife	Will repeat the husband's question to the midwife	Do not personalize husband's modesty custom; look at wife when addressing couple; speak with wife and not directly to husband, yet answer all of husband's questions and concerns
	Ultra-Orthodox and religious practice: Will not touch (i.e., shakehands) with women other than wife	Will not touch (i.e., shakehands) with a man other than husband	
	May/may not prefer to touch wife or show any type of physical affection while in the presence of others, even when permissible by Jewish law	May/may not prefer to touch husband or show any type of physical affection while in the presence of others, even when permissible by Jewish law	Need alternative assessment of husband—wife relationship because customary verbal and physical cues are not applicable
Niddah	No physical contact (including no passing of objects) with wife after any of the following: (a) bloodyshow (b) rupture of membranes(c) labor progressed to when wife needs to lie down	May or may not verbalize need for physical support and care	Provide physical and therapeutic interventions (e.g., massage) as needed. Understand husband and wife do not maintain physical contact during this time
		May need assistance with basic provisions, that is, getting a cup of water, help getting to the bathroom	Some fathers will not touch baby until blood from birth is wiped from baby
Dietary laws	Will only eat foods that are accepted by family as kosher	Same practice as husband	Israeli hospitals are all kosher; however, individuals may prefer to eat meals prepared by volunteer Orthodox Jewish organizations or only by family
	Will not eat meat and milk together		
	After eating meat meal, will wait 1 to 6 hours until dairy may be eaten (hours vary on custom)		Order special kosher meals from organization or give family members the telephone number to self-order the meals
	All utensils must be kept kosher		Do not mix meat and milk dishes
	Will only eat food on dishes that have only been used for kosher food		Ask patient and family when to serve dairy meal after last meat meal
	Uses separate dishes for meat and milk meals		Serve food as provided by kosher food service, allow woman and/or family to open double wrapping
			If possible, provide disposable dishes

Table 1   Observant Jewish Customs, Laws, and Practices During Labor, Delivery, and Postpartum (continued)

Practice	Husband	Wife	Provider's Appropriate Response
Modesty issues	Will not view actual birth or wife's perineal area at this time	Is permitted to view birth and will do so according to personal preference Will prefer to keep her hair covered Wears dresses with sleeves covering elbows, not low-cut Does not wear pants. May want to cover toes with stockings	Provide hair covering (surgical cap) if woman needs to go to OR or other cases where patient's head covering must be removed Provide patient gown that adheres to her modesty practice
Labor and birth customs	May prefer to say psalms or special prayers on behalf of wife Will or will not cut the umbilical cord	Is allowed to cut the umbilical cord but will do so according to personal preference	For the birth: Allow husband to stand by the top of the bed so he may face mother's head Allow husband to recite psalms or special prayers Do not ask father to cut the umbilical cord
	Will wait to announce baby's name until the male baby is circumcised or the female baby is named in the synagogue	Same practice as husband	Do not ask couple for baby's name
	Some Ultra-Orthodox prefer not to hold baby if blood from birth has not been wiped from baby's body	Is allowed to hold baby even when blood from birth is present	Ask father if he wants to hold the baby
Sabbath	Will try and prepare travel arrangements before the Sabbath either with a non-Jewish driver or drive himself while trying to minimize Sabbath violation	Same practice as husband	Provide safe and satisfying birth. Allow family to observe Sabbath as much as possible
	Will not turn electrical lights or appliances on or off, activate electronic devices such as doors and elevators, or use call-bell during Sabbath and holidays	Same practice as husband	Ask couple which lights they wish to leave on or off. Although couple will not ask, non-Jewish staff should anticipate the needs of the couple and operate items as needed. Determine how nurse will be called upon during Sabbath shifts
	For elevators and automatic doors, may wait for a non-Jew to activate before entering Will not write during the Sabbath		Will not sign consents. In the United States, witnessed verbal consent will suffice for medical procedures
	Will say prayers for the Sabbath	If able, will light Sabbath candles Friday night before sundown	If possible, provide room outside of proximity to oxygen where woman can light Sabbath candles. If woman cannot leave room with oxygen, allow to light electric Sabbath candles or make the blessing on an incandescent light that will remain lit till the end of the Sabbath
	May prefer to pray in synagogue if within walking distance	Same practice as husband	Provide list of closest local synagogues
	Will not touch money on the Sabbath	Same as husband	Allow prepaid vouchers to be used in cafeteria

(continued)

Table 1   Observant Jewish Customs, Laws, and Practices During Labor, Delivery, and Postpartum (continued)

Practice	Husband	Wife	Provider's Appropriate Response
	Will violate Sabbath and holidays only for emergency and absolute need to ensure a safe and satisfying delivery	Same as husband	Do not schedule induction of labor on Friday, Saturday, or Jewish holidays unless the mother or baby's life is in danger
	Will not tear paper or plastic and soon except to open/prepare food	Same as husband	Tear open wrapping to sanitary pads; if woman gives birth before Sabbath, provide her with enough pads and tissues and/or toilet paper before the Sabbath and/or holiday so that she can tear wrappings beforehand
Circumcision (male, healthy infants on 8th day after birth)	If wife's hospitalization extends to the 8th day postpartum, father may be occupied with the preparations for the ceremony	Same as husband. Mother may request to leave hospital for the ceremony	Some hospitals provide a room for the ceremony (as in all Israeli hospitals)
			Parents will invite a ritual practitioner(mohel) as opposed to using a health professional

or manipulating fire or anything deemed by Jewish law authorities to be similar (i.e., electricity) is prohibited. Many Jewish holidays (Table 2) include refraining from certain activities similar to Sabbath observance.

Although health emergencies, according to Jewish law, take precedence over Sabbath and Holiday laws, observant Jews will try to adhere to Sabbath laws as much as the situation allows. Emergency situations may necessitate the patient and family to use the telephone to call an ambulance or health care provider and travel to the hospital. In nonemergency cases, health care professionals can assist the Sabbath-observant patient and family by performing a cultural assessment pertaining to specific needs that would allow the family to maintain their cultural practices, in this case the Sabbath or holiday (see Table 3). Some examples of issues that may arise during a hospitalization of Sabbath-observant Jewish patients are the following: What lights may they leave on or off during the Sabbath hours (such as the bathroom light), alternative key option for doors that use electronic devices, arrangement for family members to stay in or near the hospital to avoid travel, kosher food availability that does not necessitate warming. Many hospitals with kosher kitchens keep food warm on the Sabbath by using hotplates (*blech, plata;* Gorga-Williams, 2003).

Additionally, some U.S. hospitals and all-Israeli hospitals maintain "Sabbath elevators". These elevators are electrically programmed to automatically stop on every hospital floor. This allows the observant patient and family to avoid pressing elevator buttons, as does keeping stairwells accessible for emergency exits per hospital policy. If a Sabbath elevator is not available, stairwells will be used. There are observant Jews who will use the stairwells, even when a Sabbath elevator is available (Broyde & Jachter, 1991; Gorga-Williams, 2003; Neustadt, 2006).

In Israel, culturally sensitive scientists in conjunction with rabbinical authorities have developed technical solutions to the Sabbath challenge of electronic devices. The technological mechanism used is referred to as *grama* and is based on indirect control and delayed automation. Religiously observant Israeli health care professionals as well as patients have the opportunity to take advantage of these innovations to minimize Sabbath violation (Broyde & Jachter, 1991). Many Israeli hospitals have incorporated these technical solutions for use on the Sabbath and Holidays. Examples of the innovations being used

by many Israeli medical institutions are the *grama*-delayed automation of such devices such as telephones, patient call bells, and electric wheelchairs. These devices are equipped to provide use of the device without desecrating the Sabbath laws.

Laws Pertaining to the Parturient (Yoledet) in Labor on the Sabbath

According to Jewish law, those caring for a dangerously ill person or a laboring woman (*yoledet*), including her husband and/or labor support attendant, are permitted to violate the Sabbath in order to ensure a safe and satisfying delivery (Neustadt, 2006). Care should be taken by the Jewish participants to minimize Sabbath violation when possible. One way to minimize Sabbath violation would be to ask a non-Jewish person to assist in those activities that require Sabbath violation. For example, arranging for a non-Jewish driver to transport the laboring woman, her husband, and/or labor support attendant to the hospital is preferable. If not possible, a Jewish person is allowed to drive the laboring woman, her husband, and attendant to the hospital. When a Jewish person must drive on the Sabbath, provisions should be taken to minimize Sabbath violation. If an observant Jewish person has transported a seriously ill person or laboring woman to the hospital, assistance by non-Jewish personnel would be helpful at time of arrival to perform nonemergency tasks such as shutting down the car or locking the doors.

Another Sabbath prohibition is carrying from public areas to private areas and vice versa. This "carrying" includes carrying by hand, in a pocket or in a bag/suitcase. Sabbath-observant Jews arriving at the hospital on the Sabbath may bring minimal items with them. They may not have, in their possession, a change of clothes, prenatal paperwork, or money. In Israel and in some American cities, there is a *halachic* (Jewish law) provision that allows for carrying certain items between public and private areas called an *eruv*. The *eruv* is a cooperative effort between the Jewish community and the municipality of a given area. The *eruv* is a symbolic boundary that transforms a public area into a private area (Siemiatycki, 2005). The observant Jewish family traveling for emergency care purposes to a health care facility, on the Sabbath, without the *eruv* provision may need assistance in carrying necessary items from the car to the hospital.

Table 2 Annual Jewish Holidays and Fast[a] Days

Name	Time of Year (Stated With Gregorian Calendar Time Frame)[a]	Time Frame	Special Observances
Rosh Hashanah (Jewish New Year)	Fall (September/October)	2 days	Daily blowing of the ram's horn (Shofar)
			Custom to eat special representative foods for a sweet year
Tzom Gedaliah	Day after Rosh Hashanah	From sunrise to sunset	Minor fast—refrain from eating and drinking. Pregnant and nursing women not obligated to fast
Yom Kippur (Day of Atonement)	10 days after RoshHashanah	1 day-from sunset night before (25 hours)	Major fast (no food or fluids)
			No leather shoes worn
			No sexual relations
			No full bathing
			No deodorant or perfumes
Sukkot (Feast of Tabernacles)	4 days after Yom Kippur	7 days in Israel 8 days outside Israel	First and last days are similar observance to Sabbath except for food preparation
			Middle (intermediate) days certain Sabbath prohibited activities permitted in keeping with the holiday spirit. Customarily to eat all meals in temporary huts built under the sky
			Customary to pray while holding the lulav and etrog (palmbranch, myrtle, willow branch, and citron)
Asarah B'Tevet	Winter (usually in December)	From sunrise to sundown	Minor fast—refrain from eating and drinking. Pregnant and nursing women not obligated to fast
Chanuka	December	8 days	Festival holiday commemorated each night with lighting of the Chanukah menorah and special foods. No Sabbath restrictions
Ta'anit Esther	March	From sunrise to sunset	Minor fast—refrain from eating and drinking. Pregnant and nursing women not obligated to fast
Purim	March	1 day	Festive holiday; reading of the Megillah (Biblical parchment scroll), food baskets exchanged, giving to the poor; festive meal. No Sabbath restrictions
Ta'anit B'chorin	One day before Passover	From sunrise to sunset	Minor fast—designated for first-born males only
Pesach (Passover)	1 month after Purim	7 days in Israel 8 days outside Israel	First and last days are similar observance to Sabbath except for food preparation
			Middle (intermediate) days certain Sabbath prohibited activities permitted in keeping with the holiday spirit
			No leavened bread. All foods are specially prepared for Passover use. Need for dishes and utensils that were not used during rest of year
Shavuot	7 weeks after Passover	1 day in Israel 2 days outside Israel	Similar observance to Sabbath except for food preparation
Sh'va Asar B'tamuz	July or August	From sunsetnight before (25 hours)	Minor fast–refrain from eating and drinking. Pregnant and nursing women not obligated to fast
Tisha B'av	3 weeks after Sheva Asar B'tamuz	1 day-from sunset-night before (25 hours)	Major fast; no eating or drinking; no leather shoes worn, certain mourning activities observed

a. Fast information (Nishmat Women's Online Information Center, 2003): (1) An active laboring woman is exempt from fasting. (2) Pregnant and nursing women are not obligated to fast on minor fasts. (3) Pregnant and nursing women are, in general, obligated to fast on the two major fasts of Yom Kippur and Tisha B'Av if there are no medical contraindications. (4) For Yom Kippur, laboring women and those that have given birth within 3 days after delivery are exempt from fasting, with exemptions easily given for postpartum days 4 to 7. The laws are more lenient with the fast of Tisha B'Av. (5) Women (or a family member) should consult with a rabbinical authority if there is any question as to the ability of the woman to fast. Leniencies are also given if the pregnant woman or fetus will be at risk, the woman had a difficult delivery, baby is ill, or there is difficulty breastfeeding.

Religious Practices of Observant Jewish Couples

Examples of religious practices of observant couples during labor, delivery, and postpartum are given in Table 1. In addition, suggested culturally appropriate responses are offered to the health professional. These examples are generalities and individuals may choose to conduct themselves differently than what is presented in Table 1.

Implications for Practice

The nurse and midwife can assist the traditionally, religious, observant Jewish couples to maintain their religious laws, customs, and practices. The religious laws, customs, and practices that will be most apparent during labor, delivery, and postpartum will be those that pertain to prayer, communication between husband and wife, dietary laws, the Sabbath, modesty issues, and labor and birth customs. Although permitted to violate tenets of religious practice in health-emergency situations, many religiously observant Jews will attempt to minimize those violations either out of absolute respect for upholding the laws or fear of Divine punishment. The culturally competent women's health care professional is behooved to follow the cues of the religious family, tailoring their health care provision in a manner that allows the family to practice their traditions in their specific designated manner while employing professionalism and creativity in providing quality patient care. The depth to which the practices and customs are presented here are not meant to overwhelm the practitioner that he or she needs to become a Jewish heritage expert. It is meant to offer a beginner's guide to the issues surrounding the labor, delivery, and postpartum experience for the observant Jewish couple. Most hospitals have access to clergy who are often used to support the professional and patient population. Clinicians can also seek information and religious items from either the institutional Jewish clergy or local Jewish community organizations.

Suggestions for a cultural assessment for this population can be found in Table 3. These suggestions include initial questions and clarification questions. This type of dialogue is more accurate than asking the patient to which branch of Judaism they prescribe (i.e., Orthodox, Conservative, Reform, Reconstructionist, Pluralistic, etc.). Even after the practitioner has prepared himself for a culturally competent patient experience, people do make individual choices during the hospitalization experience, or ask their personal Rabbi for specific guidance, and therefore, their practices may be different from the general laws and practices presented here. Asking clarification questions displays great respect to the patient. If, however, the patient and her family do not practice as they originally described, or do not explain their practice, this needs to be respected as an individual choice and verbal questioning maybe be seen as threatening or disrespectful. For example, in the recovery room, an Ultra-Orthodox man hugs his wife after the delivery. The nurse should not comment, "That is so nice, I did not know you were allowed to do that." Most religious-observant people know how to navigate their traditions and needs in the general population framework.

Below are two case studies that present cultural competence challenges to the health professional. The first case is a scenario where, though the practitioners were attempting to be culturally sensitive, because of not performing a cultural assessment, certain cues were overlooked. The second case is one in which the cultural assessment was performed, and therefore the birth experience was more tailored to the needs of the patient and her family.

Case 1

Mrs. Stein is in labor with her third child on a sunny Saturday morning in a hospital labor and delivery unit. She has been laboring for the past 5 hours with spontaneous rupture of membranes that occurred 2 hours ago. Present in the room with her are her husband, the nurse–midwife, and a labor and delivery nurse. The midwife examines Mrs. Stein and announces that she is fully dilated and needs to gather her strength to eventually begin pushing. The midwife asks Mr. Stein to please help her to move his wife up in the bed so she can be more comfortable. Mr. Stein looks down and does not respond. The midwife repeats her request a little more assertively and receives the same response from Mr. Stein. The labor and delivery nurse seeing Mr. Stein's lack of response determines that he must be nervous or embarrassed. To defuse the perceived discomfort, she sends Mr. Stein to get a cup of ice from the ice machine in the nourishment room. She reminds him of the code needed to open the electronic lock. Mr. Stein looks down again and does not respond.

Discussion

What is happening here? In using the cultural assessment tool in Table 3, the following issues would have been determined. The Steins are observant Jews. As Mrs. Stein has already ruptured membranes, she is considered to be *niddah* according to Jewish law, which prohibits the physical contact between husband and wife. Out of modesty and because of the nature of the subject, Mr. Stein does not feel comfortable to the request of the midwife. In addition, with the labor occurring on Saturday, the Sabbath, they will attempt to observe the Sabbath laws to the best of their ability especially in nonemergency situations. Both the ice machine and lock on the nourishment door have electronic mechanisms that are not used on the Sabbath by observant Jews. If, during the intake session, these data would have been obtained, other options could have been devised to meet these needs. For example, the midwife could have, originally, asked the nurse to assist her while suggesting that the husband stand close by to provide emotional support. Ice or other nourishment could have been arranged to be stored near the patient's bed in a thermos, or the staff could have provided for that need.

Case 2

Mrs. Hadad was admitted to the postpartum floor after delivering a healthy baby boy at 3:15 in the morning. The couple was exuberant, and the admitting nurse, Jane, guided them to their room and transferred Mrs. Hadad to her bed. The nurse began her admissions intake. Integrated into her intake, during the section of questions about activities of daily living, Jane asked the Hadads some specific questions about their traditions and practices. "How can I be helpful to you to keep your traditions while you are hospitalized? Is there anything I need to know about you and your baby to tailor our care to your needs?" (See Table 1 for clarification of issues that may arise during the hospitalization of the observant Jewish patient and Table 3 for more details as to questions that can be asked for clarification.) The Hadads shared with the nurse their dietary restrictions and requested a place to store the special kosher food that they eat, provided by their close-knit sect. In addition, they spoke of their concern that everyone from the staff would want to know the baby's name, and that they would not be naming the baby till the eighth day after the birth during a ceremony called a "brit" (circumcision). The nurse noted these issues and documented them in the patient's chart under cultural assessment. This came in handy for the various shifts that came in contact with the Hadads in terms of being respectful and providing for their needs.

Discussion

As a result of the foresight of the attending nurse in taking a cultural assessment, she, as well as her colleagues from the other shifts, were made aware of the Hadads' specific cultural needs and were able to

Table 3 Cultural Assessment for Jewish Clients in Labor, Delivery, and Postpartum

I. Dietary laws

   Opening question: Do you have any special dietary requirements? Is so, what?

   Clarification questions

   1. Will you eat the kosher food served in the hospital?

     1a. (If yes) The staff will order kosher food for you that will be served with a double wrapping

     1b. (If no) Do you need to order a special kosher meal? Is there anyway that we can assist you?

   2. If you are served a meat meal, please let the staff know when you can be served a dairy meal.

   3. Is there any additional food that the hospital can provide you with (i.e. uncut fruits and vegetables, soft drinks, snacks)?

   4. If you brought your own food, will you need disposable dishes and cutlery, space in the refrigerator, and patient identification stickers?

   5. Would you like us to get you a cup so you can ritually wash your hands before bread?

   6. Let me show you where the available patient nourishment area is. How can we assist you to make the microwave kosher for you to heat up your food?

II. Prayer

   Opening question: Is there anything that you or your spouse (or whoever is accompanying you) will need in regard to prayer, that is, location of the closest synagogue, prayers are said facing toward Jerusalem (e.g., in the United States, this is east)?

   Clarification questions:

   1. Will your spouse be attending prayer services during your labor? If so, can you provide his cell phone number in case we need to reach him?

   2. If your husband will pray in the labor room, we understand that we should not speak to him unless it is an urgent.

   3. Would you prefer that we leave a cup near the sink so that you can ritually wash your hands? We can also leave a basin and cup near your bedside so that you can ritually wash your hands after the birth.

III. Communication

   There are Ultra-Orthodox men who do not traditionally speak with women, or look at a woman, who is not a family member. Would your husband prefer that we direct all communication to you?

IV. Niddah (traditional physical separation between husband and wife)

   Opening question: Do you and your husband prefer that the staff provide all physical support, such as massages, help to and from the bathroom, or other comfort measures during labor?

   Clarification questions:

   1. Do you and your husband prefer that he not view the actual birth? Where would you both prefer that he stand during the birth?

   2. Would you want us to keep you covered in a way that he cannot view the birth?

   3. Does your husband prefer to hold the baby after the birth? Would he prefer that we hand him the baby after we have wiped the baby from the birth?

V. Modesty issues

   Opening question: Is there anything that we should know with regard to physical support or dress/attire to make you feel comfortable during your labor, delivery, and postpartum? (e.g., keep hair covered?)

   Clarification questions:

   1. Are there certain clothes or garments that you prefer to wear during your labor, birth, and postpartum? If yes, please tell us what garments you prefer.

   2. Will your husband be present for your labor and delivery? Will he be able to provide you with any comfort measures such as massage? How can we appropriately include your husband in your birth experience?

VI. Labor and birth customs:

   1. After the baby is born, do you want a family member to help cut the umbilical cord, or do you prefer that the birth attendant cut the cord?

   2. Do you prefer that we not ask you the baby's name?

VII. Sabbath and holidays

   Opening question: Is there anything that we should know about your traditions or preferences concerning the Sabbath and holidays?

   Clarification questions:

   1. If your birth occurs around the time of the Sabbath or holiday, do you or your husband have any special needs regarding:

     Transportation?

     Husband's sleeping arrangements?

(continued)

Table 3 Cultural Assessment for Jewish Clients in Labor, Delivery, and Postpartum *(continued)*

Use of electrical appliances including the call-bell, bathroom light (please consider the following additional electrical devices: light in the refrigerator microwave, toaster, ice machine, elevator, electronic lock to the nourishment room, etc)?

Are there any lights that you want us to leave on for the entire Sabbath/holiday?

Lighting of the Sabbath or holiday candles

There are some hospitals that provide an area for patients to light candles, if so, tell patient where that area is. If not, is there an area that the patient can light electric Sabbath candle?

Writing or signing consents—can they be done after the Sabbath or holiday?

Will you need to stay in or nearby the hospital if you are discharged on the Sabbath or holiday?

With regard to tearing paper, including toilet paper, opening sanitary napkin packages—will you need to do this before the Sabbath or holiday?

Will your family need prepaid vouchers to the cafeteria?

---

accommodate easily. The Hadads were able to feel both empowered and legitimized by the respect and flexibility of the staff. The staff felt accomplished in that they were able to meet the needs of this specific family.

Conclusion

The traditionally, religious-observant Jewish couple is committed to maintaining their religious laws, customs, and practices as much as possible throughout the labor, delivery, and postpartum periods. They view these practices as part of their value system and connection to God and their community. For observant Jews, these practices frame their year with meaning, richness, and importance and are not viewed as cumbersome. The details of each practice are seen as part and parcel of their commitment and devotion to religious processes. This article presents many of the observant Jewish practices pertaining to labor, delivery, and postpartum. Although one article cannot present an entire knowledge base, it can assist the health care professional to the provision of culturally competent care to the Jewish childbearing woman and family. Providing culturally competent care requires good communication and time management skills, patience, critical thinking, and creativity on the part of the health care professional. In turn, the patient will encounter a more individualized, holistic, and culturally appropriate hospital experience.

References

American-Israeli Cooperative Enterprise. (2008). *Context of ancient Israelite traditions.* Retrieved July 21, 2008, from //www. jewishvirtuallibrary.org/jsource/History/context.html

Andrews, M. M., & Boyle, J. S. (2003). *Transcultural concepts in nursing care.* Philadelphia: Lippincott Williams & Wilkins.

Broyde, M., & Jachter, H. (1991). The use of electricity on Shabbat and Yom Tov. *Journal of Halacha and Contemporary Society, XXI.* Retrieved April 17, 2007, from //www.daat.ac.il/daat/ english/Journal/broyde_1.htm

Callister, L. C., Semenic, S., & Foster, J. C. (1999). Cultural and spiritual meanings of childbirth: Orthodox Jewish and Mormon women. *Journal of Holistic Nursing, 17,* 280–295.

Campinha-Bacote, J. (2003). *The process of cultural competence in the delivery of healthcare services: A culturally competent model of care.* Cincinnati, OH: Transcultural C.A.R.E. Associates.

Dobrinsky, H. C. (1986). *A treasury of Sephardic laws and customs: The ritual practices of Syrian, Moroccan, Judeo-Spanish, and Spanish and Portuguese Jews of North America.* New York: Yeshiva University Press.

Galanti, G. A. (1997). *Caring for patients from different cultures: Case studies from American Hospitals.* Philadelphia: University of Pennsylvania Press.

Gorga-Williams, C. (2003, April 24). *To cater to the Orthodox community:Back to the basics.* Retrieved June 12, 2006, from //www.injersey.com/monmed/story/0,21508,726753,00.html

Kater, V. (2000). A tale of teaching in two cities. *International Nursing Review, 47,* 121–125.

Leininger, M., & McFarland, M. R. (2002). *Transcultural nursing: Concepts, theories, research and practice.* New York: McGraw-Hill.

Lewis, J. A. (2003). Jewish perspectives on pregnancy and childbearing. *American Journal of Maternal Child Nursing, 28,* 306–312.

Lutwak, R., Ney, A. M., & White, J. E. (1988). Maternity nursing and Jewish law. *MCN, American Journal of Maternal Child Nursing, 13,* 44–46.

Mattson, S. (2000). Providing culturally competent care: Strategies and approaches for perinatal clients. *Association of Women's Health, Obstetric and Neonatal Nurses, 4,* 37–39.

Nishmat Women's Online Information Center. (2003). *Fast days pregnant or nursing.* Retrieved April 28, 2008, from //www. yoatzot.org/question.php?id=762

Nishmat Women's Online Information Center. (2006). *Yoledet status & epidural.* Retrieved April 28, 2008, from //www.yoatzot.org/ question.php?id=5065

Neustadt, D. (2006). *The yoledes in halacha.* Retrieved January 30, 2007, from www.torah.org/advanced/weekly-halacha/5757/tazria.html

Orthodox Union. (2004). *How do I know it's kosher: An OU kosher primer.* Retrieved March 16, 2004, from //www.ou.org/kosher/ primer.html

Purnell, L. D., & Paulanka, B. J. (2003). *Transcultural health care: A culturally competent approach.* Philadelphia: F. A. Davis.

Schuiling, K. D., & Sampselle, C. M. (1999). Comfort in labor and midwifery art. *Image Journal of Nursing Scholarship, 31,* 77–81.

Schwartz, E. A. (2004). Jewish Americans. In J. N. Giger & R. E. Davidhizar (Eds.), *Transcultural nursing: Assessment and intervention* (pp. 545–569). St. Louis, MO: Mosby.

Selekman, J. (2003). People of Jewish heritage. In L. Purnell, & B. J. Paulanka (Eds.), *Transcultural health care: A culturally competent approach* (pp. 234–248). Philadelphia: F. A. Davis.

Shema Yisrael Torah Network. (n.d.). *Religious needs of the Orthodox Jewish patient*. Retrieved June 12, 2006, from //www. shemayisrael.co.il/burial/needs/htm

Shuzman, E. (2004). Perinatal health issues of Jewish women. In M. A. Shah (Ed.), *Transcultural aspects of perinatal health care:A resource guide*. Tampa, FL: National Perinatal Association.

Siemiatycki, M. (2005). Contesting sacred urban space: The case of the eruv. *Journal of International Migration and Integration, 6,* 255–270.

Spector, R. E. (2009). *Cultural diversity in health and illness*. Upper Saddle River, NJ: Prentice Hall Health.

Tal, R. (Ed.). (2007). *Jewish People Policy Planning Institute annual assessment* (Executive Report No. 4). Retrieved July 31, 2008, from //www.jpppi.org.il

Tendler, M. D. (1977). *Pardes rimonim: A marriage manual for the Jewish family*. Monsey, NY: Gross Brothers.

Weber, S. (1996). Cultural aspects of pain in childbearing women. *Journal of Obstetric, Gynecologic & Neonatal Nursing, 25,* 67–72.

Webster, Y. D. (1997). *The halachos of pregnancy and childbirth*. Retrieved April 8, 2009, from //www.eisheschayil.com/private/ birth/birth.htm#1

Critical Thinking

1. What Jewish customs, laws, and practices are taken into account during labor?

2. What is the purpose of a cultural assessment?

3. What are the common cultural differences between the male and female in the Jewish community?

The first and last authors were co-primary authors. The authors thank Rabbi Shalom Gold, Rabbi Morrie Wruble, and Dr. Lawrence Noble for their careful review of the manuscript. Correspondence concerning this article should be addressed to **ANITA NOBLE,** School of Nursing, P.O. Box 12000, Jerusalem 91120, Israel; e-mail: anoble@hadassah.org.il.

Understanding Transcultural Nursing

Be aware of cultural trends while respecting individual patients' preferences.

A patient's behavior is influenced in part by his cultural background. However, although certain attributes and attitudes are associated with particular cultural groups as described in the following pages, not all people from the same cultural background share the same behaviors and views.

When caring for a patient from a culture different from your own, you need to be aware of and respect his cultural preferences and beliefs; otherwise, he may consider you insensitive and indifferent, possibly even incompetent. But beware of assuming that all members of any one culture act and behave in the same way; in other words, don't stereotype people.

The best way to avoid stereotyping is to view each patient as an individual and to find out his cultural preferences. Using a culture assessment tool or questionnaire can help you discover these and document them for other members of the health care team.

Keeping the caveat about stereotyping in mind, let's take a look at how people from various cultural groups tend to perceive some common behaviors and key health care issues.

Space and Distance

People tend to regard the space immediately around them as an extension of themselves. The amount of space they prefer between themselves and others to feel comfortable is a culturally determined phenomenon.

Most people aren't conscious of their personal space requirements—it's just a feeling about what's comfortable for them—and you may be unaware of what people from another culture expect. For example, one patient may perceive your sitting close to him as an expression of warmth and caring; another may feel that you're invading his personal space.

Research reveals that people from the United States, Canada, and Great Britain require the most personal space between themselves and others. Those from Latin America, Japan, and the Middle East need the least amount of space and feel comfortable standing close to others. Keep these general trends in mind if a patient tends to position himself unusually close or far from you and be sensitive to his preference when giving nursing care.

Eye Contact

Eye contact is also a culturally determined behavior. Although most nurses are taught to maintain eye contact when speaking with patients, people from some cultural backgrounds may prefer you don't. In fact, your strong gaze may be interpreted as a sign of disrespect among Asian, American Indian, Indo-Chinese, Arab, and Appalachian patients who feel that direct eye contact is impolite or aggressive. These patients may avert their eyes when talking with you and others they perceive as authority figures.

An American Indian patient may stare at the floor during conversations. That's a cultural behavior conveying respect, and it shows that he's paying close attention to you. Likewise, a Hispanic patient may maintain downcast eyes in deference to someone's age, sex, social position, economic status, or position of authority. Being aware that whether a person makes eye contact may reflect his cultural background can help you avoid misunderstandings and make him feel more comfortable with you.

Time and Punctuality

Attitudes about time vary widely among cultures and can be a barrier to effective communication between nurses and patients. Concepts of time and punctuality are culturally determined, as is the concept of waiting.

In U.S. culture, we measure the passing and duration of time using clocks and watches. For most health care providers in our culture, time and promptness are extremely important. For example, we expect patients to arrive at an exact time for an appointment—despite the fact that they may have to wait for health care providers who are running late.

For patients from some other cultures, however, time is a relative phenomenon, and they may pay little attention to the exact hour or minute. Some Hispanic people, for example, consider time in a wider frame of reference and make the primary distinction between day and night but not hours of the day. Time may also be marked according to traditional times for meals, sleep, and other routine activities or events.

In some cultures, the "present" is of the greatest importance, and time is viewed in broad ranges rather than in terms of a fixed hour. Being flexible in regard to schedules is the best way to accommodate these differences.

Overcoming Barriers to Communication

Establishing an environment where cultural differences are respected begins with effective communication. This occurs not just from speaking the same language, but also through body language and other cues, such as voice, tone, and loudness. The Joint Commission on Accreditation of Healthcare Organizations (JCAHO) requires facilities to have interpreters available, so your facility should make a list available. But at times you'll be on your own, interacting with patients and families who don't speak English. To overcome the barriers you'll face, use these tips.

- Greet the patient using his last name or his complete name. Avoid being too casual or familiar. Point to yourself, say your name, and smile.
- Proceed in an unhurried manner. Pay attention to any effort the patient or his family makes to communicate.
- Speak in a low, moderate voice. Avoid talking loudly. Remember, we all have a tendency to raise the volume and pitch of our voice when a listener appears not to understand. But he may think that you're angry and shouting.
- Organize your thoughts. Repeat and summarize frequently. Use audiovisual aids when feasible.
- Use short, simple sentences and speak in the active voice.
- Use simple words, such as "pain" rather than "discomfort." Avoid medical jargon, idioms, and slang.
- Avoid using contractions, such as don't, can't, or won't.
- Use nouns instead of pronouns. For example, ask your patient's parent, "Does Juan take this medicine?" rather than "Does he take this medicine?"

- Pantomime words, using gestures such as pointing or drinking from a cup, and perform simple actions while verbalizing them.
- Give instructions in the proper sequence. For example, rather than saying, "Before you take the medicine, get into bed," you should say, "Get into your bed, then take your medicine."
- Discuss one topic at a time and avoid giving too much information in a single conversation. For example, instead of asking, "Are you cold and in pain?" separate your questions and gesture as you ask them: "Are you cold?" "Are you in pain?"
- Validate whether the patient understands by having him repeat instructions, demonstrate the procedure you've taught him, or act out the meaning.
- Use any appropriate words you know in the person's language. This shows that you're aware of and respect his native language.
- See if you have another language in common. For example, many Indo-Chinese people speak French, and many Europeans know three or four languages. Try Latin words or phrases, if you're familiar with the language.
- Do what you can to pick up a language that many patients in your area speak. Get phrase books from a library or bookstore, make or buy flash cards, or make a list for your bulletin board of key phrases everyone on staff can use. Your patients will appreciate your efforts, and you'll be prepared to provide better care.

Value differences also may influence someone's sense of time and priorities. For example, responding to a family matter may be more important to a patient than meeting a scheduled health care appointment. Allowing for these different values is essential in maintaining effective nurse/patient relationships. Scolding or acting annoyed when a patient is late would undermine his confidence in the health care system and might result in more missed appointments or indifference to patient teaching.

Touch

The meaning people associate with touching is culturally determined to a great degree. In Hispanic and Arab cultures, male health care providers may be prohibited from touching or examining certain parts of the female body; similarly, females may be prohibited from caring for males. Among many Asian Americans, touching a person's head may be impolite because that's where they believe the spirit resides. Before assessing an Asian American patient's head or evaluating a head injury, you may need to clearly explain what you're doing and why.

Always consider a patient's culturally defined sense of modesty when giving nursing care. For example, some Jewish and Islamic women believe that modesty requires covering their head, arms, and legs with clothing. Respect their tradition and help them remain covered while in your care.

Communication

In some aspects of care, the perspectives of health care providers, patients, and families may be in conflict. One example is the issue of informed consent and full disclosure. For example, you may feel that each patient has the right to full disclosure about his disease and prognosis and advocate that he be informed. But his family, coming from another culture may believe they're responsible for protecting and sparing him from knowledge about a serious illness. Similarly, patients may not want to know about their condition, expecting their relatives to "take the burden" of that knowledge and related decision making. If so, you need to respect their beliefs; don't just decide that they're wrong and inform the patient on your own.

You may face similar dilemmas when a patient refuses pain medication or treatment because of cultural or religious beliefs about pain or his belief in divine intervention or faith healing. You may not agree with his choice, but competent adults

Examples of Incidence of Disease among Various Cultures*

	AIDS/ HIV	Alcohol Abuse	Cancer	Cardiovascular Disorders	Endocrine Disorders	Gastrointestinal Disorders	Renal Disease
African Americans	Increased incidence		Increased incidence of lung, breast, and cervical cancer. Increased incidence of esophageal, stomach, and prostate cancer among men.	Highest mortality rates of coronary heart disease. Increased incidence of hypertension	Increased incidence of diabetes	Increased incidence of lactose intolerance	Increased incidence
American Indian		Twice the rate of whites tolerance	High incidence of nasopharyngeal cancer	Coronary heart disease causes 32% of heart-related deaths. Increased incidence of hypertension.	Increased incidence of diabetes mellitus		Increased incidence
Chinese Americans		Decreased alcohol tolerance	High incidence of nasopharyngeal, esophageal, stomach, liver, and cervical cancer			Increased incidence of lactose intolerance	
Hispanic Americans	Increased incidence			Decreased incidence of myocardial infarction	Increased incidence of diabetes mellitus	Increased incidence of lactose intolerance	Possible increased incidence
Japanese Americans		Decreased alcohol tolerance				Increased incidence of colitis. Decreased incidence of ulcers	Decreased incidence

*Because of space considerations, this chart provides only a few brief examples of increased/decreased incidence of diseases and disorders among some cultural groups.

Adapted from *Introductory Medical-Surgical Nursing,* 8th edition, B. Timby and N. Smith, Lippincott Williams & Wilkins, 2003.

have the legal right to refuse treatment, regardless of the reason. Thinking about your beliefs and recognizing your cultural bias and world view will help you understand differences and resolve cultural and ethical conflicts you may face. But while caring for this patient, promote open dialogue and work with him, his family, and health care providers to reach a culturally appropriate solution. For example, a patient who refuses a routine blood transfusion might accept an autologous one.

Holidays

People from all cultures celebrate civil and religious holidays. Get familiar with major holidays for the cultural groups your facility serves. You can find out more about various celebrations from religious organizations, hospital chaplains, and patients themselves. Expect to schedule routine health appointments, diagnostic tests, surgery, and other major procedures to avoid such holidays. If their holiday rituals aren't contradicted in the health care setting, try to accommodate them.

Diet

The cultural meanings associated with food vary widely. For example, sharing meals may be associated with solidifying social or business ties, celebrating life events, expressing appreciation, recognizing accomplishment, expressing wealth or social status, and validating social, cultural, or religious ceremonial functions. Culture determines which foods are served and when, the number and frequency of meals, who eats with whom, and who gets the choicest portions. Culture also determines how foods are prepared and served, how they're eaten (with chopsticks, fingers, or forks), and where people shop for their favorite food.

Religious practices may include fasting, abstaining from selected foods at particular times, and avoiding certain

Prohibited Foods and Beverages of Selected Religious Groups

Hinduism

All meats
Animal shortenings

Islam

Pork
Alcoholic products and beverages (including extracts containing alcohol, such as vanilla and lemon)
Animal shortenings
Gelatin made with pork, marshmallow, and other confections made with gelatin

Judaism

Pork
Predatory fowl
Shellfish and scavenger fish (shrimp, crab, lobster, escargot, catfish). Fish with fins and scales are permissible.
Mixing milk and meat dishes at same meal
Blood by ingestion (blood sausage, raw meat); blood by transfusion is acceptable.
Note: Packaged foods will contain labels identifying kosher ("properly preserved" or "fitting") and pareve (made without meat or milk) items.

Mormonism (Church of Jesus Christ of Latter-Day Saints)

Alcohol
Tobacco
Beverages containing caffeine stimulants (coffee, tea, colas, and selected carbonated soft drinks)

Seventh-Day Adventism

Pork
Certain seafood, including shellfish
Fermented beverages
Note: Optional vegetarianism is encouraged.

Biologic Variations

Along with psychosocial adaptations, you also need to consider culture's physiologic impact on how patients respond to treatment, particularly medications. Data have been collected for many years regarding different effects some medications have on persons of diverse ethnic or cultural origins. For example, because of genetic predisposition, patients may metabolize drugs in different ways or at different rates. For one patient, a "normal dose" of a medication may trigger an adverse reaction; for another, it might not work at all. (Think of how antihypertensive drugs don't work as well for African Americans as they do for white ones.) Culturally competent medication administration requires you to consider ethnicity and related factors—including values and beliefs about herbal supplements, dietary intake, and genetic factors that can affect how effective a treatment is and how well patients adhere to the treatment plan.

Environmental Variations

Various cultural groups have wide-ranging beliefs about man's relationship with the environment. A patient's attitude toward his treatment and prognosis is influenced by whether he generally believes that man has some control over events or whether he's more fatalistic and believes that chance and luck determine what will happen. If your patient holds the former view, you're likely to see good cooperation with health care regimens; he'll see the benefit of developing behavior that could improve his health. Some American Indians and Asian Americans are likely to fall into this category.

In contrast, Hispanic and Appalachian patients tend to be more fatalistic about nature, health, and death, feeling that they can't control these things. Patients who believe that they can't do much to improve their health through their actions may need more teaching and reinforcement about how diet and medications can affect their health. Provide information in a nonjudgmental way and respect their fatalistic beliefs.

Recipe for Success

Clearly, you can't take a "cookbook" approach to caring for patients based on their cultural heritage or background. Transcultural nursing means being sensitive to cultural differences as you focus on individual patients, their needs, and their preferences. Show your patients your respect for their culture by asking them about it, their beliefs, and related health care practices. They'll respond to your honesty and interest, and most will be happy to tell you more about their culture.

Critical Thinking

1. How is a patient's behavior influenced by their culture?

medications, such as pork-derived insulin. Practices may also include the ritualistic use of food and beverages. (See *Prohibited Foods and Beverages of Selected Religious Groups.*)

Many groups tend to feast, often with family and friends, on selected holidays. For example, many Christians eat large dinners on Christmas and Easter and traditionally consume certain high-calorie, high-fat foods, such as seasonal cookies, pastries, and candies. These culturally based dietary practices are especially significant when caring for patients with diabetes, hypertension, gastrointestinal disorders, and other conditions in which dietary modifications are important parts of the treatment regimen.

Test-Your-Knowledge Form

We encourage you to photocopy and use this page as a tool to assess how the articles in *Annual Editions* expand on the information in your textbook. By reflecting on the articles you will gain enhanced text information. You can also access this useful form on a product's book support website at www.mhhe.com/cls.

NAME: DATE:

_____

TITLE AND NUMBER OF ARTICLE:

_____

BRIEFLY STATE THE MAIN IDEA OF THIS ARTICLE:

_____

LIST THREE IMPORTANT FACTS THAT THE AUTHOR USES TO SUPPORT THE MAIN IDEA:

_____

WHAT INFORMATION OR IDEAS DISCUSSED IN THIS ARTICLE ARE ALSO DISCUSSED IN YOUR TEXTBOOK OR OTHER READINGS THAT YOU HAVE DONE? LIST THE TEXTBOOK CHAPTERS AND PAGE NUMBERS:

_____

LIST ANY EXAMPLES OF BIAS OR FAULTY REASONING THAT YOU FOUND IN THE ARTICLE:

_____

LIST ANY NEW TERMS/CONCEPTS THAT WERE DISCUSSED IN THE ARTICLE, AND WRITE A SHORT DEFINITION:

We Want Your Advice

ANNUAL EDITIONS revisions depend on two major opinion sources: one is our Advisory Board, listed in the front of this volume, which works with us in scanning the thousands of articles published in the public press each year; the other is you—the person actually using the book. Please help us and the users of the next edition by completing the prepaid article rating form on this page and returning it to us. Thank you for your help!

ANNUAL EDITIONS: Nursing 11/12

ARTICLE RATING FORM

Here is an opportunity for you to have direct input into the next revision of this volume.
We would like you to rate each of the articles listed below, using the following scale:

1. **Excellent: should definitely be retained**
2. **Above average: should probably be retained**
3. **Below average: should probably be deleted**
4. **Poor: should definitely be deleted**

Your ratings will play a vital part in the next revision.
Please mail this prepaid form to us as soon as possible.
Thanks for your help!

RATING	ARTICLE	RATING	ARTICLE
	1. Mary Breckinridge		30. Pediatric Hospice: Butterflies
	2. Hospitals Were for the Really Sick		31. Doing More with Less: Public Health Nurses Serve Their Communities
	3. Jane Delano		32. A New Way to Treat the World
	4. An End to Angels		33. Emergency Preparedness
	5. Delores O'Hara		34. A nursing Career to Consider: Assisted Living
	6. Shots Heard 'Round the World		35. Fetal Nutrition and Adult Hypertension, Diabetes, Obesity, and Coronary Artery Disease
	7. Linda Richards		36. Nutrition through the Life-Span. Part 1: Preconception, Pregnancy and Infancy
	8. Susie Walking Bear Yellowtail		37. Nutrition in Palliative Care
	9. Lillian Wald		38. Nutrition in the Elderly: A Basic Standard of Care and Dignity for Older People
	10. The Ethical, Legal and Social Context of Harm Reduction		39. Nutrition Management of Gastric Bypass in Patients with Chronic Kidney Disease
	11. Dialysis: Prolonging Life or Prolonging Dying? Ethical, Legal and Professional Considerations for End of Life Decision Making		40. Men in Nursing Today
	12. Being a Research Participant: The Nurse's Ethical and Legal Rights		41. Men in Nursing: Addressing the Nursing Workforce Shortage and Our History
	13. Exploring Ethical, Legal, and Professional Issues with the Mentally Ill on Death Row		42. The Male Community Nurse
	14. Covert Medication in Older Adults Who Lack Decision-Making Capacity		43. The State of the Profession: "Code White: Nurse Needed"
	15. Arresting Drug-Resistant Organisms		44. An Evaluation of Simulated Clinical Practice for Adult Branch Students
	16. Reduce the Risk of High-Alert Drugs		45. Nursing Students' Self-Assessment of their Simulation Experiences
	17. Keeping Your Patient Hemo-Dynamically Stable		46. Mentoring as a Teaching-Learning Strategy in Nursing
	18. Medication-Monitoring Lawsuit: Case Study and Lessons Learned: Another Real-Life Case in Nursing Home Litigation		47. NCLEX Fairness and Sensitivity Review
	19. Antiemetic Drugs		48. How to Read and Really Use an Item Analysis
	20. Assessment and Management of Patients with Wound-Related Pain		49. Nurses Step to the Front
	21. Acute Abdomen: What a Pain!		50. Mitigating the Impact of Hospital Restructuring on Nurses: The Responsibility of Emotionally Intelligent Leadership
	22. Pain Management: The role of the Nurse		51. Predictors of Professional Nursing Practice Behaviors in Hospital Settings
	23. Anxiety and Open Heart Surgery		52. The Winning Job Interview: Do Your Homework
	24. Respiratory Assessment in Adults		53. Meeting the Challenges of Stress in Healthcare
	25. Medical Nutrition Therapy: A Key to Diabetes Management and Prevention		54. Ethics and Advance Care Planning in a Culturally Diverse Society
	26. Understanding Hypovolaemic, Cardiogenic and Septic Shock		55. Jewish Laws, Customs, and Practice in Labor, Delivery, and Postpartum Care
	27. Tracheostomy: Facilitating Successful Discharge from Hospital to Home		56. Understanding Transcultural Nursing
	28. Are You Ready to Care for a Patient with an Insulin Pump?		
	29. Head Attack		

251

ABOUT YOU

Name   Date
_____

Are you a teacher? ❏ A student? ❏
Your school's name
_____

Department
_____

Address   City   State   Zip
_____

School telephone #
_____

YOUR COMMENTS ARE IMPORTANT TO US!

Please fill in the following information:
For which course did you use this book?
_____

Did you use a text with this ANNUAL EDITION? ❏ yes ❏ no
What was the title of the text?
_____

What are your general reactions to the Annual Editions concept?
_____

Have you read any pertinent articles recently that you think should be included in the next edition? Explain.
_____

Are there any articles that you feel should be replaced in the next edition? Why?
_____

Are there any World Wide Websites that you feel should be included in the next edition? Please annotate.
_____

May we contact you for editorial input? ❏ yes ❏ no
May we quote your comments? ❏ yes ❏ no

NOTES

NOTES